Satellite and UAV Platforms, Remote Sensing for Geographic Information Systems

Satellite and UAV Platforms, Remote Sensing for Geographic Information Systems

Editor

Alfred Colpaert

MDPI • Basel • Beijing • Wuhan • Barcelona • Belgrade • Manchester • Tokyo • Cluj • Tianjin

Editor
Alfred Colpaert
University of Eastern Finland
Finland

Editorial Office
MDPI
St. Alban-Anlage 66
4052 Basel, Switzerland

This is a reprint of articles from the Special Issue published online in the open access journal *Sensors* (ISSN 1424-8220) (available at: https://www.mdpi.com/journal/sensors/special_issues/UAV_GIS).

For citation purposes, cite each article independently as indicated on the article page online and as indicated below:

LastName, A.A.; LastName, B.B.; LastName, C.C. Article Title. *Journal Name* **Year**, *Volume Number*, Page Range.

ISBN 978-3-0365-5361-0 (Hbk)
ISBN 978-3-0365-5362-7 (PDF)

Cover image courtesy of Augustine Gbagir.

Contents

About the Editor

Alfred Colpaert

Alfred Colpaert has been a professor of physical geography at the university of Eastern Finland since 2004. He was previously affiliated with the university of Oulu, Finland. His main research interests are the use of remote sensing (UAV and satellite platforms) for the monitoring and analysis of ecological changes on the Earth's surface. He has studied, amongst other issues, the effects of reindeer herding in northern Scandinavia, the effects of cattle and wildlife on vegetation in southern Africa, and the effects of anthropogenic pressure on ecosystems.

Preface to "Satellite and UAV Platforms, Remote Sensing for Geographic Information Systems"

LAerial photography has been used since the First World War, and after the Second World War became an important tool of the Earth Sciences, as well as Biological and Geographical Sciences. Satellite Remote Sensing has been a standard research instrument since the launch of Landsat 1 in 1972. The digitalization of geographical work resulted in the first Geographical Information Systems in the 1960s. The rapid evolution of computer hardware and software has provided ever faster and better platforms for the analysis of digital geoinformation. The introduction of cheap and reliable unmanned aerial vehicles (UAV) in the last two decades filled the gap between global satellite imagery and digital aerial photogrammetry. Together with the advent of computer modelling and geo-statistical analytical capabilities, we have now an extremely powerful complex to analyze, model and predict changes in the environment. The effects of climate change, loss of biodiversity, and complex ecological questions are all possible fields benefitting from the integration of digital remotely sensed data and the analytical powers of geoinformatics.

Alfred Colpaert
Editor

Editorial

Satellite and UAV Platforms, Remote Sensing for Geographic Information Systems

Alfred Colpaert

Department of Geographical and Historical Studies, University of Eastern Finland, Yliopistokatu 7, P.O. Box 111, 80101 Joensuu, Finland; alfred.colpaert@uef.fi

Citation: Colpaert, A. Satellite and UAV Platforms, Remote Sensing for Geographic Information Systems. *Sensors* **2022**, *22*, 4564. https://doi.org/10.3390/s22124564

Received: 14 June 2022
Accepted: 15 June 2022
Published: 17 June 2022

Publisher's Note: MDPI stays neutral with regard to jurisdictional claims in published maps and institutional affiliations.

Satellite and UAV (unmanned aerial vehicle) imagery has become an important source of data for Geographic Information Systems (GISs). Enormous volumes of remote sensing and GIS data are nowadays produced, stored and analyzed in high-capacity network-based geoinformatics systems. Satellite imagery in a wide range of spatial, spectral, and temporal resolutions provides the scientific community with rapidly available global data to be used as an integral part of spatial data structures and analyses. Remote sensing platforms, such as Modis and Landsat, have a unique historical record of providing tens of years of uninterrupted global data, provided by repositories such as NASA and ESA.

For local applications, the rapid evolution of unmanned aerial vehicles and lightweight sensors has provided the scientific community with a tool for acquiring data of extremely high resolution, covering areas that vary from several hectares to hundreds of square kilometers in size. These data can be used for precision farming, forestry, and environmental monitoring.

The present Special Issue on the integration of UAV and satellite imagery with GIS contains ten articles, which can be divided into three parts: UAV applications, satellite remote sensing and methodological work using RS data.

Three articles deal with pure UAV applications, two applying UAV for agricultural crop monitoring [1,2] and one paper [3] using GIS and computer vision to analyze UAV orthomosaics. The planning of high-altitude long-endurance pseudo-satellite missions is dealt with in this paper [4].

The integration of satellite RS data in GIS systems for vegetation monitoring is used in three papers: one paper dealing with winter stress on arctic understory vegetation [5], one on the application of Copernicus (CMEMS GlobColour-Merged CHL-OC5 Satellite Observations) satellite-derived data concerning the aquatic environment [6] and one on the application of MODIS NDVI data and GIS to assess the effect of wildlife upon tropical savannah vegetation [7].

The problem of mosaicking multiple high-resolution orthoimages (UAV or satellite) is dealt with in paper [8], introducing a novel method based upon the D-LinkNet Neural Network.

Precise satellite telemetry data can be used to monitor the minute deformations occurring in the Earth's crust. The paper by Jagoda et al. describes the use of high-precision laser altimetry data to assess the deformation of the Earth's crust caused by external planetary bodies (tidal forces) [9]. The last paper uses data derived from global navigation satellite systems (GNSS) to monitor the crustal deformation of the Earth [10].

Funding: This research received no external funding.

Conflicts of Interest: The author declare no conflict of interest.

References

1. Cheng, Z.; Meng, J.; Shang, J.; Liu, J.; Huang, J.; Qiao, Y.; Qian, B.; Jing, Q.; Dong, T.; Yu, L. Generating Time-Series LAI Estimates of Maize Using Combined Methods Based on Multispectral UAV Observations and WOFOST Model. *Sensors* **2020**, *20*, 6006. [CrossRef] [PubMed]
2. Qi, G.; Zhao, G.; Xi, X. Soil Salinity Inversion of Winter Wheat Areas Based on Satellite-Unmanned Aerial Vehicle-Ground Collaborative System in Coastal of the Yellow River Delta. *Sensors* **2020**, *20*, 6521. [CrossRef] [PubMed]
3. Kentsch, S.; Cabezas, M.; Tomhave, L.; Groß, J.; Burkhard, B.; Lopez Caceres, M.; Waki, K.; Diez, Y. Analysis of UAV-Acquired Wetland Orthomosaics Using GIS, Computer Vision, Computational Topology and Deep Learning. *Sensors* **2021**, *21*, 471. [CrossRef] [PubMed]
4. Kiam, J.; Besada-Portas, E.; Schulte, A. Hierarchical Mission Planning with a GA-Optimizer for Unmanned High Altitude Pseudo-Satellites. *Sensors* **2021**, *21*, 1630. [CrossRef] [PubMed]
5. Ritz, E.; Bjerke, J.; Tømmervik, H. Monitoring Winter Stress Vulnerability of High-Latitude Understory Vegetation Using Intraspecific Trait Variability and Remote Sensing Approaches. *Sensors* **2020**, *20*, 2102. [CrossRef] [PubMed]
6. Gbagir, A.; Colpaert, A. Assessing the Trend of the Trophic State of Lake Ladoga Based on Multi-Year (1997–2019) CMEMS GlobColour-Merged CHL-OC5 Satellite Observations. *Sensors* **2020**, *20*, 6881. [CrossRef] [PubMed]
7. Gbagir, A.; Sikopo, C.; Matengu, K.; Colpaert, A. Assessing the Impact of Wildlife on Vegetation Cover Change, Northeast Namibia, Based on MODIS Satellite Imagery (2002–2021). *Sensors* **2022**, *22*, 4006. [CrossRef] [PubMed]
8. Yuan, S.; Yang, K.; Li, X.; Cai, H. Automatic Seamline Determination for Urban Image Mosaicking Based on Road Probability Map from the D-LinkNet Neural Network. *Sensors* **2020**, *20*, 1832. [CrossRef] [PubMed]
9. Jagoda, M.; Rutkowska, M.; Lejba, P.; Katzer, J.; Obuchovski, R.; Šlikas, D. Satellite Laser Ranging for Retrieval of the Local Values of the Love h2 and Shida l2 Numbers for the Australian ILRS Stations. *Sensors* **2020**, *20*, 6851. [CrossRef] [PubMed]
10. Jagoda, M. Determination of Motion Parameters of Selected Major Tectonic Plates Based on GNSS Station Positions and Velocities in the ITRF2014. *Sensors* **2021**, *21*, 5342. [CrossRef] [PubMed]

Article

Assessing the Impact of Wildlife on Vegetation Cover Change, Northeast Namibia, Based on MODIS Satellite Imagery (2002–2021)

Augustine-Moses Gaavwase Gbagir [1,*]**, Colgar Sisamu Sikopo** [2]**, Kenneth Kamwi Matengu** [3] **and Alfred Colpaert** [1,*]

[1] Department of Geographical and Historical Studies, University of Eastern Finland, Yliopistokatu 7, 80100 Joensuu, Finland

[2] Ministry of Environment, Forestry and Tourism, Windhoek 13306, Namibia; colgar.sikopo@meft.gov.na

[3] Department of Geography and Sociology, University of Namibia, Windhoek 13301, Namibia; kmatengu@unam.na

* Correspondence: augustine.gbagir@uef.fi (A.-M.G.G.); alfred.colpaert@uef.fi (A.C.)

Citation: Gbagir, A.-M.G.; Sikopo, C.S.; Matengu, K.K.; Colpaert, A. Assessing the Impact of Wildlife on Vegetation Cover Change, Northeast Namibia, Based on MODIS Satellite Imagery (2002–2021). *Sensors* **2022**, *22*, 4006. https://doi.org/10.3390/s22114006

Academic Editor: Assefa M. Melesse

Received: 6 April 2022
Accepted: 9 May 2022
Published: 25 May 2022

Publisher's Note: MDPI stays neutral with regard to jurisdictional claims in published maps and institutional affiliations.

Abstract: Human–wildlife conflict in the Zambezi region of northeast Namibia is well documented, but the impact of wildlife (e.g., elephants) on vegetation cover change has not been adequately addressed. Here, we assessed human–wildlife interaction and impact on vegetation cover change. We analyzed the 250 m MODIS and ERA5 $0.25° \times 0.25°$ drone and GPS-collar datasets. We used Time Series Segmented Residual Trends (TSS-RESTREND), Mann–Kendall Test Statistics, Sen's Slope, ensemble, Kernel Density Estimation (KDE), and Pearson correlation methods. Our results revealed (i) widespread vegetation browning along elephant migration routes and within National Parks, (ii) Pearson correlation (p-value $= 5.5 \times 10^{-8}$) showed that vegetation browning areas do not sustain high population densities of elephants. Currently, the Zambezi has about 12,008 elephants while these numbers were 1468, 7950, and 5242 in 1989, 1994, and 2005, respectively, (iii) settlements and artificial barriers have a negative impact on wildlife movement, driving vegetation browning, and (iv) vegetation greening was found mostly within communal areas where intensive farming and cattle grazing is a common practice. The findings of this study will serve as a reference for policy and decision makers. Future studies should consider integrating higher resolution multi-platform datasets for detailed micro analysis and mapping of vegetation cover change.

Keywords: vegetation monitoring; drivers of deforestation; Zambezi region; land degradation; vegetation cover change; wildlife management; TSS-RESTREND; greening and browning; MODIS; Mann–Kendall

1. Introduction

One of the persistent ongoing global environmental challenges is that of land degradation [1–3]. Land degradation is quite complex in nature and often involves the inter-play of biophysical, environmental, and socioeconomic factors [4]. There are several scientific debates on what constitutes land degradation but in this study, we adapt the general definition of land degradation by Barbier and Hochard 2018 [5], "as some measurable loss of the biological or economic productivity and complexity of rainfed cropland, irrigated cropland, or range, pasture, forest and woodlands . . . arising from human activities and habitation patterns".

Anthropogenic disturbances have been identified as a major driver of land degradation globally [6–10] and are well documented [6,11–15]. The drivers of land degradation are many, complex, and unique across regions [8,13,14,16,17], but these have been categorized as direct and indirect [16,17]. Based on this categorization, the direct causes of land degradation include: (1) infrastructure development (e.g., roads and settlements),

(2) expansion of agriculture (e.g., large- and small-scale farming and cattle grazing), and (3) wood extraction (e.g., fuelwood, pole wood, and charcoal production). While the indirect drivers of land degradation include: (a) demographic (e.g., population density, and migration/emigration), (b) economic (e.g., market growth and commercialization, and economic structures), (c) technology (e.g., agro-tech changes), (d) policy and institutional (e.g., formal policies, and property rights), and (e) cultural (e.g., public attitudes and beliefs, and individual and household behavior).

Anthropogenic activities such as wood extraction and conversion of woodland and forests into small and large-scale farming have been a major contributor to land degradation across different geographical regions [15,18]. In the tropics, particularly Africa and in Namibia, agricultural expansion, wood extraction, and infrastructure development are the key drivers of land degradation [9,12,16–18].

The impact of land degradation is quite severe in arid, semi-arid, and sub-humid regions [4,19]. Land degradation has a long history in Sub-Saharan Africa and has been well documented and researched [19–21]. In Africa, anthropogenic activities, including unsustainable land use practices (e.g., overexploitation of natural vegetation cover), are a major contributor to land degradation, in addition to other natural causes such as droughts [16,22,23]. In a country such as Namibia in south-western Africa, where about 22% of the land area is classified as desert, 70% as arid to semi-arid, and 8% as dry sub-humid [24–26], any slight change or modification in the vegetation structure could have adverse effect on the environment, social well-being, and livelihoods of the people [27].

In Namibia, the contribution of anthropogenic activities to the loss of vegetation cover has been well documented [12,24,27–29]. The loss of vegetation cover is mostly driven by the conversion of forests and woodland into agricultural farmlands [12,24]. Even though this is the case, the interaction and contribution of wildlife to vegetation cover loss is less understood and needs to be studied in more detail. On the other hand, wildlife damage and human–wildlife conflict is an on-going topic of research and discussion amongst researchers, natural resources managers and various other stakeholders [30–32]. This study will focus on one aspect of land degradation: vegetation cover loss and how anthropogenic and wildlife interaction are driving land cover change in the Zambezi region.

Amongst all the human–wildlife conflicts, the African elephant (*Loxodonta africana*) is one of the most significant wildlife species causing structural changes and damage to vegetation [33]. Elephants are herbivores and bulk feeders and require large amounts of food resources to fulfill their nutritional requirements, which they receive from trees, shrubs, and grasses [33,34]. Even though elephants' consumption of vegetation to meet their dietary needs is natural, overexploitation and mechanical damage becomes destructive, causing vegetation browning and contributing to land degradation (e.g., leaving soils bare) [33–35]. The impact of elephants on structural changes in vegetation has been documented by several studies [34,36–42]. Unlike anthropogenic activities, the impact of elephants on vegetation cover is, to a large extent, confined to locations where elephants exist mostly within protected areas [35,40,43,44]. Anthropogenic restriction of elephant movement and access to space and resources is the main factor driving the browning of vegetation cover by elephants [45,46].

In Namibia, anthropogenic activities are the primary driver of land degradation [12,24], while elephants are mainly responsible for modifying the vegetation structure [47]. As such, we will limit our discussion in this study to the elephant as the major interacting non-human factor contributing to the loss of vegetation cover in Namibia. Additionally, we will use the Zambezi region as a test case, as it is one of the best areas suitable for agriculture in the whole of Namibia and, historically, is home to a wide variety of wildlife (both large and small) [28,48]. The region has the largest savannah woodland cover in Namibia [24,27,28] and is habitat to most elephants in Namibia [30,48]. It is well documented that elephants browse, break, pull and uproot woody species, thus causing structural changes in vegetation cover [44,49–51]. The movement of elephants depends on several interrelated factors such as food, water, elevation, density and human settlements [30,49]; they have a home range of from 10 km to more than 8000 km [49]. Consequently, human–wildlife conflicts are a common occurrence in the region [32]. One of the identified reasons for these conflicts is the anthropogenic fragmentation of natural wildlife habitats [30,32,46]. Though this is the case, elephant-induced vegetation cover loss is most likely a secondary cause [51], the primary cause being the limited availability of resources driven by anthropogenic activities [25,46].

One of the solutions to these conflicts has been the establishment of wildlife reserves and national parks and the construction of fences and other barriers to keep wildlife at bay from human settlements [25]. While this has largely worked, the carrying capacity of these wildlife reserves is often not sufficient to sustain large herds of herbivores [25], thus putting pressure on available resources and causing the loss of vegetation cover [25].

Currently, in Namibia, specifically in the Zambezi region, there are projects to combat land degradation [48], conserve wildlife and manage human–wildlife conflicts [51–53]. In this study, we will use land degradation to mean the loss of vegetation cover with contextual meaning [54]. Additionally, we will use the terms greening and browning to refer to vegetation increase and decrease, respectively.

Presently, there are multiple satellites that provide datasets that can be used for different research purposes [55–57]. Satellite remote sensing is widely used in environmental monitoring, mapping of vegetation, and assessment of different land use and land cover changes [58–61]. Remote sensing is a popular mode of research as it is the cheapest and most efficient way to assess land use and land cover change [62]. Land use and land cover change assessment is still one of the most important areas of research because of the direct, immediate, and long-term impact of anthropogenic activities on the environment [3,60,63]. Thus, finding long-lasting and sustainable approaches to address land degradation is essential [3,64]. Additionally, it is important to understand that land degradation is contextual in nature and this should be taken into account during discussions [65].

In the Zambezi region, many studies have successfully used satellite remote sensing to assess and map changes in vegetation cover [24,28,29,66]. Although this is the case, to our understanding and best knowledge, assessing and characterizing the impact of wildlife on vegetation cover using remote sensing in the Zambezi region has not been attempted before.

Thus, improving our understanding of the dynamics and impact of wildlife on land degradation within the Zambezi region and beyond is important. Better understanding will provide better insight and tools to improve the management of wildlife and natural land resources in the region. One major challenge of implementing more effective wildlife conservation and natural resource management is continuous access to historical and up-to-date land use and satellite data. Fortunately, the availability of historical satellite remote sensing data and the increasing improvements in analytical software provide opportunities to assess and map changes in vegetation cover and structure. In a 2019 study, the authors successfully applied remote sensing data to characterize regional vegetation cover change in the Zambezi region [24]. In that study, they used eight km resolution Global Inventory Monitoring and Modelling Studies (GIMMS) from the Advanced Very High-Resolution Radiometer (AVHRR) [67]. In the study, only results on a regional scale were obtained due to the coarse resolution of the data [24].

Although higher resolution satellite observations exist, there are drawbacks in using this data, such as (a) the exponentially increasing amount of data result in high computational costs for a long time series, and (b) the low temporal resolution and higher impact of cloudiness (especially in the tropics) [68]. The MODIS 250 m resolution dataset provides over 20 years of continuous global daily imagery, which has been resampled into a monthly NDVI nearly cloudless dataset. This monthly NDVI dataset allows us to use sophisticated geospatial trend analysis techniques.

In this study, we will assess and characterize the vegetation cover change in the last nineteen years (2002–2021).

The specific objectives of this study are to:

(i)   Assess the impact of wildlife (elephants and other large herbivores) on the vegetation cover change (greening and browning) in the last 19 years (2002–2021).

(ii) Assess the effects of anthropogenic activities on wildlife migration and vegetation cover change (greening and browning).

To assess the vegetation cover change, we will use historical remote sensing data from the 250-meter Moderate Resolution Imaging Spectroradiometer (MODIS) Terra satellite instrument. We will use time series and geostatistics, as well as geo-spatial analytical methods. Specifically, we will use the Time Series Segmented Residual Trend (TSS-RESTREND) method, which will allow us to separate human-induced land degradation from that caused by natural climatic factors [24,69–71]. A similar approach was used by the authors in their 2019 study [24]. Additionally, we will use the Mann–Kendall non-parametric test and Sen's Slope measure of direction and magnitude of vegetation change [72]. In addition, we will use the Kernel Density Analysis [73] to cluster the elephant GPS tracking data and correlate the results with the trajectory of vegetation cover change.

2. Materials and Methods

2.1. Study Area

The Zambezi region ((formerly Caprivi Strip) (Figure 1)) is part of the Kavango-Zambezi Transfrontier Conservation Area (KAZA TFCA), stretching across five countries: Namibia, Angola, Zambia, Zimbabwe, and Botswana [48], forming the second largest conservation area in the world [51]. The land area of the Zambezi region is 14,785 km^2 [51], with a total population of 98,849 (2011 Census) [24]. Most of the vegetation in the region is woodland savanna and open grasslands [27,28,51]. The region contains three large National Parks, Bwabwata, Mudumu and Nkasa Lupala (formerly Mamili) [48,51]. Conservancies in the region include: Kwandu, Mayuni, Salambala, Sibbinda and Linyanti [25,29]. A conservancy is a legally defined area set aside and managed by local communities who have rights to live within, use, and manage wildlife and other natural resources for personal and tourism purposes (including trophy hunting) [74,75]. The region is an important migratory route and home to a high density of elephants [25,76], buffalos, and antelopes [48]. Additionally, the region is an important agricultural area, due to good soils and high rainfall [77,78].The major soil types in the region are poor ferralic arenosols containing high iron contents and fertile eutric fluvisols with high base saturation [77,78]. The yearly amount of rainfall in the region is the highest in Namibia (500–700 mm/year) [77,78] when compared to the national mean of <250 mm/year [19] and <50 mm/year in the southwestern and coastal areas [77]. The wet season in the region starts in November and ends in April. The average summer and winter temperatures in the region are 35 °C and 5 °C, respectively [78]. The region has a yearly evaporation rate of about 2500 mm [78]. The region shares common borders with Angola, Botswana, Zimbabwe, and Zambia [48]. The perennial Kwando (Cuando) River flows along the border between Angola and Zambia through the Zambezi region (with Bwabwata National Park on the west and the Mudumu National Park and the six conservancies on the east) south towards the swampy areas around Nkasa Lupala National Park. East of Nkasa Lupala is the Linyanti River that flows east through the seasonal Lake Liambezi into the Chobe River. The Chobe River flows eastward into the perennial Zambezi River, one of Africa's major and longest river systems. The Zambezi River flows from Zambia and forms the border between Zambia and Namibia in the Zambezi region.

2.2. Satellite and UAV Field Data and Image Pre-Processing

We downloaded and processed the monthly 250 m Moderate Resolution Imaging Spectroradiometer (MODIS) Terra satellite instrument NDVI datasets. We resampled the NDVI index to a common 250 m grid (UTM-35S). Because the 2001 data set is incomplete, we used only raster images from 2002 to 2021.

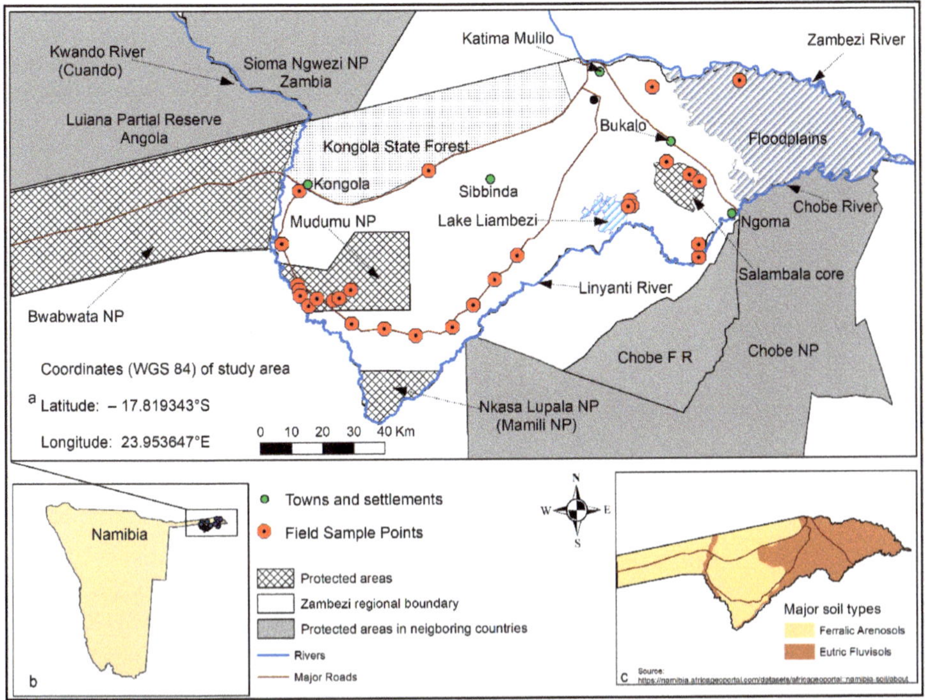

Figure 1. Study area, main roads, rivers, conservation areas, and field work way points for high resolution (2 cm) UAV data in (**a**). In (**b**), the map of Namibia, and (**c**) map of major soil types.

The time series for precipitation and temperature data were monthly ERA5, available from 1979 (ERA documentation). The monthly 0.25 × 0.25 degree resolution data was downloaded from 1999 to 2021 (https://cds.climate.copernicus.eu (accessed on 6 May 2022)). Both temperature and precipitation were resampled to a common 250 m grid and re-projected to the UTM reference system. The temperature and precipitation data must start two years before the NDVI time series because this information is required by the processing algorithms during analysis to calculate the maximum rainfall accumulation months. Finally, we used 265 gridded monthly temperature and precipitation raster layers.

2.3. Field Sample Locations and Elephant Tracking Data

During our field work, we used a hand-held GPS instrument [(Garmin GPSMAP 62ST), Garmin Finland] to collect the latitudes and longitudes of sample points. At each sample point (Figure 1), we recorded the location coordinates and geographical name. We used a DJI Mavic Pro Platinum drone to document the vegetation characteristics by taking aerial photos and videos at every sample point. At each sample location, we flew the drone at a height of 40–90 m and recorded a 360° view of the surrounding vegetation (pictures and videos).

The Government of the Republic of Namibia (Ministry of Environment Tourism and Forestry) provided the elephant tracking data (2010–2020). These data are a transboundary hourly GPS-collar tracking dataset, covering Namibia, Botswana, Angola, and Zambia. The data consisted of 31 individual elephants over a period of eight (8) years (2010–2020). The GPS-collared elephant data were collected by the Africa Wildlife Tracking company (https://awt.co.za/ (accessed on 6 May 2022)), based in Pretoria, South Africa. The GPS collars were put on the elephants by first using a tranquilizer dart from a helicopter to immobilize them. The brand of GPS collar used was the Iridium Satellite (IR-Sat) that collects and transmits continuous near real-time data. The data transmission and receiver of the IR-Sat covers a few hundred meters to multiple kilometers [79]. Table 1 presents a breakdown of the number individual elephants tracked and during which period. We downloaded additional crowd sourced wildlife observations, elephant and buffalo observations (one kilometer grid), using the Monad (1 km × 1 km) reference grid data from the Environmental Information Service Namibia (http://www.the-eis.com (accessed on 6 May 2022)).

Table 1. The tracked elephant data used in this study. The periods correspond to the 12 calendar months of the year. The total length of tracking period (in months) is shown in brackets.

Year	No. of Individual Elephants Tracked	Period
2010	8	10–12 (3)
2011	7	1–12 (12)
2012	7	1–8 (8)
2016	5	3–12 (8)
2017	13	1–12 (12)
2018	20	1–12 (12)
2019	16	1–12 (12)
2020	6	1–12 (12)

2.4. Data Analysis

To assess the vegetation changes, browning (decrease) or greening (increase), we used the Time Series Segmentation and Residual Trend analysis (TSS-RESTREND) method [69] to perform a pixel wise analysis. To achieve this, we created and used an R-script that iterates over each pixel across the image stack of the complete the time series. TSS-RESTREND is an improved method of the Residual Trends algorithm (RESTREND) [71] that incorporates the functionalities of Break For Additive Season and Trend (BFAST) algorithm [80,81] to look for break points in the time series.

RESTREND uses an Ordinary Least Squares Linear Regression model, fitted on the residual and time [1,24]. The equation is of the form:

$$y_i = \beta_0 + \beta_1 x, \tag{1}$$

where x is time in years, β_0 is the intercept and β_1 is the slope.

BFAST fits a linear piecewise harmonic model using the ordinary least squares moving sum (OLS-MOSUM) to test for structural changes within time series data [24,82].

The decomposition model takes the following form:

$$Y_t = T_t + S_t + e_t, \tag{2}$$

where Y_t is the original observed data (TS) at time t, T_t is the trend, S_t is the seasonal, and e_t is the remaining unexplained variation within the TS, respectively [24,80].

TSS-RESTREND fits a multivariate regression between the VPR-Residual (vegetation precipitation) and a dummy variable (0 before a break point and 1 after). The model is of the form:

$$y_i = \beta_0 + \beta_1 x + \beta_2 z_i + \beta_3 x_i z_i \tag{3}$$

where x is time in years, β_0 is the intercept, β_1 is the slope, β_2 is the offset at breakpoint position, β_3 is the change in slope at the breakpoint position and z is the dummy variable (0 or 1) [24,69].

In addition, we performed a pixel wise Mann–Kendall statistics test of the NDVI time series to determine the trend of total vegetation change in the Zambezi, and the Sen's Slope to determine the magnitude of the change [72]. Mann–Kendall is a non-parametric test and does not rely on a particular data distribution but rather on the relative magnitude of the sample data [83,84].

The Mann–Kendall statistics is of the form:

$$S = \sum_{k=1}^{n-1} \sum_{j=k+1}^{n} \text{sign}\left(x_j - x_k\right) \tag{4}$$

where:

$$\text{sign}\left(x_j - x_k\right) = \begin{cases} 1 \text{ if } x_j - x_k > 0 \\ 0 \text{ if } x_j - x_k = 0 \\ -1 \text{ if } x_j - x_k < 0 \end{cases} \tag{5}$$

x_j and x_k are the annual values in years j and k, respectively [29].

The non-parametric Sens Slope time series analysis was performed using the same pixel-wise moving window method to obtain the linear rate of change in the time series. During all pixel-wise analysis (TSS.RESTREND, RESTREND, Mann–Kendall and Sen's slope), we used a *p*-value parameter of 1 during the analysis, thus we obtained the change in every pixel, irrespective of the *p*-value. We took this approach because it provides for a synoptic spatial overview, showing gradual changes between distinct areas of significant degradation and vegetation increase, and areas of no change; the latter is associated with non-significant *p*-values. We observed that the changes that the algorithm interpreted as non-significant contain important information, e.g., areas of no change. This approach also provides a much more homogeneous and easier to understand cartographic map product. Although the general result of the different methods conforms very well, local differences are noticeable when comparing the results of different algorithms, therefore we made an additional ensemble analysis by combining the TSS.RESTREND, RETREND and Mann–Kendall algorithms and calculating the mean value of the results.

We computed a Kernel Density Estimation (KDE) of the elephant tracking point data in ArcMap 10.5.1. We then used the KDE and the ensemble mean to calculate a simple Pearson correlation analysis between the presence of elephants and vegetation changes. Before calculating the Pearson correlation, we used ArcMap 10.5.4 to create a grid of 1600 points over the whole Zambezi area where elephants are present (Figures 2 and 3). Additionally, we created an additional l600 points in two sub-sample grids, west of the Kwando River in Bwabwata park and around the Mudumu park (Figure 3B,C). We then extracted the values of the ensemble mean at these points and we excluded data points where the KDE was zero (no presence). We were left with 1506 points 1386 points in sub-sample locations one (Bwabwata) and 1482 points in two (Mudumu). We then used these points to compute a simple Pearson correlation in Microsoft Excel (version 2202).

To validate our results, we compared the UAV data we collected during our 2019 field trip with the outcome of the time series analysis. We used R (R Core Team, 2022) and ArcMap (version 10.5.4) for data analysis and to produce graphics. The R-code was run on the cPouta cloud services of CSC using 24 and 48 cores Ubuntu Virtual Machines (https://research.csc.fi (accessed on 6 May 2022)). We implemented Google Earth Pro for visual interpretation and verification of results.

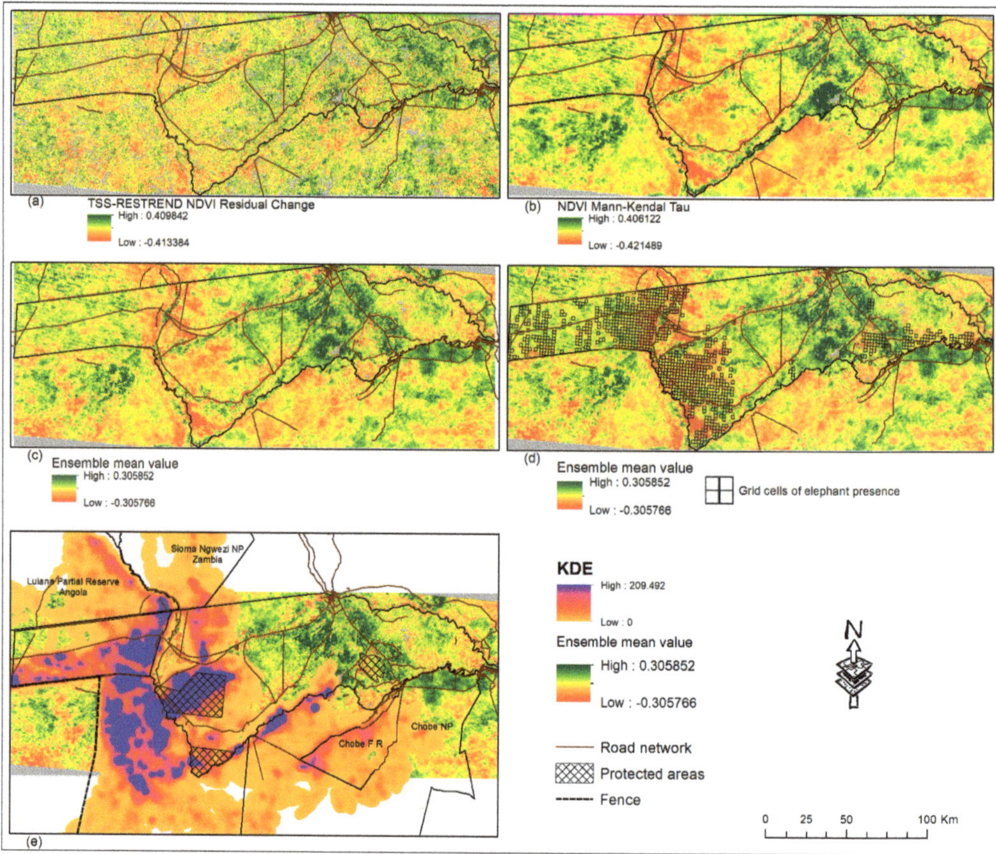

Figure 2. (**a–e**) show the output of the TSS.RESTREND, Mann–Kendall, ensemble, and kernel density estimate (KDE) analysis over the whole Zambesi region. In (**a**), the Residual Change of the TSS.RESTREND and (**b**) is the Mann–Kendall Tau. In (**c**) is the ensemble of the mean values of RESTREND, TSS.RESTREND, (**d**) the spatial distribution of elephant sightings overlayed over the ensemble mean value from (**b**) above and (**e**) the overlay of the Kernel Density Estimate (KDE), and Mann-Kendal Tau from (**b**) above. No data and background values are displayed as grey.

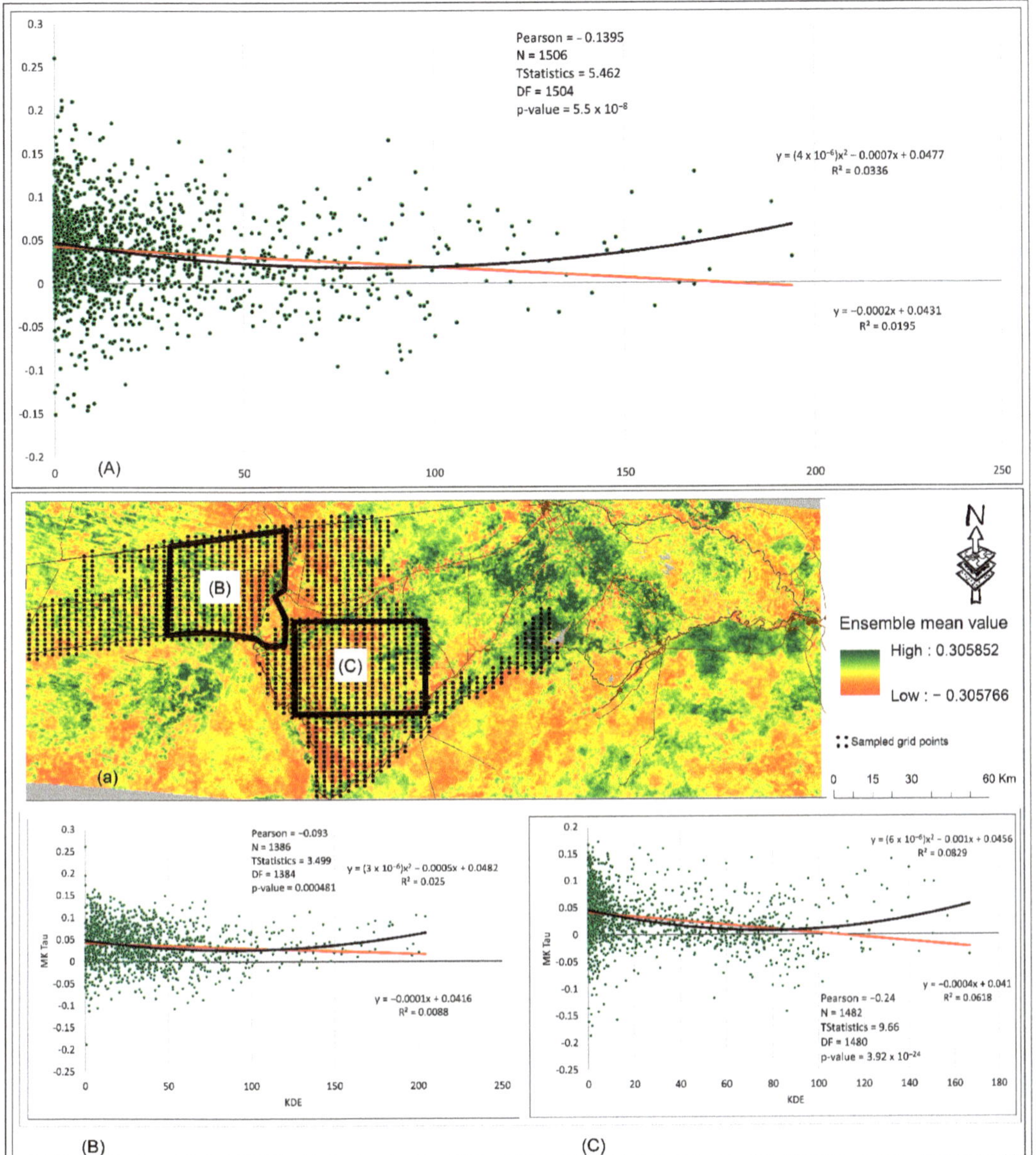

Figure 3. Shows the locations of the sampled grid points in the Zambezi region of the Kernel Density Estimate and the ensemble. The grid points are overlayed on the ensemble (a) used previously in Figure 2. The results of the Pearson correlation in (**A**) correspond to all the sampled points in the Zambezi while (**B**), and (**C**) correspond to the sub-sampled areas (black polygons) above.

3. Results

Pattern of Vegetation Trend: 2002–2021

Based on remote sensing and GIS data, this study analyzed the human–wildlife interaction in the Zambezi region. Figure 2 shows the pixel-by-pixel vegetation change pattern in the Zambezi region during the period 2002–2020. The TSS-RESTREND residual change

(Figure 2a) highlights a mixed pattern of positive (greening) and negative (browning) vegetation changes. Similarly, the ensemble and Mann–Kendall Tau (Figure 2b,c) both clearly show a mixed pattern of positive and negative values. The positive and negative pixel values indicate vegetation change increase and decrease, respectively. The observed negative pattern of the TSS-RESTREND and RESTREND is attributed to factors other than climatic aspects because the variability associated with climate was removed during the analysis. Most of the browning pixels (land degradation) are along the Kwando River (Figures 2a, 3 and 4A), which is a major migration route for wildlife, specifically elephants [25]. On the eastern part of the Kwando River are two National Parks Mudumu and Nkasa Lupala, and six conservancies (Kwandu, Mayuni, Mashi, Balyerwa, Wuparo, and Malengalenga), (Figures 1, 2a and 3) [25]. On the western part of the Kwando River is the Bwabwata National Park, a part of the home range for large herds of elephants [32], and also contains a buffalo core area [25]. Most of the vegetation browning we observed was also taking place within Mudumu and Nkasa Lupala National Parks (Figure 2). Within Bwabwata National Park (west of the Kwando River), browning is mainly close to and along the Kwando River, while elsewhere most of the pixels show greening (Figure 2).

Figure 3 shows the Pearson correlation between the Mann–Kendall Tau and the Kernel Density Estimate of the elephant data. Over the whole Zambezi (Figure 3A), as well as in the western (Figure 3B) and eastern areas (Figure 3C), there is a clear negative trend (browning) indicated by the red straight line. In the whole Zambezi area (Figure 3A) and in both sub-sampled locations (Figure 3B,C), the negative trend (browning) is significant (p-values = 5.5×10^{-8}, 0.0005, and 3.93×10^{-24}). Additionally, the results (Figure 3A–C) show that as the density of elephants decreases away from the KDE core areas, vegetation greening begins to occur (black polynomial line, Figure 3A–C). It is noteworthy that the polynomial line increase related to high KDE densities probably indicates that large herds are attracted to abundant vegetation resources. Moving further to the eastern part, we observed browning around and within the Salambala core area, located east of Lake Liambezi (Figure 2c). The Salambala core area is also home to elephant herds (Figure 2e). In addition, we also observed browning close to and around roads (Figures 2–4). These roads are locations of high-density human settlements where cattle grazing and extensive agricultural activities are a common practice. On the other hand, we also observed some relatively high greening, mostly around Lake Liambezi, communal areas, as well as some locations of the floodplains (Figures 2–4).

The browning pattern we observed is not only confined to the Zambezi region but extends across the border into neighboring countries (Figure 2). For example, the browning along the Kwando River continues across the Namibian border into Luina Partial Reserve (Angola) and Sioma Ngwwezi National Park (Zambia). We also observed a similar pattern across the border into Botswana, where we can see a clear difference along the 135 km veterinary fence which was constructed between 1991 and 1997 [25], (Figure 2c,e). On the eastern side of the fence, we observed high levels of degradation, while to the west we see relatively high greening values (Figure 2c).

We also ran the Mann–Kendall test on temperature and precipitation but did not see any significant trend, so the result was not shown here. An enlarged graphics of the Mann–Kendall NDVI Tau is shown in Figure 5 (Additional resources).

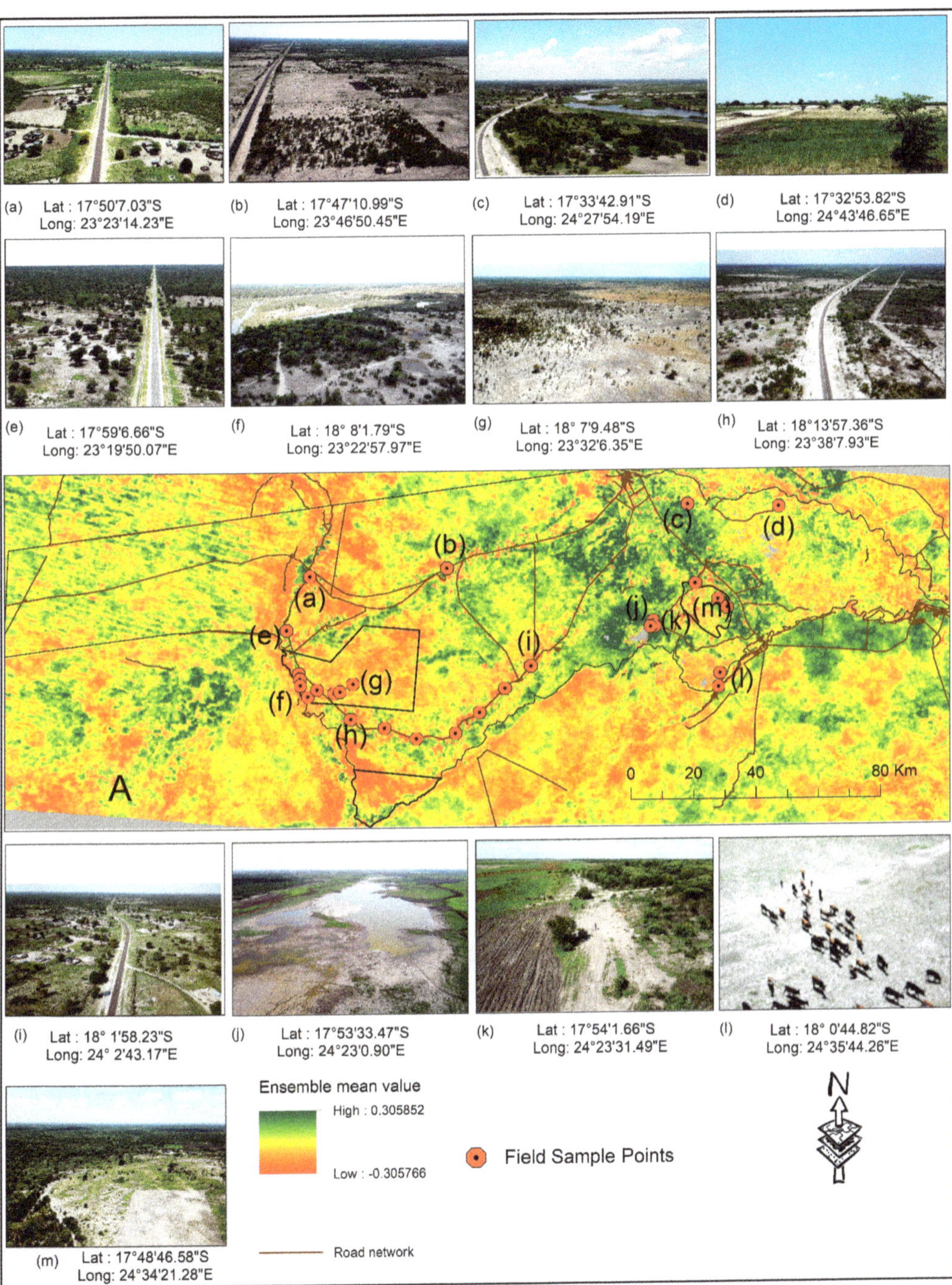

Figure 4. Shows aerial images from our 2019 field survey in the Zambezi region (**a–m**). In (A), the labels (**a–m**) correspond to the images shown with their respective GPS coordinates at the sample locations (red circle with black dot in the middle). Additionally, in (A), we use the same ensemble as in Figure 2b above for reference purposes.

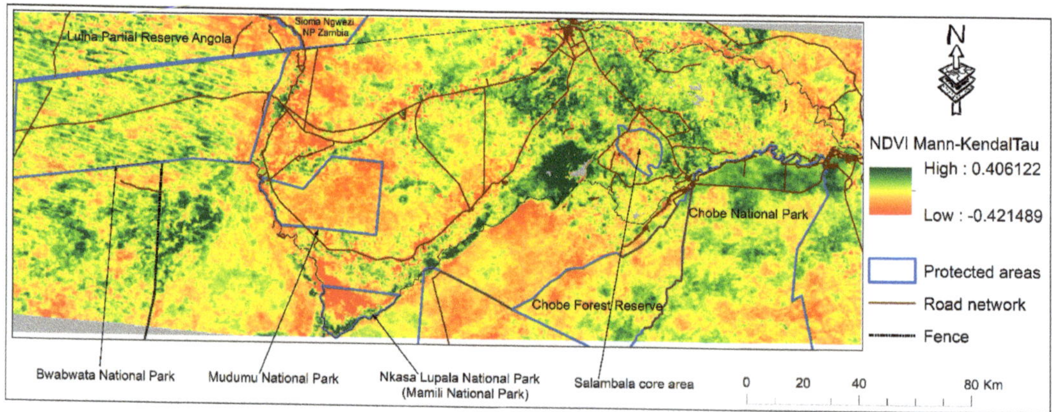

Figure 5. Additional resources.

4. Discussion

Potential Impact of Wildlife on Vegetation Cover and Land Degradation: 2002–2021

This study found that human–wildlife interaction is driving vegetation cover change in the Zambesi region. Specifically, anthropogenic restriction of space and resources for wildlife is driving the observed accelerated vegetation removal by large wildlife herbivores, in this case elephants. Consequently, this interaction is a potential contributor to land degradation in the Zambezi region. Previous studies in the Zambezi region by Gbagir et al. 2019 [24] revealed that land degradation is driven by the interaction of multiple direct and indirect factors. These factors include: demographic [24,27], ecological (e.g., floods) [27,85], and environmental factors (e.g., topography) [24,28]. Specifically, subsistence farming, infrastructure expansion (e.g., roads), including settlements, and legal and illegal wood extraction for firewood were identified as the main drivers of land degradation in the region [24]. However, the studies by Gbagir et al. 2019 [24] were not able to clearly establish more specific causes for land degradation due to the nature of the data used [24].

In the present study, a more detailed pattern and trend of vegetation cover change was revealed and the impact of wildlife on land degradation was clearly established. The contribution of wildlife on land degradation in the region corresponds to previous base line studies and ongoing statistics from the region. Reports indicate that elephant populations in the Zambezi region are stable or growing [51,86]. In addition to elephants, other wildlife populations, e.g., buffalo, are present (Figure 2d), but since elephants clearly damage trees, our results and discussion are focused on these.

Baseline studies on the presence of elephants in Namibia were carried out in the early 1980s and 1990s [32,87]. Based on these studies, several subsequent surveys have shown that the population of elephants has been increasing steadily in Namibia [25,48] from 600 to 1000 in 1934 to 22,754 in 2016 [48]. Probably the elephant population can be assumed to be even larger in 2022. The Zambezi region (Namibian KAZA) hosts most of the elephants in Namibia and is the most important migration route [30,48]. Reports show that the number of elephants within the Zambezi region has more than tripled since 1995/1996 [48]. A current estimate of elephants in the region is reported as 12008 [88]. According to Chase and Griffin (2009) [25], the elephant population in the Zambezi region was 1468 in 1989, while in 1994, these figures were reported as 5804 [25], 7950 and 5556 [87,88].The differences between the figures are mainly due to the different sampling techniques used in those studies [25,88]. By 1998, these figures were down from 5804 to 4576 [25], but in 2005 the numbers had again increased to 5242 [89]. The decline in the elephant population was thought to be the result of the civil unrest in Angola [25,89] and the construction of the veterinary fence between Namibia and Botswana in western Zambezi [25,48,76], consequently restricting and cutting off the migration of wildlife (elephants, buffalos, wildebeests, zebras, etc.) [25,48,76].

Most of the browning we observed was along the Kwando River (Figure 2), an important migration corridor for elephants [32]. The negative impact (browning) on the surrounding vegetation is clearly visible compared to elsewhere in the Zambezi region, as elephants browse on and de-back and break down trees, causing structural changes to the surrounding vegetation [47,90]. The Kwando River and the Mudumu and Nkasa Lupala parks contain medium to high numbers of elephants [32]. The studies by O'Connell-Rodwell et al. 2000 [32] put the number of elephants present on the western Kwando River as 3000, while 400 and 600 were reported for Mudumu and Nkasa Lupala, respectively. Within Bwabwata national park (west of the Kwando River), browning is limited to the riverbanks and migration route, while elsewhere, there is less indication of high levels of browning. We also observed a similar pattern of browning within the Salambala core area, while most of the greening is occurring elsewhere in the area (Figure 2c). A 2019 survey of elephants reported the current population in the Salambala conservancy as 507 [88].

Other important factors contributing to land degradation in the Zambezi region are of anthropogenic origin, as has been established by previous studies [24]. However, in this study, we now see clearly how these anthropogenic activities have contributed to wildlife-induced vegetation browning in the region. The expansion of settlements and roads and the construction of artificial barriers (e.g., fences) has diminished the habitat of elephants and reduced their access to food and water resources [25,45,47,91]. As a result, more pressure is put on the remaining resources, propagating vegetation browning in the region [25,45,47,91]. Additionally, the shrinking of elephants' habitats has modified the behavioral pattern of these animals [46] and increased human–wildlife conflicts [53,66]. Most of the visible browning was observed within the protected areas (e.g., Mudumu, Nkasa Lupala, Salambala core) due to the high density of elephants concentrated within small restricted areas. Restricting the habitat of wildlife, specifically elephants, has impacted and contributed to the observed browning [45,91]. The Pearson correlation results (Figure 3A,B) also confirmed that the concentration of elephants within a certain restricted area was contributing to the observed gradual browning. However, the right side of the curve (Figure 3A,B) show that areas of vegetation browning do not sustain high animal populations; hence, the curve rises as the green areas attract large herds, thus indicating that the available space and resources may be beyond the carrying capacity of the current number of elephants with the protected areas. Additionally, taking into consideration that the arenosols soils are poor in nutrients [77,78], any slight modification in the vegetation cover will have a visible impact, which in this case is browning.

In addition to the elephant populations in this area, along the Kwando River there are also six conservancies (Kwandu, Mayuni, Mashi, Balyerwa, Wuparo, and Malengalenga) [25]. The presence of these settlements has given rise to clearing of land for farming and cattle grazing and increases in the road network. The presence of human settlements increases human–wildlife competition for land resources, making these areas hotspots for human–wildlife conflicts. Just like the six conservancies along the Kwando River, the Salambala core is also surrounded by several villages where farming and cattle grazing is a common practice [24,29,92,93], which could explain the high levels of vegetation browning (Figure 2c).

Apart from browning, there is also greening within the Zambezi region, most of the greening is occurring within the communal areas of Lake Liambezi and the Chobe River floodplains farther east. These greening areas have high human population densities and are locations of intensive farming and cattle grazing [48]. This we were able to verify during our field visit in December 2019 (Figure 4).

The greening of the Lake Liambezi is mostly due to the present drying (Figure 4j) which opens land for vegetation growth, farming, and grazing activities (Figure 4k). Large numbers of cattle are grazing on the eastern floodplains (Figure 4l) [48]. Recent reports in 2019 estimated the total number of cattle in the Zambezi region to be 135,878 animals [88].

The pattern of vegetation browning observed along the Kwando migratory route continues into neighboring Zambia (Sioma Ngwezi National Park) and Angola (Liuana

Partial Reserve) (Figures 1 and 2). Additionally, this pattern of vegetation browning applies to the southern border with Botswana around the Chobe National Park (Figures 2–4). The Chobe National Park is where the majority of the 200,000 migratory elephant population is located [24,94,95]. Furthermore, it is quite clear that greening is mainly on the opposite side (south) of the northern buffalo fence, where access by elephants is restricted [9,35] (Figure 2c,e) [59].

Similar studies elsewhere have also linked elephants to the loss of vegetation in protected areas. Examples include: Samburu and Buffalo Springs National Reserves [39], Aberdare National Park [44] in Kenya, Addo Elephant National Park, Eastern Cape, South Africa [38], and the Serengeti National Park, Tanzania [40,42]. We anticipate that the results of this study will provide increased understanding of the interaction between wildlife and land degradation in the Zambezi region. This new additional information could potentially improve and inform policy formulation and decision-making regarding wildlife and natural resources conservation and management in the region and elsewhere in Namibia.

Future studies should consider detailed and micro-analysis, classification and mapping of vegetation cover change by combining and integrating higher resolution remote sensing datasets [56,61,63,96–98]. This form of information will provide even better data to improve the current integrated sustainable land use and management practices.

5. Conclusions

This study assessed the impact of wildlife populations, specifically elephants, on vegetation browning in the Zambezi region during the last 19 years (2002–2021). Our analysis reveals that vegetation browning is mostly in locations with a high density of elephants. Most of the browning is along the migration corridor of elephants within national parks and conservation areas as a result of exclusion and harassment in areas with human settlements. Obviously, brown vegetation areas do not sustain high population densities of animals.

We also found that the expansion of settlements and the construction of artificial barriers (e.g., fences) has affected the movement and migration pattern (behavior) of wildlife populations, specifically elephants, in the region, which has led to the concentration of game animals within confined national parks.

Furthermore, the limited amount of space and resources for wildlife populations could potentially be a major factor contributing to vegetation browning in the region. This assumption is supported by the high incidence of ongoing human–wildlife conflicts within the region. On the other hand, our study found that most of the greening was occurring in areas with intensive farming; for example, around the shrinking Lake Liambezi, and within communal areas.

Author Contributions: Conceptualization, A.-M.G.G. and A.C.; methodology, A.-M.G.G. and A.C.; software, A.-M.G.G. and A.C.; formal analysis, A.-M.G.G. and A.C.; investigation, A.-M.G.G. and A.C.; resources, A.C.; data curation, A.-M.G.G., K.K.M., C.S.S. and A.C.; writing—original draft preparation, A.-M.G.G.; writing—review and editing, A.-M.G.G., K.K.M., C.S.S. and A.C.; visualization, A.-M.G.G. and A.C.; supervision, A.C.; project administration, A.C.; funding acquisition, A.C. All authors have read and agreed to the published version of the manuscript.

Funding: We acknowledge the Government of Namibia (Ministry of Environment, Tourism and Forestry), especially the, University of Eastern Finland and the Kone Foundation for funding. CSC–IT Centre for Science, Finland (urn:nbn:fi:research-infras-2016072531) and the Open Geospatial Information Infrastructure for Research (Geoportti, urn:nbn:fi:research-infras-2016072513) for computational resources and support.

Institutional Review Board Statement: Not applicable.

Informed Consent Statement: Not applicable.

Data Availability Statement: Not applicable.

Conflicts of Interest: The authors declare no conflict of interest.

References

1. Wessels, K.J.; Prince, S.D.; Malherbe, J.; Small, J. Can Human-Induced Land Degradation be Distinguished from the Effects of Rainfall Variability ? A Case Study in South Africa. *J. Arid Environ.* **2007**, *68*, 271–297. [CrossRef]
2. IPBES. Summary for Policymakers of the Thematic Assessment Report on Land Degradation and Resporation of the Intergovernmental Platform on Biodiversity and Ecosystem Services. *J. Rural. Plan. Assoc.* **2018**, *36*, 13–16.
3. Song, C.; Kim, W.; Kim, J.; Gebru, B.M.; Adane, G.B.; Choi, Y.E.; Lee, W. Spatial Assessment of Land Degradation using MEDALUS Focusing on Potential Afforestation and Reforestation Areas in Ethiopia. *Land Degrad. Dev.* **2022**, *33*, 79–93. [CrossRef]
4. Lewińska, K.E.; Hostert, P.; Buchner, J.; Bleyhl, B.; Radeloff, V.C. Short-Term Vegetation Loss Versus Decadal Degradation of Grasslands in the Caucasus Based on Cumulative Endmember Fractions. *Remote Sens. Environ.* **2020**, *248*, 111969. [CrossRef]
5. Barbier, E.B.; Hochard, J.P. Land Degradation and Poverty. *Nat. Sustain.* **2018**, *1*, 623–631. [CrossRef]
6. Jiang, L.; Jiapaer, G.; Bao, A.; Li, Y.; Guo, H.; Zheng, G.; Chen, T.; De Maeyer, P. Assessing Land Degradation and Quantifying its Drivers in the Amudarya River Delta. *Ecol. Indic.* **2019**, *107*, 105595. [CrossRef]
7. Nkonya, E.; Srinivasan, R.; Anderson, W.; Kato, E. Economics of Land Degradation and Improvement in Bhutan. In *Economics of Land Degradation and Improvement—A Global Assessment for Sustainable Development*; Springer International Publishin AG: Cham, Switzerland, 2015; pp. 327–383.
8. Sklenicka, P. Classification of Farmland Ownership Fragmentation as a Cause of Land Degradation: A Review on Typology, Consequences, and Remedies. *Land Use Policy* **2016**, *57*, 694–701. [CrossRef]
9. Bossio, D.; Geheb, K.; Critchley, W. Managing Water by Managing Land: Addressing Land Degradation to Improve Water Productivity and Rural Livelihoods. *Agric. Water Manag.* **2010**, *97*, 536–542. [CrossRef]
10. Prăvălie, R.; Patriche, C.; Bandoc, G. Quantification of Land Degradation Sensitivity Areas in Southern and Central Southeastern Europe. New Results Based on Improving DISMED Methodology with New Climate Data. *Catena* **2017**, *158*, 309–320. [CrossRef]
11. Röder, A.; Pröpper, M.; Stellmes, M.; Schneibel, A.; Hill, J. Assessing Urban Growth and Rural Land use Transformations in a Cross-Border Situation in Northern Namibia and Southern Angola. *Land Use Policy* **2015**, *42*, 340–354. [CrossRef]
12. De Blécourt, M.; Röder, A.; Gröngröft, A.; Baumann, S.; Frantz, D.; Eschenbach, A. Deforestation for Agricultural Expansion in SW Zambia and NE Namibia and the Impacts on Soil Fertility, Soil Organic Carbon- and Nutrient Levels. *Biodivers. Ecol.* **2018**, *6*, 242–250. [CrossRef]
13. Batunacun; Wieland, R.; Lakes, T.; Yunfeng, H.; Nendel, C. Identifying Drivers of Land Degradation in Xilingol, China, between 1975 and 2015. *Land Use Policy* **2019**, *83*, 543–559. [CrossRef]
14. Karamesouti, M.; Detsis, V.; Kounalaki, A.; Vasiliou, P.; Salvati, L.; Kosmas, C. Land-use and Land Degradation Processes Affecting Soil Resources: Evidence from a Traditional Mediterranean Cropland (Greece). *Catena* **2015**, *132*, 45–55. [CrossRef]
15. Perovic, V.; Kadovic, R.; Durdevic, V.; Pavlovic, D.; Pavlovic, M.; Cakmak, D.; Mitrovic, M.; Pavlovic, P. Major Drivers of Land Degradation Risk in Western Serbia: Current Trends and Future Scenarios. *Ecol. Indic.* **2021**, *123*, 107377. [CrossRef]
16. Geist, H.J.; Lambin, E.F. Proximate Causes and Underlying Driving Forces of Tropical Deforestation. *Bioscience* **2002**, *52*, 143. [CrossRef]
17. Tegegne, Y.T.; Lindner, M.; Fobissie, K.; Kanninen, M. Evolution of Drivers of Deforestation and Forest Degradation in the Congo Basin Forests: Exploring Possible Policy Options to Address Forest Loss. *Land Use Policy* **2016**, *51*, 312–324. [CrossRef]
18. Brink, A.B.; Bodart, C.; Brodsky, L.; Defourney, P.; Ernst, C.; Donney, F.; Lupi, A.; Tuckova, K. Anthropogenic Pressure in East Africa—Monitoring 20 Years of Land Cover Changes by Means of Medium Resolution Satellite Data. *ITC J.* **2014**, *28*, 60–69. [CrossRef]
19. NAPCOD. Third National Action Programme for Namibia to Implement the United Nations Convention to Combat Desertification 2014–2024. 2014, pp. 1–80. Available online: https://www.unccd.int/sites/default/files/naps/Namibia-2014-2024-eng.pdf (accessed on 8 May 2022).
20. Bojö, J. The Costs of Land Degradation in Sub-Saharan Africa. *Ecol. Econ.* **1996**, *16*, 161–173. [CrossRef]
21. Hoffman, M.T.; Todd, S. A National Review of Land Degradation in South Africa: The Influence of Biophysical and Socio-Economic Factors. *J. South. Afr. Stud.* **2000**, *26*, 743–758. [CrossRef]
22. Ibrahim, Y.Z.; Balzter, H.; Kaduk, J.ö.; Tucker, C.J. Land Degradation Assessment using Residual Trend Analysis of GIMMS NDVI3g, Soil Moisture and Rainfall in Sub-Saharan West Africa from 1982 to 2012. *Remote Sens.* **2015**, *7*, 5471–5494. [CrossRef]
23. Weinzierl, T.; Wehberg, J.; Böhner, J.; Conrad, O. Spatial Assessment of Land Degradation Risk for the Okavango River Catchment, Southern Africa. *Land Degrad. Dev.* **2016**, *27*, 281–294. [CrossRef]
24. Gbagir, A.M.G.; Tegegne, Y.T.; Colpaert, A. Historical Trajectory in Vegetation Cover in Northeastern Namibia Based on AVHRR Satellite Imagery (1982–2015). *Land* **2019**, *8*, 160. [CrossRef]
25. Chase, M.J.; Griffin, C.R. Elephants Caught in the Middle: Impacts of War, Fences and People on Elephant Distribution and Abundance in the Caprivi Strip, Namibia. *Afr. J. Ecol.* **2009**, *47*, 223–233. [CrossRef]
26. Dirkx, E.; Hager, C.; Tadross, M.; Bethune, S.; Curtis, B. *Climate Change Vulnerability & Adaptation Assessment Namibia Final Report*; Developed by Desert Research Foundation of Namibia & Climate Systems Analysis Group for the Ministry of Environment and Tourism; Desert Research Foundation of Namibia Climate System Analysis Group: Windhoek, Namibia, 2008; pp. 1–167.
27. Kamwi, J.M.; Chirwa, P.W.C.; Manda, S.O.M.; Graz, P.F.; Kätsch, C. Livelihoods, Land use and Land Cover Change in the Zambezi Region, Namibia. *Popul. Environ.* **2015**, *37*, 207–230. [CrossRef]

28. Kamwi, J.M.; Kaetsch, C.; Graz, F.P.; Chirwa, P.; Manda, S. Trends in Land use and Land Cover Change in the Protected and Communal Areas of the Zambezi Region, Namibia. *Environ. Monit. Assess.* **2017**, *189*, 242. [CrossRef]

29. Wingate, V.R.; Phinn, S.R.; Kuhn, N.; Bloemertz, L.; Dhanjal-Adams, K. Mapping Decadal Land Cover Changes in the Woodlands of North Eastern Namibia from 1975 to 2014 using the Landsat Satellite Archived Data. *Remote Sens.* **2016**, *8*, 681. [CrossRef]

30. Purdon, A.; Mole, M.A.; Chase, M.J.; van Aarde, R.J. Partial Migration in Savanna Elephant Populations Distributed Across Southern Africa. *Sci. Rep.* **2018**, *8*, 11331. [CrossRef]

31. Stoldt, M.; Göttert, T.; Mann, C.; Zeller, U. Transfrontier Conservation Areas and Human-Wildlife Conflict: The Case of the Namibian Component of the Kavango-Zambezi (KAZA) TFCA. *Sci. Rep.* **2020**, *10*, 7964. [CrossRef]

32. O'connell-Rodwell, C.E.; Rodwell, T.; Rice, M.; Hart, L.A. Living with the Modern Conservation Paradigm: Can Agricultural Communities Co-Exist with Elephants? A Five-Year Case Study in East Caprivi, Namibia. *Biol. Conserv.* **2000**, *93*, 381–391. [CrossRef]

33. Thornley, R.; Spencer, M.; Zitzer, H.R.; Parr, C.L. Woody Vegetation Damage by the African Elephant during Severe Drought at Pongola Game Reserve, South Africa. *Afr. J. Ecol.* **2020**, *58*, 658. [CrossRef]

34. Mwambeo, H.M.; Maitho, T. Factors Influencing Elephants to Destroy Forest Trees especially Olea Africana: The Case of Ngare Ndare Forest Reserve in Meru County, Kenya. *Ethiop. J. Environ. Stud. Manag.* **2015**, *8*, 398–407. [CrossRef]

35. Guldemond, R.; Van Aarde, R. A Meta-Analysis of the Impact of African Elephants on Savanna Vegetation. *J. Wildl. Manag.* **2008**, *72*, 892–899. [CrossRef]

36. Robinson, J.A.; Lulla, K.P.; Kashiwagi, M.; Suzuki, M.; Nellis, M.D.; Bussing, C.E.; Long, W.J.L.; McKenzie, L.J. Conservation Applications of Astronaut Photographs of Earth: Tidal-Flat Loss (Japan), Elephant Effects on Vegetation (Botswana), and Seagrass and Mangrove Monitoring (Australia). *Conserv. Biol.* **2001**, *15*, 876–884. [CrossRef]

37. Midgley, J.J.; Balfour, D.; Kerley, G.I. Why do Elephants Damage Savanna Trees? *S. Afr. J. Sci.* **2005**, *101*, 213–215. [CrossRef]

38. Vincent, K.; Janis, S.; Graham, K. A Temporal Analysis of Elephant-Induced Thicket Degradation in Addo Elephant National Park, Eastern Cape, South Africa FULL TEXT Introduction. *Rangel. Ecol. Manag.* **2015**, *68*, 461–469. [CrossRef]

39. Ihwagi, F.W.; Vollrath, F.; Chira, R.M.; Douglas-Hamilton, I.; Kironchi, G. The Impact of Elephants, *Loxodonta africana*, on Woody Vegetation through Selective Debarking in Samburu and Buffalo Springs National Reserves, Kenya. *Afr. J. Ecol.* **2010**, *48*, 87–95. [CrossRef]

40. Rugemalila, D.M.; Anderson, T.M.; Holdo, R.M. Precipitation and Elephants, Not Fire, Shape Tree Community Composition in Serengeti National Park, Tanzania. *Biotropica* **2016**, *48*, 476–482. [CrossRef]

41. Trollope, W.S.W.; Trollope, L.A.; Biggs, H.C.; Pienaar, D.; Potgieter, A.L.F. Long-Term Changes in the Woody Vegetation of the Kruger National Park, with Special Reference to the Effects of Elephants and Fire. *Koedoe* **1998**, *2*, 103–112.

42. Holdo, R.M.; Holt, R.D.; Fryxell, J.M. Grazers, Browsers, and Fire Influence the Extent and Spatial Pattern of Tree Cover in the Serengeti. *Ecol. Appl.* **2009**, *19*, 95–109. [CrossRef]

43. Morrison, T.A.; Holdo, R.M.; Anderson, T.M.; Gilliam, F. Elephant Damage, Not Fire Or Rainfall, Explains Mortality of Overstorey Trees in Serengeti. *J. Ecol.* **2016**, *104*, 409–418. [CrossRef]

44. Morrison, J.; Higginbottom, T.; Symeonakis, E.; Jones, M.; Omengo, F.; Walker, S.; Cain, B. Detecting Vegetation Change in Response to Confining Elephants in Forests using MODIS Time-Series and BFAST. *Remote Sens.* **2018**, *10*, 1075. [CrossRef]

45. Cassidy, L.; Fynn, R.; Sethebe, B. Effects of Restriction of Wild Herbivore Movement on Woody and Herbaceous Vegetation in the Okavango Delta Botswana. *Afr. J. Ecol.* **2013**, *51*, 513–527. [CrossRef]

46. Buchholtz, E.K.; Spragg, S.; Songhurst, A.; Stronza, A.; Mcculloch, G.; Fitzgerald, L.A. Anthropogenic Impact on Wildlife Resource use: Spatial and Temporal Shifts in Elephants' Access to Water. *Afr. J. Ecol.* **2021**, *59*, 614. [CrossRef]

47. Watson, L.H.; Cameron, M.J.; Iifo, F. Elephant Herbivory of Knob-thorn (*Senegalia nigrescens*) and Ivory Palm (*Hyphaene petersiana*) in Bwabwata National Park, Caprivi, Namibia: The Role of Ivory Palm as a Biotic Refuge. *Afr. J. Ecol.* **2020**, *58*, 14–22. [CrossRef]

48. Colpaert, A.; Matengu, K.; Polojärvi, K. Land use Practices in Caprivi 's Changing Political Environment. *J. Stud. Hum. Soc. Sci.* **2013**, *2*, 141–162.

49. Mlambo, L.; Shekede, M.D.; Adam, E.; Odindi, J.; Murwira, A. Home Range and Space use by African Elephants (*Loxodonta africana*) in Hwange National Park, Zimbabwe. *Afr. J. Ecol.* **2021**, *59*, 842. [CrossRef]

50. Bakker, E.S.; Gill, J.L.; Johnson, C.N.; Vera, F.W.M.; Sandom, C.J.; Asner, G.P.A.; Svenning, J.C. Combining Paleo-Data and Modern Exclosure Experiments to Assess the Impact of Megafauna Extinctions on Woody Vegetation. *Proc. Natl. Acad. Sci. USA* **2016**, *113*, 847–855. [CrossRef]

51. Meyer, M.; Klingelhoeffer, E.; Naidoo, R.; Wingate, V.; Börner, J. Tourism Opportunities Drive Woodland and Wildlife Conservation Outcomes of Community-Based Conservation in Namibia's Zambezi Region. *Ecol. Econ.* **2021**, *180*. [CrossRef]

52. Klintenberg, P.; Seely, M. Land Degradation Monitoring in Namibia: A First Approximation. *Environ. Monit. Assess.* **2004**, *99*, 5–21. [CrossRef]

53. Schnegg, M.; Kiaka, R.D. Subsidized Elephants: Community-Based Resource Governance and Environmental (in) Justice in Namibia. *Geoforum* **2018**, *93*, 105–115. [CrossRef]

54. De Jong, R.; Verbesselt, J.; Zeileis, A.; Schaepman, M. Shifts in Global Vegetation Activity Trends. *Remote Sens.* **2013**, *5*, 1117–1133. [CrossRef]

55. Heck, E.; de Beurs, K.M.; Owsley, B.C.; Henebry, G.M. Evaluation of the MODIS Collections 5 and 6 for Change Analysis of Vegetation and Land Surface Temperature Dynamics in North and South America. *ISPRS J. Photogramm. Remote Sens.* **2019**, *156*, 121–134. [CrossRef]

56. Fortin, J.A.; Cardille, J.A.; Perez, E. Multi-Sensor Detection of Forest-Cover Change across 45 years in Mato Grosso, Brazil. *Remote Sens. Environ.* **2020**, *238*, 111266. [CrossRef]
57. Moon, M.; Zhang, X.; Henebry, G.M.; Liu, L.; Gray, J.M.; Melaas, E.K.; Friedl, M.A. Long-Term Continuity in Land Surface Phenology Measurements: A Comparative Assessment of the MODIS Land Cover Dynamics and VIIRS Land Surface Phenology Products. *Remote Sens. Environ.* **2019**, *226*, 74–92. [CrossRef]
58. Xie, Y.; Sha, Z.; Yu, M. Remote Sensing Imagery in Vegetation Mapping: A Review. *J. Plant Ecol.* **2008**, *1*, 9–23. [CrossRef]
59. Basommi, P.L.; Guan, Q.; Cheng, D. Exploring Land use and Land Cover Change in Themining Areas of Wa East District, Ghana usingSatellite Imagery. *Open Geosci.* **2015**, *7*, 618–626. [CrossRef]
60. Ren, Y.; Li, Z.; Li, J.; Ding, Y.; Miao, X. Analysis of Land use/Cover Change and Driving Forces in the Selenga River Basin. *Sensors* **2022**, *22*, 1041. [CrossRef]
61. Afrin, S.; Gupta, A.; Farjad, B.; Ahmed, M.R.; Achari, G.; Hassan, Q.K. Development of Land-use/Land-Cover Maps using Landsat-8 and MODIS Data, and their Integration for Hydro-Ecological Applications. *Sensors* **2019**, *19*, 4891. [CrossRef]
62. Tang, X.; Woodcock, C.E.; Olofsson, P.; Hutyra, L.R. Spatiotemporal Assessment of Land use/Land Cover Change and Associated Carbon Emissions and Uptake in the Mekong River Basin. *Remote Sens. Environ.* **2021**, *256*, 112336. [CrossRef]
63. Xu, L.; Herold, M.; Tsendbazar, N.; Masiliūnas, D.; Li, L.; Lesiv, M.; Fritz, S.; Verbesselt, J. Time Series Analysis for Global Land Cover Change Monitoring: A Comparison across Sensors. *Remote Sens. Environ.* **2022**, *271*, 112905. [CrossRef]
64. Cowie, A.L.; Orr, B.J.; Castillo Sanchez, V.M.; Chasek, P.; Crossman, N.D.; Erlewein, A.; Louwagie, G.; Maron, M.; Metternicht, G.I.; Minelli, S.; et al. Land in Balance: The Scientific Conceptual Framework for Land Degradation Neutrality. *Environ. Sci. Policy* **2018**, *79*, 25–35. [CrossRef]
65. Warren, A. Land Degradation is Contextual. *Land Degrad. Dev.* **2002**, *13*, 449–459. [CrossRef]
66. Khumalo, K.E.; Yung, L.A. Women, Human-Wildlife Conflict, and CBNRM: Hidden Impacts and Vulnerabilities in Kwandu Conservancy, Namibia. *Conserv. Soc.* **2015**, *13*, 232–243. [CrossRef]
67. Detsch, F. *Download and Process GIMMS NDVI3g Data*; R Package Version 1.2.1; R Core Team: Vienna, Austria, 2021; pp. 1–18.
68. Ruan, Z.; Kuang, Y.; He, Y.; Zhen, W.; Ding, S. Detecting Vegetation Change in the Pearl River Delta Region Based on Time Series Segmentation and Residual Trend Analysis (TSS-RESTREND) and MODIS NDVI. *Remote Sens.* **2020**, *12*, 4049. [CrossRef]
69. Burrell, A.L.; Evans, J.P.; Liu, Y. Detecting Dryland Degradation using Time Series Segmentation and Residual Trend Analysis (TSS-RESTREND). *Remote Sens. Environ.* **2017**, *197*, 43–57. [CrossRef]
70. Wessels, K.J.; Prince, S.D.; Frost, P.E.; Van Zyl, D. Assessing the Effects of Human-Induced Land Degradation in the Former Homelands of Northern South Africa with a 1 km AVHRR NDVI Time-Series. *Remote Sens. Environ.* **2004**, *91*, 47–67. [CrossRef]
71. Evans, J.; Geerken, R. Discrimination between Climate and Human-Induced Dryland Degradation. *J. Arid Environ.* **2004**, *57*, 535–554. [CrossRef]
72. Ashraf, M.S.; Ahmad, I.; Khan, N.M.; Zhang, F.; Bilal, A.; Guo, J. Streamflow Variations in Monthly, Seasonal, Annual and Extreme Values using Mann-Kendall, Spearmen's Rho and Innovative Trend Analysis. *Water Resour Manag.* **2020**, *35*, 243–261. [CrossRef]
73. Fan, X.; Xuan, C.; Zhang, M.; Ma, Y.; Meng, Y. Estimation of Spatial-Temporal Distribution of Grazing Intensity Based on Sheep Trajectory Data. *Sensors* **2022**, *22*, 1469. [CrossRef]
74. Mufune, P. Community Based Natural Resource Management (CBNRM) and Sustainable Development in Namibia. *J. Land Rural Stud.* **2015**, *3*, 121–138. [CrossRef]
75. NACSO. *Namibia's Communal Conservancies: A Review of Progress and Challenges in 2011*; NASCO: Windhoek, Namibia, 2013; p. 59.
76. Chase, M. Status of Wildlife Populations and Land Degradation in Botswana's Forest Reserves and Chobe District. 2013. Available online: https://library.wur.nl/ojs/index.php/Botswana_documents/article/view/15966 (accessed on 5 April 2022).
77. Mendelsohn, J.; Jarvis, A.; Roberst, C.; Robertson, T. *Atlas of Namibia: A Portrait of the Land and Its People.*; Published for the Ministry of Environment and Tourism by David Philip: Cape Town, South Africa, 2002; pp. 1–201.
78. Mendelsohn, J.; Roberts, C. *An Environmental Profile and Atlas of Caprivi*; Directorate of Environment and Tourism Affairs, Ministry of Enviornment and Tourism: Windhoek, Namibia, 1997; pp. 1–57.
79. AWT. Africa Wildlife Tracking Tag User Manual:Version 02. (n.d). *AWT Tag User Manual.pdf.* Available online: africawildlifetracking.com (accessed on 5 April 2022).
80. Verbesselt, J.; Hyndman, R.; Newnham, G.; Culvenor, D. Detecting Trend and Seasonal Changes in Satellite Image Time Series. *Remote Sens. Environ.* **2010**, *114*, 106–115. [CrossRef]
81. Verbesselt, J.; Zeileis, A.; Hyndman, R. Breaks for Additive Season and Trend (BFAST). Technical Report. 2012. Available online: http://r-forge.r-project.org/projects/bfast (accessed on 5 April 2022).
82. Jong, R.; Verbesselt, J.; Schaepman, M.E.; Bruin, S. Trend Changes in Global Greening and Browning: Contribution of Short-Term Trends to Longer-Term Change. *Global Change Biol.* **2012**, *18*, 642–655. [CrossRef]
83. Mann, H.B. Nonparametric Tests Against Trend. *Mann Source: Econom.* **1945**, *13*, 245–259. [CrossRef]
84. Hamed, K.H. Trend Detection in Hydrologic Data: The Mann–Kendall Trend Test Under the Scaling Hypothesis. *J. Hydrol.* **2008**, *349*, 350–363. [CrossRef]
85. Long, S.; Fatoyinbo, T.E.; Policelli, F. Flood Extent Mapping for Namibia using Change Detection and Thresholding with SAR. *Environ. Res. Lett.* **2014**, *9*, 35002. [CrossRef]

86. Thouless, C.R.; Dublin, H.T.; Blanc, J.J.; Skinner, D.P.; Daniel, T.E.; Taylor, R.D.; Maisels, F.; Frederick, H.L.; Bouché, P. *African Elephant Status Report 2007: An Update from the African Elephant Database*; Occassional Paper Series of the IUCN Species Survival Commission, no.60 IUCN/SSC Africa Elephant Specialist Group; IUCN: Gland, Switzerland, 2016.

87. Rodwell, T. Wildlife Resources in the Caprivi, Namibia: The Results of an Aerial Census in 1994 and Comparisons with Past Surveys; Research Discussion Paper Number 9. 1995. Available online: https://www.researchgate.net/publication/267681244_Wildlife_Resources_in_the_Caprivi_Namibia_The_Results_of_an_Aerial_Census_in_1994_and_Comparisons_with_Past_Surveys (accessed on 5 April 2022).

88. Craig, G.C.; St, D.C.; Gibson. *Aerial Survey of North-East Namibia- Elephants and Other Wildlife in Zambezi Region September/October 2019*; Ministry of Environment & Tourism: Windhoek, Namibia, 2019.

89. Chase, M.J.; Griffin, C.R. Elephants of South-East Angola in War and Peace: Their Decline, Re-Colonization and Recent Status. *Afr. J. Ecol.* **2011**, *49*, 353–361. [CrossRef]

90. Mapaure, I.; Moe, S.R. Changes in the Structure and Composition of Miombo Woodlands Mediated by Elephants (*Loxodonta africana*) and Fire Over a 26-Year Period in North-Western Zimbabwe. *Afr. J. Ecol.* **2009**, *47*, 175–183. [CrossRef]

91. Fullman, T.J.; Child, B. African Journal of Ecology—2012—Fullman—Water Distribution at Local and Landscape Scales Affects Tree Utilization By. *Afr. J. Ecol.* **2012**, *51*, 235–243. [CrossRef]

92. Kamwi, J.M.; Chirwa, P.; Graz, F.P.; Manda, S.; Mosimane, A.W.; Kätsch, C. Livelihood Activities and Skills in Rural Areas of the Zambezi Region, Namibia: Implications for Policy and Poverty Reduction. *Afr. J. Food Agric. Nutr. Dev.* **2018**, *18*, 13074–13094. [CrossRef]

93. Laamanen, R.; Michael, O. *Ministry of Environment and Tourism Directorate of Forestry Forest Management Plan for the Salambala Conservancy Core Area Namibia-Finland Forestry Programme Forest Management Plan for the Salambala Conservancy Core Area-Draft*; Ministry of Environment and Tourism Directorate of Forestry: Windhoek, Namibia, 2002; pp. 1–53.

94. Pricope, N.G. Variable-Source Flood Pulsing in a Semi-Arid Transboundary Watershed: The Chobe River, Botswana and Namibia. *Environ. Monit. Assess.* **2013**, *185*, 1883–1906. [CrossRef]

95. Skarpe, C.; Aarrestad, P.A.; Andreassen, H.P.; Dhillion, S.S.; Dimakatso, T.; du Toit, J.T.; Duncan; Halley, J.; Hytteborn, H.; Makhabu, S.; et al. The Return of the Giants: Ecological Effects of an Increasing Elephant Population. *Ambio* **2004**, *33*, 276–282. [CrossRef] [PubMed]

96. Herrero, H.V.; Southworth, J.; Bunting, E.; Kohlhaas, R.R.; Child, B. Integrating Surface-Based Temperature and Vegetation Abundance Estimates into Land Cover Classifications for Conservation Efforts in Savanna Landscapes. *Sensors* **2019**, *19*, 3456. [CrossRef]

97. Bunting, E.; Southworth, J.; Herrero, H.; Ryan, S.; Waylen, P. Understanding Long-Term Savanna Vegetation Persistence across Three Drainage Basins in Southern Africa. *Remote Sens.* **2018**, *10*, 1013. [CrossRef]

98. Nguyen, L.H.; Joshi, D.R.; Clay, D.E.; Henebry, G.M. Characterizing Land Cover/Land use from Multiple Years of Landsat and MODIS Time Series: A Novel Approach using Land Surface Phenology Modeling and Random Forest Classifier. *Remote Sens. Environ.* **2020**, *238*, 111017. [CrossRef]

Article

Soil Salinity Inversion of Winter Wheat Areas Based on Satellite-Unmanned Aerial Vehicle-Ground Collaborative System in Coastal of the Yellow River Delta

Guanghui Qi [1,2], Gengxing Zhao [2,*] and Xue Xi [2]

[1]   College of Information Science and Engineering, Shandong Agricultural University, Tai'an 271018, China; qghui@sdau.edu.cn
[2]   National Engineering Laboratory for Efficient Utilization of Soil and Fertilizer Resources, College of Resources and Environment, Shandong Agricultural University, Tai'an 271018, China; 2018110265@sdau.edu.cn
*   Correspondence: zhaogx@sdau.edu.cn; Tel.: +86-053-8824-3939

Received: 7 October 2020; Accepted: 12 November 2020; Published: 14 November 2020

Abstract: Soil salinization is an important factor affecting winter wheat growth in coastal areas. The rapid, accurate and efficient estimation of soil salt content is of great significance for agricultural production. The Kenli area in the Yellow River Delta was taken as the research area. Three machine learning inversion models, namely, BP neural network (BPNN), support vector machine (SVM) and random forest (RF) were constructed using ground-measured data and UAV images, and the optimal model is applied to UAV images to obtain the salinity inversion result, which is used as the true salt value of the Sentinel-2A image to establish BPNN, SVM and RF collaborative inversion models, and apply the optimal model to the study area. The results showed that the RF collaborative inversion model is optimal, $R^2 = 0.885$. The inversion results are verified by using the measured soil salt data in the study area, which is significantly better than the directly satellite remote sensing inversion method. This study integrates the advantages of multi-scale data and proposes an effective "Satellite-UAV-Ground" collaborative inversion method for soil salinity, so as to obtain more accurate soil information, and provide more effective technical support for agricultural production.

Keywords: sentinel-2A image; UAV image; remote sensing; soil salinity

---

1. Introduction

Soil salinization is a form of soil degradation. Soil salinization will not only cause a series of problems such as ecological deterioration, but also have a negative impact on the growth of crops [1]. Therefore, it is of great practical significance for agricultural production to carry out research on soil salinization in coastal winter wheat planting areas and grasp the spatial distribution of salinization.

Traditional soil salt information is obtained mainly through field survey sampling and chemical analysis method. This method is relatively accurate, but it is time-consuming and labor-intensive, and has obvious limitations in terms of spatial globalization and effectiveness. In addition, field sampling can also cause damage to winter wheat and other crops. Satellite remote sensing data can make up for these shortcomings. Because of its high timeliness, economy and large-area simultaneous observation ability, it has become an important method for quantitative extraction of saline soil information in a large area. A large number of scholars have carried out related research and achieved excellent results. Most of them have realized a large-scale quantitative inversion of soil salt content based on multi-source satellite images [2–11], and they inverted soil salinity by studying the quantitative relationship between soil salinity and vegetation index. This is mainly because the vegetation cover and growth can reflect the status of soil salinity. In addition, the satellite image detects mainly the vegetation coverage information, so it is hard to directly use the spectrum of bare soil to monitor the soil salinity in the vegetation coverage area on a large scale [12–15]. When constructing soil salinity inversion model based on satellite imagery, most studies use traditional linear regression methods and machine learning algorithms. After comparison, it is found that most machine learning methods are more accurate and therefore more widely used [16–19].

However, satellite imagery is greatly affected by factors such as fixed orbit, time phase, weather, etc., especially the low spatial resolution and imaging quality, and it often fails to meet the demand of high-precision and real-time monitoring of soil salinization [4]. In recent years, UAV technology has rapidly developed, due to its advantages of high image accuracy, simple operation and flexibility, it has been applied to soil salt monitoring by scholars [20–22]. However, UAV technology has some limitations in monitoring soil salinization on a large scale due to its limited observation range. Therefore, it is necessary to make use of the complementarity of "Satellite-UAV-Ground" multi-scale data to carry out collaborative inversion of soil salinity.

At present, scholars have carried out certain researches on the inversion of surface-related parameters by combining data from two platforms such as Satellite-Ground and UAV-Ground. For example, An et al. [23] built soil salinity inversion model by simulating Landsat8 band through actual hyperspectral measurements on the ground. Schut et al. [24] combined the vegetable index of UAV and satellite with the crop growth model to evaluate the yield and fertilizer response in the field of heterogeneous smallholders, which would improve the accuracy of yield and crop production assessment. Kattenborn et al. [25] built a random forest model based on the spectrum, texture information and canopy structure obtained from UAV data, and upscaled it to Sentinel satellite data to draw a large range distribution map of tree species; Hu et al. [26] found that the inversion accuracy of soil salinization using the combination of UAV hyperspectral data and GF-2 multi-spectral data was better than that of GF-2 multi-spectral data inversion. Daryaei et al. [27] used Sentinel-2 and UAV data to conduct fine-scale monitoring of vegetation in semi-arid mountainous areas focusing on riparian landscapes, and the accuracy has been improved. Zhang et al. [2] established the reflectance relationship between satellite and UAV data and applied the constructed UAV high-precision model to satellite imagery to realize the SPAD worth inversion during the recovery period of winter wheat in a large area. Studies have shown that the combination of data from the two platforms can effectively improve the accuracy of the inversion, but most of these studies are based on the spectral information relationship between the data samples of different platforms for data fusion to achieve collaborative inversion. Due to the large spatial scale and spectral differences between the samples, it will cause the spectrum information loss and errors, which will affect the accuracy of inversion results. How to realize the accurate and efficient connection of data from the three platforms of "Satellite-UAV-Ground", and then the high-precision and large-scale monitoring of ground surface information, especially soil salinization information, needs further research and exploration.

Therefore, this study selects typical coastal areas in the Yellow River Delta, and uses Sentinel-2, UAV and ground multi-platform data to explore its efficient and accurate collaborative method to carry out "Satellite-UAV-Ground" collaborative inversion of soil salinity in winter wheat planting areas. Thus, a rapid and accurate method for obtaining soil salt in the coastal area was proposed, which provided a scientific basis for the production and management of winter wheat.

2. Materials and Methods

2.1. Study Area

The study area is located in Kenli District (37°24′–38°10′ N, 118°15′–119°19′ E), which is the core area of the Yellow River Delta. It has a warm temperate continental monsoon climate, with sufficient sunlight but little precipitation and evaporation large, uneven droughts and floods, and obvious seasonal alternating wet and dry. The annual mean precipitation, evaporation and air temperature are 511.6 mm, 1928.2 mm and 12.4 °C, respectively [28]. The terrain of the study area is low and flat, decreasing from southwest to northeast. The source of surface water is natural precipitation and water from the Yellow River. The groundwater level is relatively shallow and the salinity is high. The main soil type is gleyic solonchaks with a high sand proportion and high salinization. Soil texture is light, organic matter is generally lack, nitrogen and phosphorus are less, soil overall nutrient is poor, the pH value of the soil is greater than 7 [29]. The main crops are winter wheat, corn, rice and cotton, but the overall management is extensive and the yield is low. Wheat planting areas are mainly distributed in the higher terrain area in the southwest and the Yellow River coast area in the northeast. The variation of soil salinization is obvious, which is an ideal area for this study. Based on the investigation of Kenli area, landform, soil and crop distribution, two test areas A and B (Figure 1) were selected for the concentrated distribution of winter wheat in the southwest and northeast of the study area respectively to carry out field soil salinity measurement and UAV flight test. Test area A was a square area of 200 m × 200 m, and a 100 m × 50 m area was selected to arrange sampling points. Test area B was a rectangular area of 200 m × 100 m, and a 50 m × 50 m area was selected to arrange sampling points. An investigation suggests that planting time, farming methods, and fertilization conditions are all basically the same in A and B test areas, while the growth of winter wheat there is obviously different, with all soil salt content levels distributed, making the experimental area more typical and representative.

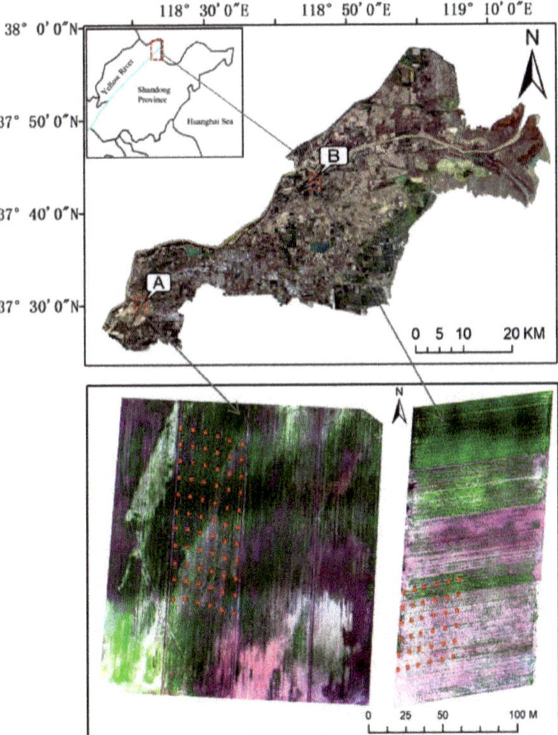

Figure 1. Location map of the study area and test areas (**A**, **B**), the red dots are the sampling points in the test area. The study area is true color image of Sentinel-2A, and the test area (**A**, **B**) are false color images of UAV.

2.2. Data Acquisition and Preprocessing

2.2.1. Acquisition of Ground Soil Salinity Data

There is less rainfall in spring in Kenli District, and the surface salinization is obvious and stable. Winter wheat is at the reviving stage, while other major crops are not sown, which is convenient for extracting spectral characteristics [30]. A field survey in the study area was conducted from April 10 to 16, 2019. To ensure uniform distribution, three samples were pre-arranged in the study area every 5 km × 5 km grid, and finally 77 winter wheat sample data were collected. At the same time, for the two test areas, a ground card was placed every 10m at the outer boundary of the sampling point area of the A and B test areas, and connect them with a measuring rope to form a 10 m × 10 m grid of sample points, taking the intersection of the grid as sampling points. This 10 m × 10 m grid is the same as the pixel size of Sentinel-2A. A total of 102 sample points was collected, including 66 in the A test area and 36 in the B test area. Four outliers in the sampling points were eliminated, and the remaining 98 samples were used to construct and verify the soil salinity inversion model of winter wheat. An EC110 portable salinity meter (Spectrum Technologies, Inc., Aurora, USA) equipped with a 2225FST series probe (conductivity temperature correction was performed) was used to make multiple measurements of the electrical conductivity (EC) of the soil surface layer 10 cm below the plant at each sample point and make a record after stabilization. The average of the measured values is taken as the EC value of each sample point, in ds/m. According to the results of earlier research which in our laboratory, the measured EC data were converted into soil salt content (SSC) in g/kg by using the formula SSC = 2.18 EC + 0.727 which was obtained by chemical analysis of soil in the same study area in spring [31]. At the same time, the orientation, topography, soil, wheat growth and other relevant information were recorded.

2.2.2. Acquisition and Processing of UAV Imagery

A multispectral camera (Parrot Sequoia, Parrot Inc., Paris, France) was mounted on a Dajiang Matrice 600 Pro UAV (loaded mass: 5.5 kg; flying time: 18 min) (SZ DJI Technology Co., Ltd. Shenzhen, Guangdong Province, China). The camera can receive a total of 4 bands of information, which are green light (G), red light (R), red edge (RE), near infrared (NIR). The wavelengths are 550 nm, 660 nm, 735 nm, and 790 nm, and the band widths are 40 nm, 40 nm, 10 nm and 40 nm. The Sequoia multi-spectral camera is mounted on the head of the UAV, and the radiation sensor is mounted on the top of the UAV to write the radiometric correction data into the image during flight.

The data collection time was from 11:00 to 15:00 on 14 April 2019. The weather was clear and cloudless with low wind force when the UAV was flying. Before takeoff, the Sequoia multispectral camera and radiation sensor were calibrated, and the ground standard whiteboard image was collected. The flight height was 50 m, the flight speed was 5 m/s, and the image acquisition interval was 1.5 s. After the data are collected, they will be imported into Pix4D Mapper software (Pix4D, S.A., Prilly, Switzerland) for splicing, radiation correction and other processing to obtain the high-resolution orthophoto image of the test area, with a spatial resolution of 4–5 cm. Finally, in ENVI5.3, the decision tree method is used to remove the soil background. In order to eliminate the random error caused by the reflectance of a single point, a 5 × 5 pixels image is taken with the sampling point as the center, and the average reflectance value is taken as the reflectance data of the sampling point.

2.2.3. Acquisition and Processing of Sentinel-2A Satellite Data

Sentinel-2 satellite is a multispectral imaging satellite with high resolution, revisit rate and update rate. It includes two small satellites A and B. The revisit period is 5 days. The main payload is MSI multispectral imager, covering 0.4–2.4 μm spectral range, including 10 m (four bands), 20 m (six bands), 60 m (three bands) ground resolution, which can monitor the growth, coverage and health of land vegetation, and obtain information on crop planting, land use changes, etc. In this paper, the Sentinel-2A products were downloaded from the ESA Copernicus data sharing website (https://scihub.copernicus.eu/). Considering the acquisition time of ground and UAV data and the quality of the image, the Sentinel-2A Level-1C multispectral image on 17 April 2019 was selected for modeling and inversion of the soil salt content, and the images of 3 November 2018 and 26 June 2019 were used for extraction of winter wheat planting areas.

The downloaded L1C data are orthophoto with geometric precision correction, without radiometric calibration and atmospheric correction. First, radiometric calibration and atmospheric correction are carried out by using Sen2cor, a plug-in published by ESA. Then, the data are re-sampled by the Sentinel Application Platform (SNAP) software to generate 10m spatial resolution images, and the data are exported in ENVI format. Finally, the Sentinel-2A true color images of the research area were obtained by splicing the images in ENVI 5.3 (Exelis Visual Information Solutions, Inc., Colorado, USA) and clipping the images using kenli District administrative boundary vector documents (Figure 1). In order to be consistent with the UAV image band, the green (G), red (R), red edge (RE), and near infrared (NIR) of Sentinel-2A image are selected in this study. The wavelengths are 560 nm, 665 nm, 740 nm, 865 nm, and the band widths are 45 nm, 38 nm, 18 nm and 33 nm.

2.3. Methods

2.3.1. Calculation and Optimization of Vegetation Indices

Studies have shown that different levels of soil salinization have an impact on vegetation growth and morphology, plasma membrane permeability, photosynthetic pigments of leaves, gas exchange parameters, chlorophyll fluorescence characteristics, etc. [32]. Therefore, there are differences in spectral information of vegetation at different levels of soil salinization, which can indirectly reflect the level of soil salinization [33,34]. The vegetation index can highlight the characteristics of vegetation and effectively reflect the health and growth of vegetation. When the soil salt content increases, the reflectivity of visible red light of the salt-sensitive vegetation will increase, and the near-infrared reflectance will decrease [35].In order to better reflect the vegetation conditions, 8 vegetation indexes related to red light and near-infrared are selected in this study, including normalized difference vegetation index (NDVI), normalized difference red edge index (NDRE), optimized soil adjusted vegetation index (OSAVI), green normalized difference vegetation index (GNDVI), triangle vegetation index (TVI), difference vegetation index (DVI), Improved chlorophyll absorption vegetation index based on PROSPECT and SAILH radiation transfer model (MCARI2) and renormalized difference vegetation index (RDVI).

The multi-band spectrum collected by the UAV is used to calculate the 8 vegetation indexes, and the formulas are shown in Table 1. Then the correlation coefficient between each vegetation index and soil salinity was calculated, and the variance inflation factor (VIF) between vegetation indexes was calculated by using the formula VIF = 1/(1-r*r) (r is the correlation coefficient between vegetation indexes) [12], excluding the low correlation or VIF > 10, which is the parameter that cannot be diagnosed by collinearity. The sensitive vegetation indexes are selected for soil salt modeling.

Table 1. Formulas and corresponding citation for vegetation indexes.

No.	Vegetation Index	Formula	Reference
1	NDVI	$(NIR - R)/(NIR + R)$	
2	NDRE	$(NIR - RE)/(NIR + RE)$	[36]
3	OSAVI	$(1 + 0.16)(NIR - R)/(NIR + R + 0.16)$	
4	MCARI2	$[3[(RE - R) - 0.2(RE - G)(RE/R)]]/RE/R$	
5	TVI	$\sqrt{(NIR - R)/(NIR + R) + 0.5}$	[37]
6	DVI	$NIR - R$	
7	GNDVI	$(NIR - G)/(NIR + G)$	[38]
8	RDVI	$(NIR - R)/\sqrt{(NIR + R)}$	

2.3.2. Construction and Optimization of UAV-Ground Collaborative Inversion Model of Soil Salinity in Wheat Field Based on UAV Images

The 98 samples were sorted from small to large, and the modeling set and the validation set were sampled at equal intervals in a ratio of 2:1 to ensure the same range and uniform distribution of the model samples and the validation samples. 68 samples were selected for modeling and 30 samples for validation.

Taking the sensitive vegetation index as the input variable of the model, three methods were used to construct the winter wheat soil salinity inversion model, namely, BPNN, SVM and RF. BPNN is a multi-layer feedforward neural network trained according to the error back propagation algorithm, and it has also been applied to the salt inversion problem [39,40]. SVM is a new machine learning method from linear separable to linear nonseparable based on the principle of minimizing structural risk according to the statistical theory. It has been widely applied in image recognition and classification and has also been applied in regression problems in recent years [41,42]. The RF algorithm is an integrated learning algorithm obtained by combining the bagging algorithm with the decision tree algorithm. In recent years, many scholars have applied it to remote sensing technology [43,44]. All the three methods were implemented in MATLAB R2016b (MathWorks, Inc., Natick, MA, USA). The BPNN sets the number of training iterations to 1000, the accuracy to 0.003, the learning rate to 0.01; The SVM method was set as V-SVR, Gaussian kernel function was selected, the best penalty factor C and kernel parameter gamma were selected through network search and cross validation, and the SVM model was trained. The RF method called MATLAB random forest toolbox, and the parameter leaf node Leaf = 5 and the number of trees Ntrees = 200 were finally determined by Bayesian optimization.

The accuracy of model modeling and verification was evaluated by the coefficient of determination (R^2) and root mean square error (RMSE) [45–47]. R^2 is used to measure the fitting degree of the model, and RMSE reflects the deviation between measured value and predicted value. The closer R^2 is to 1, the smaller the RMSE, which means the higher the accuracy of the model, the better the effect. The model with the best accuracy and effect was selected for soil salinity inversion of winter wheat. The degree of soil salinization is divided into five grades according to relevant criteria [48], non-salinization (<1 g/kg), mild salinization (1–2 g/kg), and moderate salinization (2–4 g/kg), severe salinization (4–6 g/kg) and saline soil (>6 g/kg), and we get the distribution map of soil salinity grade.

2.3.3. Information Extraction of Winter Wheat Planting Area

The planting area of winter wheat in the study area was extracted by using the time series features composed of the NDVI of the Sentinel-2A images on 3 November 2018, 17 April 2019 and 26 June 2019. According to the investigation and analysis of various vegetation types in the study area, only winter wheat was in the seedling stage in early November, its growth reached the peak stage in late April of the second year, and entered the maturity stage at the end of June. Its NDVI time series curve showed a rapid rise from early November to late April, while other vegetation types showed little change in NDVI, the NDVI of winter wheat declined rapidly from late April to the end of June, while the NDVI time series curves of other vegetations were in varying degrees of rising stages, as shown in Figure 2. The valley-peak-valley in the NDVI time series at the beginning of November, late April and the end of June is a significant feature that distinguishes winter wheat from other vegetation. Therefore, by calculating the changes in NDVI from early November to late April, and late April to the end of June, a decision tree is established to extract this feature. Combined with the training sample data, the following decision rules are established:

$$\text{Types of ground objects} = \begin{cases} \text{Winter wheat, } b_2 > 0.1 \text{ and } b_2 - b_1 > 0.1 \text{ and } b_3 - b_2 < 0 \\ \text{Other ground objects} \end{cases} \tag{1}$$

where b_1, b_2, and b_3 are the NDVI values of Sentinel-2A images on 3 November 2018, 17 April 2019, and 26 June 2019, respectively.

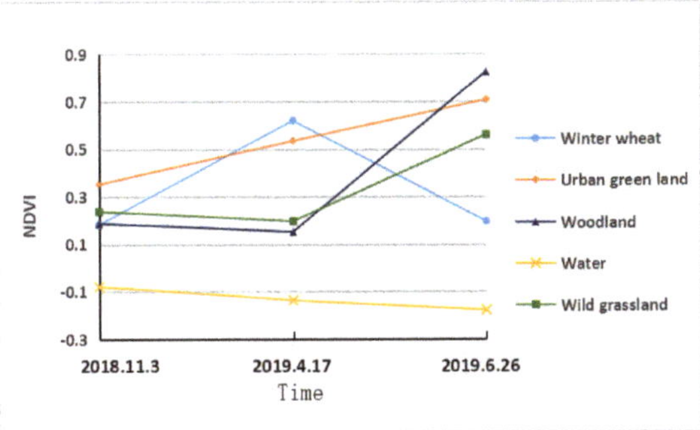

Figure 2. Changes of NDVI of different vegetation types in the study area with time.

Thus, the distribution of winter wheat planting areas on 17 April 2019 was obtained.

2.3.4. Construction of the Satellite-UAV-Ground Collaborative Inversion Model of Soil Salinity in Wheat Area Based on Sentinel-2 Images

Sentinel-2A images have mixed pixels, and it is difficult to accurately obtain the measured data of soil salinity within the corresponding range of pixels. However, the resolution of UAV images is up to centimeter level, so the measured data of soil salinity within the corresponding range of pixels are easy to obtain and accurate. Therefore, the measured ground salt data and the UAV image vegetation index are used to construct a high-precision inversion model to obtain the soil salt content of the test areas A and B, then the salt values corresponding to sentinel-2 image pixels in the test area were calculated and taken as the "salt true value" of the Sentinel-2A image construction inversion model. It is combined with the Sentinel-2A image vegetation index to construct three inversion models including BPNN, SVM and RF.

First, the 10 m × 10 m surface vector data corresponding to the size positions of Sentinel-2A image pixels were successively established in the test area A and the test area B. In order to ensure the objectivity of the data, in the formed surface vector data grid, every three rows and three columns of surface vector data are used as a unit to extract the surface vector data at the center of the unit. If the center position is not in the unit, the surface vectors at other positions in the unit are extracted. Figure 3 is the distribution map of the area vector data extracted from the test area A and a total of 75 surface vector data are extracted. The vegetation index corresponding to the Sentinel-2A image pixel of the surface vector was counted and entered into the attribute information, and the combination of the vegetation indexes were used as the input variable of the model. Secondly, A vector surface corresponds to 40,000 image UAV pixels, and the average salt value of 40,000 UAV pixels is calculated as the salt "truth value" of the surface vector data. The soil salt content of 75 surface vector data was recorded into the attribute information as the output variable of the model. Finally, BPNN, SVM and RF were used to construct soil salt content collaborative inversion model based on Sentinel-2A image. All three methods were implemented in MATLAB R2016b.The BPNN sets the number of training iterations to 1000, the accuracy to 0.003, and the learning rate to 0.02; The SVM method was set as V-SVR, and the training set cross-validation and network search method were used to optimize the parameters. According to the principle of minimum variance, the penalty coefficient was determined as C = 10000, γ = 0.01; the RF method called MATLAB random forest toolbox, and the parameter leaf node, Leaf = 5, and the number of trees, Ntrees = 300, was finally determined by Bayesian optimization.

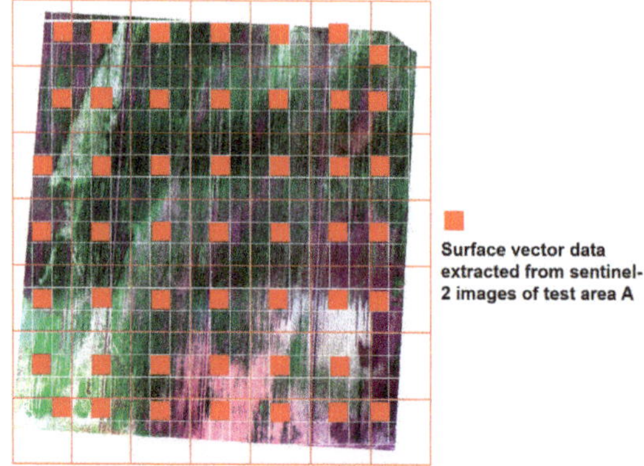

Figure 3. Distribution map of the extracted surface vector data of test area A.

2.3.5. Soil Salinity Inversion Results and Accuracy Analysis in Wheat Area

Among the three machine learning models constructed, the optimal model was selected according to R^2 and RMSE. The optimal model was used to invert the soil salinity of the winter wheat planting area in Kenli District, and the soil salinity distribution map of winter wheat in Kenli District was obtained.

At the same time, this method was compared with the direct inversion method of satellite remote sensing, namely, based on Sentinel-2A images and ground-measured salt data, the optimal model was directly constructed and selected to invert the winter wheat soil salt distribution in Kenli District. In order to verify the accuracy of the two methods, the R^2 and RMSE of the inversion results and the data of 77 winter wheat soil sampling points in Kenli District were calculated for quantitative evaluation.

3. Results and Analysis

3.1. Screening of Soil Salt-Sensitive Vegetation Index

Correlation analysis was conducted between UAV image vegetation index and measured soil salinity content respectively, as is shown in Table 2. In the correlation matrix, the DVI, MCARI2, and TVI of the eight vegetation indices had low correlations with soil salinity. The VIF value of OSAVI and NDVI was greater than 10, and there was strong multicollinearity. Therefore, the combination of vegetation index NDVI, NDRE, GNDVI and RDVI was selected as independent variables for modeling.

Table 2. Correlation coefficient between vegetation index and soil salinity.

r	SS	NDVI	NDRE	GNDVI	OSAVI	RDVI	DVI	MCARI2	TVI
SS	1								
NDVI	−0.730 **	1							
NDRE	−0.669 **	0.709 **	1						
GNDVI	−0.710 **	0.928 **	0.691 **	1					
OSAVI	−0.699 **	0.961 **	0.666 **	0.931 **	1				
RDVI	−0.631 **	0.916 **	0.674 **	0.857 **	0.883 **	1			
DVI	−0.546 **	0.769 **	0.689 **	0.741 **	0.718 **	0.768 **	1		
MCARI2	−0.584 **	0.850 **	0.678 **	0.828 **	0.862 **	0.819 **	0.842 **	1	
TVI	−0.511 **	0.462 **	0.314 **	0.452 **	0.440 **	0.395 **	0.328 **	0.371 **	1

Significance levels: ** 0.01.

3.2. UAV-Ground Collaborative Inversion Model of Soil Salinity Based on UAV Images

Soil salinity inversion model was established by taking 4 vegetation indexes of 68 modeling samples as independent variables and soil salinity content as dependent variables, and 30 validation samples were used to verify the model. The results are shown in Table 3.

Table 3. Accuracy of three inversion models based on UAV images.

Modeling Methods	Modeling Set ($n = 68$)		Validation Set ($n = 30$)	
	$R_m{}^2$	$RMSE_m$	$R_v{}^2$	$RMSE_v$
BPNN	0.789	0.671	0.667	0.689
SVM	0.608	0.891	0.601	0.813
RF	0.878	0.511	0.827	0.473

It can be seen that all three inversion models show good stability, with their $R_m{}^2$ and $R_v{}^2$ both exceeding 0.6, and there is no "overfitting phenomenon". The highest modeling $R_m{}^2$ of the RF algorithm is 0.878, which is 0.089 and 0.27 higher than the $R_m{}^2$ of the BPNN and SVM algorithms, and the $RMSE_m$ is the lowest of 0.511, which is 0.159 and 0.38 lower than the $RMSE_m$ of the BPNN and SVM algorithms, respectively. From the validation results, the highest $R_v{}^2$ of the RF algorithm is 0.827, which is 0.16 and 0.226 higher than the $R_v{}^2$ of the BPNN and SVM algorithms, and the $RMSE_v$ is the lowest 0.473, which is 0.216 and 0.340 lower than the $RMSE_v$ of the BPNN and SVM algorithms, respectively. The inversion model of the RF algorithm has the highest accuracy. Therefore, The RF inversion model was selected to conduct soil salinity inversion for UAV images in the test areas A and B, and the salinity levels were divided according to five levels to obtain the distribution diagram (Figure 4a,b). Figure 4a,b are the interpolation maps of the measured soil salinity in the sampling area. Generally speaking, the inversion results of the wheat field soil salinity are basically the same as the interpolation results of measured data, and the change trend of the area proportions of each grade is roughly the same, but in comparison, the reflection of the inversion results on spatial distribution of soil salinity is more refined. Therefore, the soil salinity inversion model of winter wheat based on RF is better and has the best stable effect.

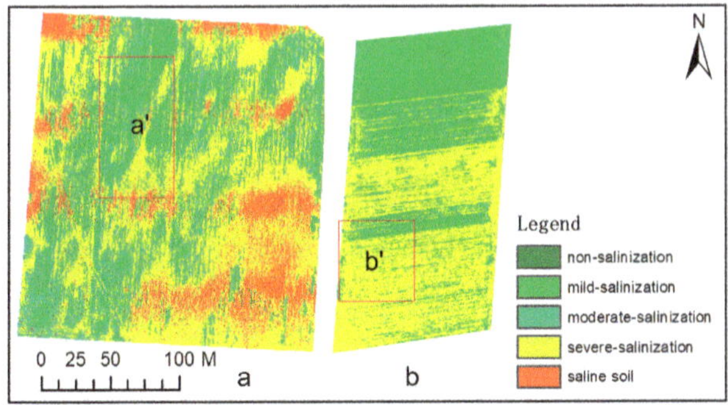

Figure 4. *Cont.*

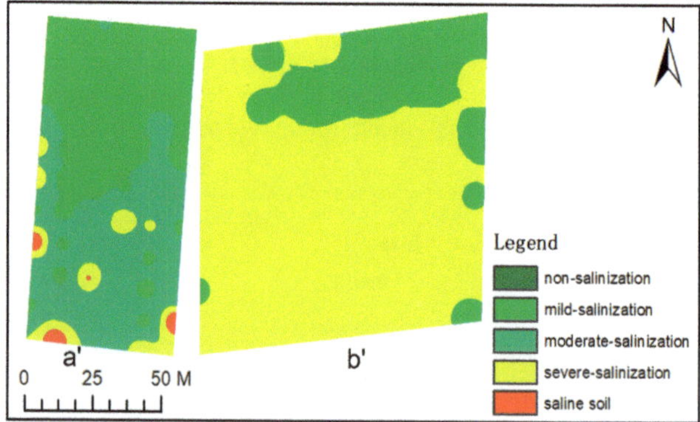

Figure 4. Salinity inversion map (**a**,**b**) and regional interpolation map (**a′**,**b′**) of sampling points in the test area.

3.3. Collaborative Inversion of Soil Salinity Based on "Satellite-UAV-Ground"

Based on the 75 surface vector data of test areas, the NDVI, NDRE, GNDVI and RDVI of the Sentinel-2A images of 52 modeling samples were taken as independent variables, and the soil salt content estimated by UAV was taken as dependent variable to establish soil salt inversion model, and 23 validation samples were used to verify the model. Figure 5 shows the accuracy of the modeling set and validation set of the three models. By comparison, RF model has the highest accuracy, modeling set $R^2 = 0.886$, RMSE = 0.456, validation set $R^2 = 0.850$, RMSE = 0.505, followed by SVM model, the modeling set and validation set R^2 are 0.796 and 0.649 respectively, and BPNN model has the lowest accuracy, the modeling set and validation set R^2 are 0.734 and 0.630 respectively. Therefore, RF model is the optimal model for collaborative inversion of soil salinity based on "Satellite-UAV-Ground".

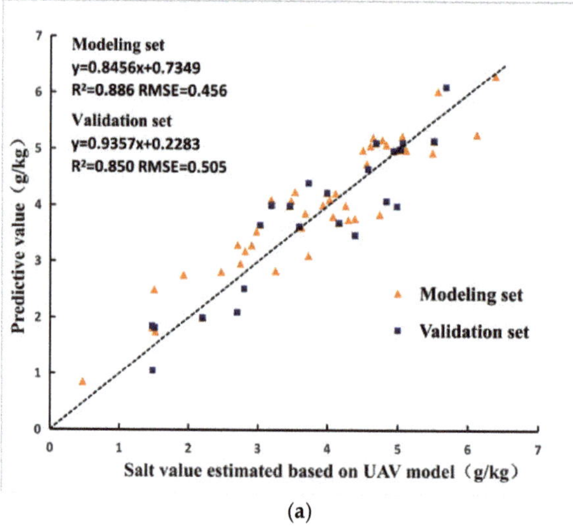

(**a**)

Figure 5. *Cont.*

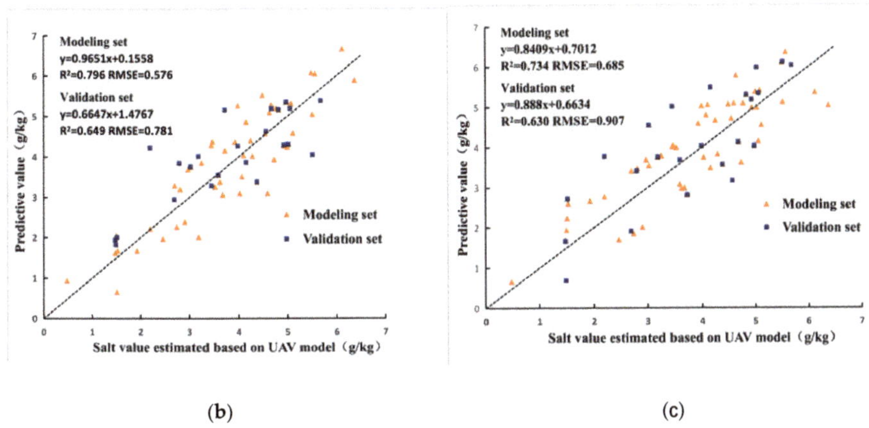

(b)   **(c)**

Figure 5. Accuracy of the three collaborative inversion models, (**a**) RF model, (**b**) SVM model, and (**c**) BPNN model.

3.3.1. Analysis of Inversion Results of Soil Salinity in Wheat Area

A large-scale inversion of the soil salinity in the winter wheat planting area in Kenli district was carried out through the optimal RF model of the "Satellite-UAV-Ground" collaborative inversion, and the soil salinity level distribution map was obtained, as shown in Figure 6. Table 4 shows the area ratio of each salinity level. It can be seen that the distribution of non-saline soil in the wheat planting area of the study area is very small, accounting for only 0.05%. The proportion of winter wheat soil area shows a trend of decreasing with the increase of salinization, which is consistent with the actual situation of the survey. Among them, mildly salinized soil in wheat area is widely distributed, accounting for 61.32% of the total area. It is concentrated in the southwest where the terrain is relatively high and the northeast area affected by the fresh water of the Yellow River. The moderately salinized soil area was the second, accounting for 19.53%, which was distributed in all wheat regions. Severe salinized soil and salinized soil accounted for 19.1% of the total area and were scattered in wheat area.

Table 4. Statistics of soil salinity grade area in wheat area (Unit: %).

Soil Salinity Level	Non-Saline	Mild Salinization	Moderate Salinization	Severe Salinization	Saline Soil
Proportion of inversion result	0.05	61.32	19.53	12.28	6.82

3.3.2. Accuracy Comparison between "Satellite-UAV-Ground" Collaborative and "Satellite-Ground" Direct Inversion

Figure 7 shows the soil salinity inversion results of the RF model constructed directly based on Sentinel-2A images and measured salinity data. The R^2 of the model is 0.56, which is far lower than the RF model of collaborative inversion. From the comparison of Figures 6 and 7, it can be seen that the trend of soil salinity distributions obtained by the two inversion methods are basically consistent. However, Figure 7 is generally lower than the salt value in the survey. This is mainly due to the influence of the mixed pixels during the directly satellite remote sensing inversion method. Figure 6 has fewer non-saline areas and more saline soil areas, which is consistent with the field survey. As a result, using the UAV inversion result as the medium data can effectively reduce the influence of mixed pixels on the inversion result.

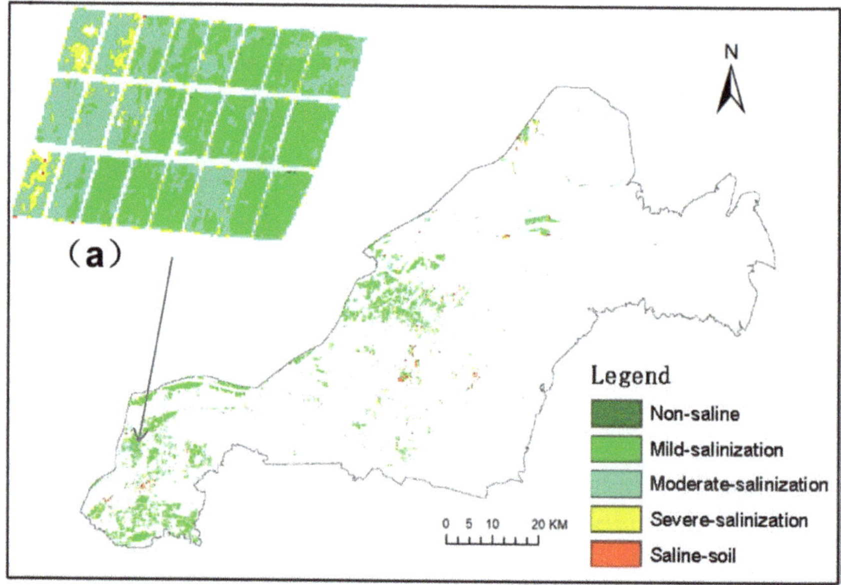

Figure 6. Inversion results based on "Satellite-UAV-Ground", and (a) is a partial enlarged view.

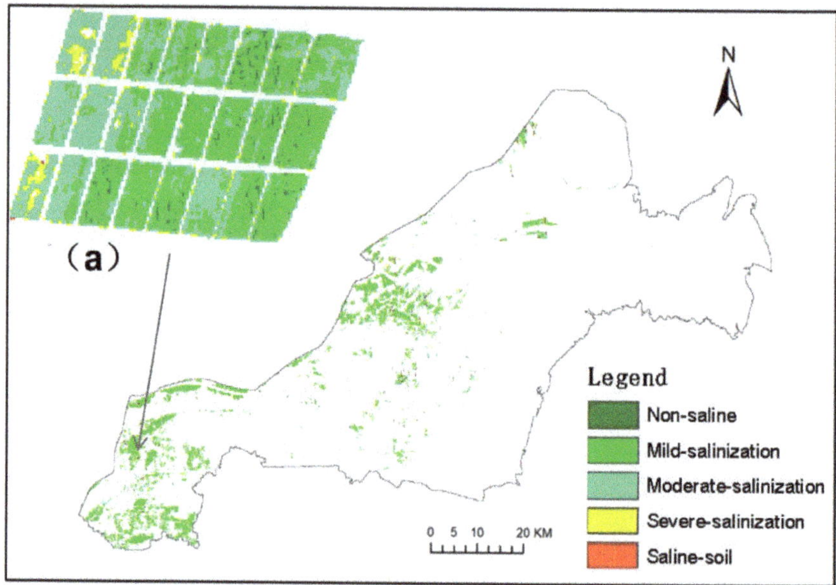

Figure 7. Inversion results based on satellite remote sensing, and (a) is a partial enlarged view.

In order to further compare the inversion results of the "Satellite-UAV-Ground" collaborative method and the direct satellite remote sensing method, the salt values of 77 sampling points in Kenli District at the two inversion results were extracted and compared with the measured data. As is shown in Figure 8. The $R^2 = 0.741$ and RMSE = 0.776 of the "Satellite-UAV-Ground" collaborative inversion results and the measured values, while the $R^2 = 0.591$ and RMSE = 0.831 of the satellite remote sensing method and the measured values, indicating that the results of the "Satellite-UAV-Ground" collaborative inversion are highly consistent with the measured salinity, and the direct satellite remote sensing inversion results have a large deviation.

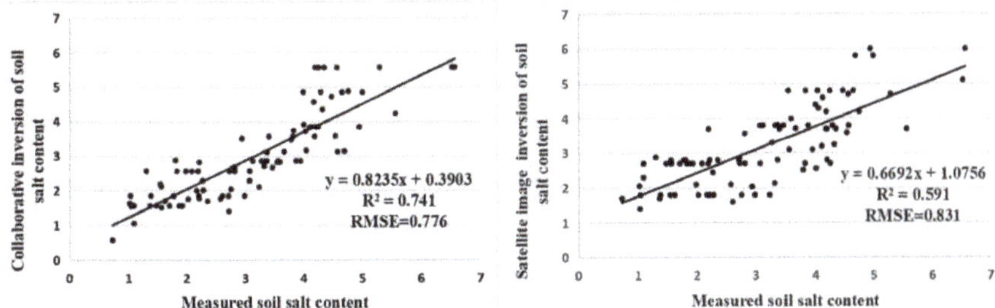

Figure 8. Scatter plots of measured sample points and soil salinity inversion results by two methods in Kenli District.

Therefore, "Satellite-UAV-Ground" collaborative inversion effectively improves the accuracy of soil salt inversion in winter wheat planting areas and obtains a more reliable soil salt distribution.

4. Discussion

The measured data on the ground is the basis for quantitative analysis of soil, UAV near-earth remote sensing is the link between the satellite and the ground, and satellite remote sensing is the platform for large regional inversion. Combining the three will be an important way to obtain soil salt information at present and in the future. Therefore, in this study, the three platforms of "Satellite-UAV-Ground" data were used to perform soil salinity inversion. Compared with the direct satellite image inversion results, the inversion quality was significantly improved. The inversion results of two models are verified with ground-based data. The results show that the R^2 based on the "Satellite-UAV-Ground" collaborative inversion model has been significantly improved, and RMSE and distribution pattern is also improved, but not so obviously. It might be because the salinity inversion result of UAV image is used as an intermediate bridge, its scope is limited and it is hard to fully express the situation of the entire study area. We will continue to improve research on collaborative approach, models, vegetation spectra, data sampling, etc. to improve the performance in terms of scale, accuracy, and temporal and spatial resolution.

The selected test areas in this study were located in two concentrated winter wheat planting areas in the southwest and northeast of Kenli District. The growth of winter wheat was significantly different, and the soil salt content was distributed in all grades, which was typical and representative. In order to improve the spectral and salinity data accuracy of winter wheat, the grid intersection method was used to determine the location of ground data collection, which could ensure the uniformity and accuracy of sample points more than the traditional random sampling point determination method. In order to ensure the representativeness of modeling samples, the samples are sorted and sampled at equal intervals at the ratio of 2:1 between the calibration set and the validation set. Compared with the traditional method of random partition modeling set and validation set [5,49], the data is more uniform and effective, thus ensuring the strong universality and high stability of the model.

The soil salt contents in the study area vary greatly, so the smaller the pixel and the closer the sampling point, the more accurate the information reflected. In this study, an inversion model was built based on the measured ground data and UAV images. The salinity value obtained from the inversion was regarded as the "salinity truth value", one pixel of Sentinel-2A corresponds to 40,000 UAV pixels, the average value of the soil salinity corresponding to these 40,000 UAV pixels was calculated, which was used as the salt value corresponding to the pixel of Sentinel-2A. Compared with the salt value measured on the ground, the "true salt value" is a more comprehensive representation of the true salt status in the Sentinel-2A pixel. Most of the previous studies have realized the up-scaling inversion method of UAV model by establishing the relationship of spectral information between UAV image and satellite image [5,50]. The method in this paper is a collaborative inversion model based on satellite images which is constructed by taking the UAV "salt truth value" as the bridge of "Satellite-UAV-Ground" collaborative inversion. It can better maintain the information content and spatial structure characteristics of the original remote sensing data, and better guarantee the accuracy of salt inversion. It is an effective way to realize the integrated inversion of "Satellite-UAV-Ground".

Previous research results show that the machine learning inversion model is superior to the statistical model [51–54]. The machine learning method can show strong nonlinear fitting ability and excellent data mining ability and can better simulate the complex nonlinear relationship between soil salinity and remote sensing image characteristics. Therefore, three machine learning modeling methods were directly adopted in this study, and the comparison found that the RF model was particularly effective, especially the "Satellite-UAV-Ground" collaborative inversion model, with R^2 and RMSE up to 0.886 and 0.456, respectively. Therefore, if there is a non-linear relationship between the predictor variable and the response variable, a non-linear model such as RF will usually have a better fitting effect and produce excellent estimation accuracy.

The growth of winter wheat in coastal salinized areas is mainly affected by soil salinity, while other factors such as soil texture, pH value, climate, water, nutrients, and fertility have a more balanced impact on crop growth, which is a systematic error. Therefore, salt content is mainly considered for the impact on vegetation index. The vegetation index can be used to invert soil salinity indirectly, which has been confirmed by previous research [11,55,56]. However, the relationship between the vegetation index and soil salinity is different under different environmental conditions. Consequently, the constructed model is only applicable to wheat fields in coastal saline soil in spring. In order to make the obtained soil salinity better support crop production, the next step is to perform salinity inversion for different crops in the study area in different seasons, and use satellite data at different times to verify the model to further improve the reliability of the model.

When salinity inversion is carried out by using data from different remote sensing platforms, matching between data is very important. The predecessors usually take the average of multiple measurements within the ground range corresponding to the satellite pixel as the salt value corresponding to the pixel [57,58]. In this paper, the salt value of the corresponding range of Sentinel-2A pixels is obtained by calculating the average value of 40,000 pixels of the corresponding UAV. Compared with the previous methods, the accuracy of salt value corresponding to satellite pixel is improved, thus the spatio-temporal matching accuracy of the data is improved. Due to the uncertainty of remote sensing data, the band response functions of different sensors are different. When building a collaborative inversion model, it is necessary to fully consider the radiation characteristics of the data and select similar bands to reduce the impact caused by radiation and improve the accuracy of data spectrum matching.

5. Conclusions

In this study, satellite images, UAV images and measured soil salt data were combined to build an inversion model to realize the "Satellite-UAV-Ground" collaborative inversion of soil salt in the coastal area of winter wheat. The main conclusions are as follows:

(1)   The correlation between 8 vegetation indexes based on the multi-spectral bands of UAV images in the winter wheat test area and soil salinity was all greater than 0.5. According to the correlation coefficient and variance expansion factor, four sensitive vegetation indices, including NDVI, RDVI, GNDVI, and NDRE, were selected for modeling, and three machine learning inversion models, BPNN, SVM and RF were constructed. The RF model modeling set has $R^2 = 0.878$, RMSE = 0.511, and its accuracy is higher than BPNN and SVM. It is the best salt estimation model. The inversion results are in good agreement with the actual distribution of soil salt in the test

area. The model has good predictive ability and applicability for the estimation of soil salinity of winter wheat in spring in coastal salinization areas.

(2) The soil salinity in the test area obtained from the inversion of the best model of UAV imagery is used as the "true value of salinity" for satellite image modeling, and three collaborative inversion models are constructed. The RF inversion model has $R^2 = 0.885$, which is significantly better than the BPNN and SVM models. The model is applied to the study area to obtain a large-scale distribution map of soil salinity in the winter wheat area. The soil in the winter wheat planting area of the study area is dominated by light and moderate salinization, accounting for 80.85% of the area, mainly distributed in the southwest and northeast regions. The area of heavily salinized and saline soil is smaller, only accounting for 19.1%, and is scattered in the wheat area.

(3) The result of "Satellite-UAV-Ground" collaborative inversion method and the satellite remote sensing direct inversion method were quantitatively compared and evaluated by using the measured salinity data of 77 sample points in the wheat field in the study area. The results show that the $R^2 = 0.741$ and RMSE = 0.776 of the "Satellite-UAV-Ground" collaborative inversion result and the measured value, while the $R^2 = 0.591$ and RMSE = 0.831 of the satellite remote sensing method and the measured value. Therefore, the "Satellite-UAV-Ground" collaborative inversion method can effectively improve the accuracy of the soil salt inversion results and obtain more accurate spatial distribution of winter wheat soil salt in spring in coastal salinization areas.

This study makes full use of the advantages of satellites, UAV images, and ground data to construct soil salinity inversion model, using UAV soil salinity inversion results as intermediate data, and machine learning modeling methods to obtain a grade distribution map of winter wheat soil salinity in the study area which is consistent with the actual distribution of soil salt. The study proposed an effective "Satellite-UAV-Ground" collaborative inversion method for soil salinity, which provides a scientific basis for accurately grasping the distribution of winter wheat soil salinity levels and guiding agricultural production in the study area.

Author Contributions: Conceptualization, G.Z. and G.Q.; methodology, G.Q. and X.X.; software, G.Q. and X.X.; validation, G.Z., G.Q. and X.X.; formal analysis, X.X.; investigation, G.Q. and X.X.; resources, G.Z.; data curation, X.X.; writing—original draft preparation, G.Q.; writing—review and editing, G.Z.; visualization, X.X.; supervision, G.Q. and X.X.; project administration, G.Z.; funding acquisition, G.Z. All authors have read and agreed to the published version of the manuscript.

Funding: This work was financially supported by the National Natural Science Foundation of China(41877003), the Shandong Province Key Science and Technology Innovation Project(2019JZZY010724), and the Funds of Shandong "Double Tops" Program(SYL2017XTTD02).

Conflicts of Interest: There are no conflict of interest exits in the submission of this manuscript.

References

1. Tian, C.Y.; Zhou, H.F.; Liu, G.Q. Research suggestions on the regulation of soil salinization and sustainable agricultural development in Xinjiang in the 21st century. *Arid Land Geogr.* **2000**, *23*, 178–181.
2. Zhang, S.; Zhao, G.; Lang, K.; Su, B.; Chen, X.; Xi, X.; Zhang, H. Integrated Satellite, Unmanned Aerial Vehicle (UAV) and Ground Inversion of the SPAD of Winter Wheat in the Reviving Stage. *Sensors* **2019**, *19*, 1485. [CrossRef] [PubMed]
3. Allbed, A.; Kumar, L.; Aldakheel, Y.Y. Assessing soil salinity using soil salinity and vegetation indices derived from IKONOS high-spatial resolution imageries: Applications in a date palm dominated region. *Geoderma* **2014**, *230*, 1–8. [CrossRef]
4. Yao, Y.; Ding, J.L.; Zhang, F. Regional soil salinization monitoring model based on hyperspectral index and electromagnetic induction technology. *Spectrosc. Spect. Anal.* **2013**, *6*, 1658–1664.
5. Zhang, S.; Zhao, G. A Harmonious Satellite-Unmanned Aerial Vehicle-Ground Measurement Inversion Method for Monitoring Salinity in Coastal Saline Soil. *Remote Sens.* **2019**, *11*, 1700. [CrossRef]
6. Wang, D.Y.; Chen, H.Y.; Wang, G.F.; Cong, J.Q.; Wang, X.F.; Wei, X.W. Research on UAV Multispectral Inversion of the Salt in the Severely Saline Soil of the Yellow River Estuary. *Sci. Agric. Sin.* **2019**, *52*, 1698–1709.

7.   Zhang, T.-T.; Qi, J.; Gao, Y.; Ouyang, Z.; Zeng, S.-L.; Zhao, B. Detecting soil salinity with MODIS time series VI data. *Ecol. Indic.* **2015**, *52*, 480–489. [CrossRef]
8.   Zhang, T.R.; Zhao, G.X.; Gao, M.X.; Chang, C.Y.; Wang, Z.R. Salinity estimation and remote sensing inversion of winter wheat planting areas in the Yellow River Delta based on near-Earth multispectral and OLI images: Taking Kenli County and Wudi County in Shandong Province as examples. *J. Nat. Resour.* **2016**, *31*, 1051–1060.
9.   Chen, H.; Zhao, G.; Li, Y.; Wang, D.; Ma, Y. Monitoring the seasonal dynamics of soil salinization in the Yellow River delta of China using Landsat data. *Nat. Hazards Earth Syst. Sci.* **2019**, *19*, 1499–1508. [CrossRef]
10.   Abbas, A.; Khan, S.; Hussain, N.; Hanjra, M.A.; Akbar, S. Characterizing soil salinity in irrigated agriculture using a remote sensing approach. *Phys. Chem. Earth Parts A/B/C* **2013**, *55*, 43–52. [CrossRef]
11.   Scudiero, E.; Skaggs, T.H.; Corwin, D.L. Regional-scale soil salinity assessment using Landsat ETM + canopy reflectance. *Remote Sens. Environ.* **2015**, *169*, 335–343. [CrossRef]
12.   Chen, H.Y.; Zhao, G.X.; Chen, J.C.; Wang, R.Y.; Gao, M.X. Remote sensing inversion of saline soil salinity in the mouth of the Yellow River based on improved vegetation index. *Trans. Chin. Soc. Agric. Eng.* **2015**, *31*, 107–114.
13.   Yao, Y.; Ding, J.L.; Lei, L. Monitoring spatial variability of soil salinity in dry and wet seasons in the North Tarim Basin using remote sensing and electromagnetic induction instruments. *Acta Ecol. Sin.* **2013**, *33*, 5308–5319. [CrossRef]
14.   Zhang, X.L.; Zhang, F.; Zhang, H.W.; Li, Z.; Hai, Q.; Chen, L.H. Optimization of hyperspectral index soil salinity inversion model based on spectral transformation. *Trans. Chin. Soc. Agric. Eng.* **2018**, *34*, 110–117.
15.   Guo, P.; Li, H.; Chen, H.Y. Quantitative spectral estimation of soil salinity based on spectral index optimization. *Bull. Soil Water Conserv.* **2018**, *38*, 193–199.
16.   Wang, F.; Yang, S.T.; Ding, J.L. Selection of environmentally sensitive variables and machine learning algorithm to predict oasis soil salinity. *Trans. Chin. Soc. Agric. Eng.* **2018**, *34*, 102–110.
17.   Jia, P.P.; Sun, Y.; Shang, T.H.; Zhang, J.H. Estimation models of soil water-salt based on hyperspectral and Landsat-8 OLI image. *J. Ecol.* **2020**. [CrossRef]
18.   Wang, M.K.; Mo, H.W.; Chen, H.Y. Study on the Modeling Method of Retrieving Soil Salinity Based on Multispectral Image. *Soil Bull.* **2016**, *47*, 1036–1041.
19.   Zhang, S.M.; Zhao, G.X.; Wang, Z.R.; Xiao, Y.; Lang, K. Remote sensing inversion and dynamic monitoring of soil salinity in coastal saline area. *J. Agric. Resour. Environ.* **2018**, *35*, 349–358.
20.   Lin, F.; Zhao, G.X.; Chang, C.Y.; Wang, Z.R.; Li, H. Area extraction and growth analysis of winter wheat based on adjacent orbit image. *J. Agric. Resour. Environ.* **2016**, *33*, 384–389.
21.   Schut, A.G.T.; Traore, P.C.S.; Blaes, X.; De By, R.A. Assessing yield and fertilizer response in heterogeneous smallholder fields with UAVs and satellites. *Field Crops Res.* **2018**, *221*, 98–107. [CrossRef]
22.   Ivushkin, K.; Bartholomeus, H.; Bregt, A.K.; Pulatov, A.; Franceschini, M.H.D.; Kramer, H.; Van Loo, E.N.; Roman, V.J.; Finkers, R. UAV based soil salinity assessment of cropland. *Geoderma* **2019**, *338*, 502–512. [CrossRef]
23.   An, D.; Zhao, G.; Chang, C.; Wang, Z.; Li, P.; Zhang, T.; Jia, J. Hyperspectral field estimation and remote-sensing inversion of salt content in coastal saline soils of the Yellow River Delta. *Int. J. Remote Sens.* **2016**, *37*, 455–470. [CrossRef]
24.   Said, N.; Henning, B.; Joachim, H. Estimation of soil salinity using three quantitative methods based on visible and near-infrared reflectance spectroscopy: A case study from Egypt. *Arab J Geosci* **2015**, *8*, 5127–5140. [CrossRef]
25.   Kattenborn, T.; Lopatin, J.; Förster, M.; Braun, A.C.; Fassnacht, F.E. UAV data as alternative to field sampling to map woody invasive species based on combined Sentinel-1 and Sentinel-2 data. *Remote Sens. Environ.* **2019**, *227*, 61–73. [CrossRef]
26.   Hu, J.; Peng, J.; Zhou, Y.; Xu, D.; Zhao, R.; Jiang, Q.; Fu, T.; Wang, F.; Shi, Z. Quantitative Estimation of Soil Salinity Using UAV-Borne Hyperspectral and Satellite Multispectral Images. *Remote Sens.* **2019**, *11*, 736. [CrossRef]
27.   Ardalan, D.; Hormoz, S.; Clement, A. Fine-scale detection of vegetation in semi-arid mountainous areas with focus on riparian landscapes using Sentinel-2 and UAV data. *Comput. Electron. Agric.* **2020**, *177*, 105686.
28.   Zhao, Q.; Bai, J.; Gao, Y.; Zhao, H.; Zhang, G.; Cui, B. Shifts in the soil bacterial community along a salinity gradient in the Yellow River Delta. *Land Degrad. Dev.* **2020**, 2255–2267. [CrossRef]

29. Wen, Y.; Guo, B.; Zang, W.; Ge, D.; Luo, W.; Zhao, H. Desertification detection model in Naiman Banner based on the albedo-modified soil adjusted vegetation index feature space using the Landsat8 OLI images. *Geomat. Nat. Hazards Risk* **2020**, *11*, 544–558. [CrossRef]

30. Revill, A.; Florence, A.; MacArthur, A.; Hoad, S.P.; Rees, R.W.; Williams, M. The Value of Sentinel-2 Spectral Bands for the Assessment of Winter Wheat Growth and Development. *Remote Sens.* **2019**, *11*, 2050. [CrossRef]

31. Wang, Z.R.; Zhao, G.X.; Gao, M.X.; Chang, C.Y.; Jiang, S.Q.; Jia, J.C.; Li, J. Spatial variability of soil water and salt in summer and soil salt microdomain characteristics in Kenli County, Yellow River Delta. *Acta Ecol. Sin.* **2016**, *36*, 1040–1049.

32. Li, J.L. Study on the Mechanism of the Influence of Salt in Coastal Saline Soil Cotton Field on Cotton Yield and Quality. Ph.D. Thesis, Nanjing Agricultural University, Nanjing, China, 2013.

33. Jia, J.C.; Zhao, G.X.; Gao, M.X.; Wang, Z.R.; Chang, C.Y.; Jiang, S.Q.; Li, J. Study on the relationship between winter wheat sowing area changes and soil salinity in typical area of the Yellow River Delta. *J. Plant Nutr. Fertil.* **2015**, *21*, 1200–1208. [CrossRef]

34. Zhang, T.R.; Zhao, G.X.; Gao, M.X.; Wang, Z.R.; Jia, J.C.; Li, P.; An, D.Y. Estimation of soil salinity in typical areas of the Yellow River Delta based on near-surface multi-spectrum. *Spectrosc. Spect. Anal.* **2016**, *36*, 248–253.

35. Tilley, D.R.; Ahmed, M.; Son, J.H.; Badrinarayanan, H. Hyperspectral Reflectance Response of Freshwater Macrophytes to Salinity in a Brackish Subtropical Marsh. *J. Environ. Qual.* **2007**, *36*, 780–789. [CrossRef] [PubMed]

36. Dong, C.; Zhao, G.X.; Su, B.W.; Chen, X.N.; Zhang, S.M. Research on the Decision Model of Variable Nitrogen Application in Winter Wheat's Greening Period Based on UAV Multispectral Image. *Spectrosc. Spect. Anal.* **2019**, *39*, 3599–3605.

37. Fang, X.R.; Gao, J.F.; Xie, C.Q.; Zhu, F.L.; Huang, L.X.; He, Y. Survey of detection techniques and methods for crop canopy spectral information. *Spectrosc. Spect. Anal.* **2015**, *35*, 1949–1955.

38. Dong, J.J.; Niu, J.M.; Zhang, Q.; KangSa, R.L.; Han, F. Remote sensing yield estimation of typical grassland based on multi-source satellite data. *Chin. J. Grassl.* **2013**, *35*, 64–69.

39. Yao, R.J.; Yang, J.S.; Zou, P.; Liu, G.M. BP neural network model for spatial distribution of regional soil water and salinity. *Acta Pedol. Sin.* **2009**, *46*, 788–794.

40. Zheng, Z.; Zhang, F.; Chai, X.; Zhu, Z.; Ma, F. SPATIAL ESTIMATION OF SOIL MOISTURE AND SALINITY WITH NEURAL KRIGING. In *Stochastic Differential Systems*; Springer Science and Business Media LLC: Berlin/Heidelberg, Germany, 2009; Volume 294, pp. 1227–1237.

41. Diao, S.; Liu, C.; Zhang, T. Extraction of remote sensing information for lake salinity level based on SVM: A case from Badain Jaran desert in Inner Mongolia. *Remote Sens. Land Resour.* **2016**, *28*, 114–118.

42. Liang, D.; Guan, Q.S.; Huang, W.J. Remote sensing inversion of leaf area index based on support vector machine regression in winter wheat. *Trans. CSAE* **2013**, *29*, 117–123.

43. Song, R.J.; Ning, J.F.; Chang, Q.R. Kiwifruit orchard mapping based on wavelet textures and random forest. *Trans. Chin. Soc. Agric. Mach.* **2018**, *49*, 222–231.

44. Yuan, H.; Yang, G.; Li, C.; Wang, Y.; Liu, J.; Yu, H.; Feng, H.; Xu, B.; Zhao, X.; Yang, X. Retrieving Soybean Leaf Area Index from Unmanned Aerial Vehicle Hyperspectral Remote Sensing: Analysis of RF, ANN, and SVM Regression Models. *Remote Sens.* **2017**, *9*, 309. [CrossRef]

45. Zhou, X.H.; Zhang, F.; Zhang, H.W. A Study of Soil Salinity Inversion Based on Multispectral Remote Sensing Index in Ebinur Lake Wetland Nature Reserve. *Spectrosc. Spect. Anal.* **2018**, *39*, 1229–1235.

46. Xu, K.; Su, Y.; Liu, J.; Hu, T.; Jin, S.; Ma, Q.; Zhai, Q.; Wang, R.; Zhang, J.; Li, Y.; et al. Estimation of degraded grassland aboveground biomass using machine learning methods from terrestrial laser scanning data. *Ecol. Indic.* **2020**, *108*, 105747. [CrossRef]

47. Xie, Y.; Sha, Z.; Yu, M.; Bai, Y.; Zhang, L. A comparison of two models with Landsat data for estimating above ground grassland biomass in Inner Mongolia, China. *Ecol. Model.* **2009**, *220*, 1810–1818. [CrossRef]

48. Bao, S.D. *Soil Agrochemical Analysis*; China Agriculture Press: Beijing, China, 2000.

49. Xi, X.; Zhao, G.X. Retrieval and Monitoring of Chlorophyll Content in Winter Wheat Based on UAV Multi-spectral Remote Sensing. *Chin. Agric. Sci. Bull.* **2020**, *36*, 119–126.

50. Chen, J.Y.; Wang, X.T.; Zhang, Z.T.; Han, J.; Yao, Z.H.; Wei, G.F. Soil salinization monitoring method based on UAV-satellite remote sensing upscaling. *Trans. Chin. Soc. Agric. Mach.* **2019**, *50*, 161–169.

51. Zhang, Z.T.; Wei, G.F.; Yao, Z.H.; Tan, C.X.; Wang, X.T.; Han, J. Research on Soil Salt Inversion Model Based on UAV Multispectral Remote Sensing. *Trans. Chin. Soc. Agric. Mach.* **2019**, *50*, 151–160.

52. Li, Y.L.; Zhao, G.X.; Chang, C.Y.; Wang, Z.R.; Wang, L.; Zheng, J.R. Soil salt inversion model based on the fusion of OLI and HSI images. *Trans. Chin. Soc. Agric. Eng.* **2017**, *33*, 173–180.

53. Yao, Z.H.; Chen, J.Y.; Zhang, Z.T.; Tan, C.Y.; Wei, G.F.; Wang, X.T. Effect of plastic film mulch ing on soil salinity inversion by using UAV multispectral remote sensing. *Trans. Chin. Soc. Agric. Eng.* **2019**, *35*, 89–97.

54. Qiu, Y.L.; Chen, C.; Han, J.; Wang, X.T.; Wei, S.Y.; Zhang, Z.T. Satellite Remote Sensing Estimation Model of Soil Salinity in Jiefangzha Irrigation under Vegetation Coverage. *Water Sav. Irrig.* **2019**, *10*, 108–112.

55. Wu, W.C. The Generalized Difference Vegetation Index (GDVI) for Dryland Characterization. *Remote Sens.* **2014**, *6*, 1211–1233. [CrossRef]

56. Fernández-Buces, N.; Siebe, C.; Cram, S.; Palacio, J.L. Mapping soil salinity using a combined spectral response index for bare soil and vegetation: A case study in the former lake Texcoco, Mexico. *J. Arid Environ.* **2006**, *65*, 644–667. [CrossRef]

57. Wang, X.; Zhang, F.; Ding, J.; Kung, H.-T.; Latif, A.; Johnson, V.C. Estimation of soil salt content (SSC) in the Ebinur Lake Wetland National Nature Reserve (ELWNNR), Northwest China, based on a Bootstrap-BP neural network model and optimal spectral indices. *Sci. Total Environ.* **2018**, *615*, 918–930. [CrossRef] [PubMed]

58. Yue, J.; Yang, G.; Tian, Q.; Feng, H.; Xu, K.; Zhou, C. Estimate of winter-wheat above-ground biomass based on UAV ultrahigh-ground-resolution image textures and vegetation indices. *ISPRS J. Photogramm. Remote Sens.* **2019**, *150*, 226–244. [CrossRef]

Publisher's Note: MDPI stays neutral with regard to jurisdictional claims in published maps and institutional affiliations.

Article

Analysis of UAV-Acquired Wetland Orthomosaics Using GIS, Computer Vision, Computational Topology and Deep Learning

Sarah Kentsch [1,2,*], Mariano Cabezas [3], Luca Tomhave [2], Jens Groß [2], Benjamin Burkhard [2], Maximo Larry Lopez Caceres [1], Katsushi Waki [4] and Yago Diez [4,*]

1   Faculty of Agriculture, Yamagata University, Tsuruoka 997-8555, Japan; larry@tds1.tr.yamagata-u.ac.jp
2   Faculty of Natural Sciences, Leibniz Universität, 30167 Hannover, Germany; luca.tomhave@kabelmail.de (L.T.); gross@phygeo.uni-hannover.de (J.G.); burkhard@phygeo.uni-hannover.de (B.B.)
3   Brain and Mind Centre, University of Sydney, Sydney 2015, Australia; mariano.cabezas@sydney.edu.au
4   Faculty of Science, Yamagata University, Yamagata 990-8560, Japan; waki@sci.kj.yamagata-u.ac.jp
*   Correspondence: sarah@tds1.tr.yamagata-u.ac.jp (S.K.); yago@sci.kj.yamagata-u.ac.jp (Y.D.)

Abstract: Invasive blueberry species endanger the sensitive environment of wetlands and protection laws call for management measures. Therefore, methods are needed to identify blueberry bushes, locate them, and characterise their distribution and properties with a minimum of disturbance. UAVs (Unmanned Aerial Vehicles) and image analysis have become important tools for classification and detection approaches. In this study, techniques, such as GIS (Geographical Information Systems) and deep learning, were combined in order to detect invasive blueberry species in wetland environments. Images that were collected by UAV were used to produce orthomosaics, which were analysed to produce maps of blueberry location, distribution, and spread in each study site, as well as bush height and area information. Deep learning networks were used with transfer learning and unfrozen weights in order to automatically detect blueberry bushes reaching True Positive Values (TPV) of 93.83% and an Overall Accuracy (OA) of 98.83%. A refinement of the result masks reached a Dice of 0.624. This study provides an efficient and effective methodology to study wetlands while using different techniques.

Keywords: ArcGIS; big data; blueberries; deep learning; image analysis; orthomosaics; segmentation refinement; UAVs

Citation: Kentsch, S.; Cabezas, M.; Tomhave, L.; Groß, J.; Burkhard, B.; Lopez Caceres, M.L.; Waki, K.; Diez, Y. Analysis of UAV-Acquired Wetland Orthomosaics Using GIS, Computer Vision, Computational Topology and Deep Learning. *Sensors* **2021**, *21*, 471. https://doi.org/10.3390/s21020471

Received: 27 November 2020
Accepted: 7 January 2021
Published: 11 January 2021

Publisher's Note: MDPI stays neutral with regard to jurisdictional claims in published maps and institutional affiliations.

1. Introduction

Recent changes in global climate conditions influence species composition and accelerating the presence of invasive plant species in natural environments. Species that spread outside their native habitat and rapidly and effectively adapt to new environments are known as invasive species [1]. The spread of invasive species often benefits from ecosystem changes and habitat disturbances that weaken the natural species and open an ecological niche for invaders. Hence, invasive species can influence the biodiversity, thus limiting the growth of natural plant species due to a higher occurrence of an invasive species, which could lead to ecosystem degradation [2]. The fast adaption to multiple stress factors in environments could also lead to a replacement of native species and it may increase economic costs due to production losses in agriculture and forestry [3]. In Europe, 11% of the 12,000 identified species have caused damage to the economy, society, and the environment [4]. Reference [5] states that hundreds of invasive species find their pathways through horticulture, agriculture, etc., and the linearly increasing trend of invasive species numbers (from 1970 to 2007) indicates higher impacts of invasive species in the future. Reference [6] pointed out that not only invasive species have an impact on native plants, since several factors often interact with the environment that influence species distributions. In recent years, the need to precisely understand the ecological impacts of invasive species in ecosystems has become a key in designing and prioritizing natural resource management approaches [2], since the behaviour and impact of invasive species is still not well understood [3,7]. Furthermore, a high number of invasive plant are species spreading in natural environments, which increases the demand of management practices [5].

In the past two decades, an explosive spread of North American blueberry hybrids (*Vaccinium corymbosum* x *angustifolium*) has been observed in several moors in the northern German geest area, endangering the natural development of these protected raised bog areas [8]. The starting point of the spread has been almost exclusively located at existing or former commercial blueberry plantations [9,10], which can be found near or in the immediate vicinity of bogs or former peat extraction areas, due to good soil and local climatic conditions. Most of the recipient habitats are pine forests and bogs in various stages of de- and regeneration. Because of these characteristics, the American Blueberry (*Vaccinium angustifolium* x *corymbosum*) has been classified as a potentially invasive neophyte by the *German Federal Agency for Nature Conservation* [11]. After the degradation of wetland areas due to anthropogenic activities, protection programs, called "Moorschutzprogramme", were established by the state government of Lower Saxony for the conservation and the development of rare animal and plant communities in these areas [12]. Furthermore, activities that could threaten the goals of the protection program are prohibited, which increases the difficulty of conducting relevant field studies [12]. However, maintenance and development measures are needed in order to rehabilitate protected and relatively sensitive wetland areas, into which the invasive blueberry species *Vaccinium corymbosum* x angustifolium has migrated. In 2011, 20 counties in Lower Saxony reported stands of spontaneously growing blueberry bushes [13]. The potential area that is occupied by spontaneously growing blueberry bushes can reach several square kilometres within a few years [14]. Previous studies, as presented in [10,14], used grids in the field in order to plot the distribution of blueberry bushes within wetlands. Both of the studies focused on sites near blueberry cultivation areas, as the biggest spread was found in close proximity to blueberry plantations [9]. Those studies show the limitations in the studied area and lack an overview. It is still unclear how far the blueberry bushes have already spread and in which areas they occur. In order to implement effective measures in these areas and minimise the disturbance of sensitive biotopes, it is necessary to locate the individual blueberry bushes as accurately and early as possible. In addition, the following questions arise: does a displacement of natural species occur and where should what measures be taken against a continuing invasion? According to [14,15], relatively simple counter measures can lead to good results and prevent further spread, especially when invasive blueberry bushes are identified early. Therefore, a suitable and non-invasive method for recording stock development and distribution is needed. A simple tool is needed to rapidly, cost-effectively, and precisely detect invasive species in wetlands to counteract their rapid reproduction. Wetlands are protected environments with limited ground accessibility making UAVs particularly appropriate for data collection. UAVs offer the possibility to cover large areas with high resolution images, and they have proved their usefulness in a variety of studies in agriculture [16,17] and forestry [18–20]. Still, UAV images present challenges like the pre-pocessing of the data. Large amounts of data need to be processed, labeled, and annotated by experts, which is usually time consuming, before the data can be further analysed. The use of deep-learning techniques reduces the amount of time that is needed to extract information from the data, which increases the benefits for several applications.

Images that are acquired by UAVs can be analysed while using computer vision and GIS techniques. Important results can then be obtained by reducing the complexity contained in the images (using different image interpretation strategies) and the findings can be presented in elaborate visualisations [21]. Persistent homology, a tool of topological data analysis, can help to understand complex datasets by analysing their large scale geometric features [22]. In our study, we have used persistent homology to measure the spread of invasive species. Processing images and creating orthomosaics allow for a fast analysis of large amounts of data. Deep learning techniques additionally automatize classification and localisation processes, and make it possible to incorporate expert knowledge into the automatic image processing pipeline. This has the potential to increase the scale of the resulting studies to reach large regions that are significant in terms of country-level invasive species detection and management.

In this study, the incorporation of all the mentioned techniques from remote sensing, GIS, computer vision, computational topology, and artificial intelligence allows for us to study invasive species on a large scale, with minimum disturbances and the incorporation of expert knowledge. To the best of our knowledge, this is the first study that includes different techniques and UAV gathered data to increase the understanding of an invasive blueberry species in wetlands. Furthermore, locating and studying small bushes in large areas and at the single-bush level was done for the first time. Therefore, UAV-supported methods offer an efficient possibility to discover individual young plants on a large scale and detect propagation hotspots at an early stage.

The following objectives guided this study:

(I)   To use UAV data in order to provide allometric statistics (height, area, and number) of invasive blueberry in large areas at bush level.
(II)   To use clustering techniques and persistent homology to quantitatively and qualitatively assess the spread of blueberry invasions.
(III)   To assess the potential of Deep learning to automatically segment blueberry bushes, initiating the possibility for even larger-scale studies.

State of the Art

In recent years, remote sensing techniques have been used in various natural environments with the goal of reducing the need for in situ measurements [23]. Low-cost data gathering, time saving, and larger study areas are the benefits. Furthermore, data can be directly used and processed within Geographical Information Systems (GIS) [23]. This has been done successfully, for environmental studies [24–26]. Closely related to the current work, Reference [27] proposed using GIS as a synthesising tool in invasive species management approaches. Reference [28] used satellite images of 1992 and 2002 in order to identify stress indicators and change detection in a wetland in Sri Lanka to quantify the conditions of the complex. The authors state that the inventory, mapping, and monitoring are needed to understand interactions in the ecosystem. Classification with the Maximum Likelihood Algorithm were performed, mapping and spatial analysis were used, and finally refined and verified with ground data. Their approach ([28]) reached 86% accuracy and provided detailed analysis. GIS in combination with remote sensing data was found to be an effective methodology for investigating wetlands. Reference [29] evaluated vegetation change detection while using the NDVI of remote sensing data and applied GIS in order to visualise the results. Landsat and Shuttle Radar Topography Missions were used to capture the Vellore District. Image interpretations were carried out using ERGDAS IMAGINE software in order to classify and detect changes in the vegetation. The differences in the NDVI values were used to analyse data sets of the years 2001 to 2006. The study provided information about the lowest decrease in the forest area by 6%, while agriculture land increased the most by 19%.

In comparison to satellite images UAV images provide a higher resolution and appear to be more suitable for wetland investigations, especially when focusing on invasive species. Higher resolution images allow for higher accuracies of image interpretations and feature extractions [30]. Several studies using UAVs in wetlands have already been carried out [31–34]. Reference [34]developed a method for detecting and mapping invasive species with UAVs. The authors acknowledged UAVs as suitable for monitoring eradication efforts in wetlands. Reference [33] realized the higher efficiency in gathering valuable and accurate information in comparison to field studies, when using UAVs and computer vision techniques to enhance classifications and health assessments in wetlands. Gandhi et al. [31] used UAV imagery for detecting invasive species and mapping their distribution and spread. This study also compared data from two years (2009 and 2011) and detected an increase of 19.07%, which was confirmed by field studies with a total agreement of 94% and shows the suitability of UAV imaging for this kind of application. Reference [31] lanalysed the spread of *Spartina alterniflora* in Beihai in the years 2009 and 2011, using high resolution images acquired by UAVs. They captured images at a flight height of 800 m, generated orthomosiacs, performed multi-resolution segmentation by grouping homogenous pixels, and classified them. The target species was extracted by their pixel values. In a final step the accuracy was assessed and verified with field data by comparing three sample plots (a total of 166 samples) with the image results. A total accuracy of 94.0% could be achieved and, hence, provided information regarding an increasing spread of 19.07% from 2009 to 2011. The total infected area was, in 2011, 357.2 ha. Moreover, the image analysis provided the opportunity to identify areas with different levels of densities. Reference [35] collected UAV images of *Harrisia pomanensis* in the Limpopo province of South Africa. An area of 87 ha was captured by images that were taken at a height of 800 to 817 m. Orthomosaics generated with Agisoft Photoscan and pixel as well as object-based classifiers were used. The classification results of supervised and unsupervised classifiers were assessed. The supervised classification outperformed the unclassified one, and the object-based approach outperformed the pixel-based one. The best accuracy achieved was 86.1%.

Given the sizes of the data sets used and the need to detect and classify parts of the images that only trained experts can confidently tell apart the use of deep learning in problems that are closely related to the current work is of common use. For example, a recent study [36] used CNN (convolutional neutral network) in order to classify images from wetlands in an area of 700 km^2. They used a fully trained and fine-tuned CNN with a limited amount of data. A comparison of the CNN with random forest was performed to evaluate the capacity and classification accuracies of CNNs. Canadian wetlands were captured with two RapidEye images with 5 m resolution in 2015. The validation data were sampled in 2015 and 2016 and four wetland classes were identified using 1000 samples. The network was trained with patches and 30,000 iterations and then tested in a second step. The CNN outperformed the random forests and an overall accuracy of 94.82% was achieved, varying between 76.65% and 98.74%. Deep convolutional neural networks were also used by [37] in order to classify AUV-acquired wetland images. The 677 m × 518 m study area was located in Southern Florida. The authors used processed orthomosaics and multi-view images and then compared them with the performances of random forests and support vector machines. Image segmentation was done with Trimble's eCognition by first segmenting objects, then extracting features, and finally training a classifier. The results of the study show the advantages of deep CNN, reaching an accuracy of 82.02%, when multi-view images were used and with lower accuracy when orthomosaics were used (71.69%). A similar approach was used in [38]. 3800 images of the Brazilian national forest (Kaggle dataset) were used in order to identify *hydrangea* in the images. The dataset contained two-thirds of images, including the invasive species, a smaller fraction, where the invasive species appeared only in parts of the images, and a third small fraction that included no plants at all. The authors used three models: VGGNet, DenseNet, and Inception, which were pre-trained with ImageNet. Data augmentation was used and accuracies of 97.6% were reached.

2. Materials and Methods

2.1. Study Area

The study area "Lichtenmoor" is located in a wetland region about 60 km northwest of Hanover in Lower Saxony, Germany (52°43′06.2″ N 9°20′41.5″ E) (Figure 1).The total size of the moor area is 38 km^2 [39]. Post industrial peat cutting characterises the central area of the studied wetland. As a subsequent use, some former peat cutting areas have been rewetted with the aim of regenerating raised bogs. In the surrounding area, parts of the Lichtenmoor have been designated as nature reserves. Former hand peat cutting sites are located at the edges and in parts of the nature reserves. Agricultural areas dominate the remaining areas, mostly grassland, dry to moist moorland forests, scrubby heather and moorland degeneration stages, pioneer stages of moorland rewetting, and peat extraction areas under cultivation. The case study area covers a total area of 62 ha and it is located in the central part of the Lichtenmoor region. To the northeast, it borders a pine plantation, while at its southwest border current peat extraction areas are located. Blueberry plantations can be found throughout the Lichtenmoor, especially on the outskirts of the localities Lichtenhorst, Heemsen, Sonnenborstel, and Steimbke.

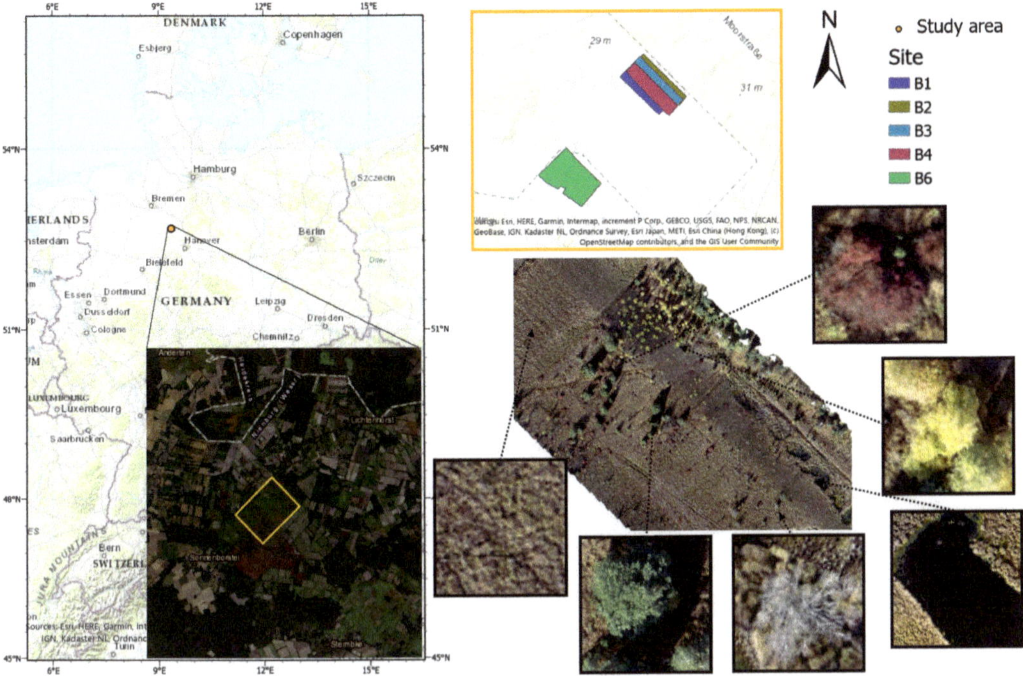

Figure 1. Location of the study area in the north of Germany. In the upper right part of the figure, the study sites are marked with different colours. The bottom right part an example orthomosaic is shown with detail of the different classes.

2.2. Characteristics of the Blueberry Species

Blueberries have been cultivated in commercial plantations in Lower Saxony since the early 1930's [14]. Since then, the area under cultivation has steadily increased. In 2005, the area with blueberry cultivation in Lower Saxony was approximately 1400 hectares [13]. In nature, blueberries are mainly distributed by birds and small mammals, who spread the seeds. Once established, plants can spread in a vicinity by clonal growth and the high regeneration potential of blueberries favours a strong spread [10]. The species prefers acidic locations, such as pine forests or wetlands. Especially, raised bogs in their degeneration stage provide ideal habitat conditions for invasive blueberries [14]. Thus, blueberries are growing increasingly wild in neighbouring areas. References [9,14] found a correlation between the density of blueberries and the distance to blueberry plantations; a maximum distance of 1700 m was recorded. Near cultivated areas, the feral blueberries form dense shrub stands with height of up to 2–3 m and have a ground coverage of up to >60%. With increasing distance, the degree of coverage decreases rapidly [8,9,13]. Since these studies were already conducted in the last millennium, a larger distribution must be expected today. Reference [10] describes a distribution of blueberry bushes over 4 ha in the "Krähenmoor", where the maximum distance to the plantation was identified at 100 m. Blueberry bushes are low and can reach a height of 60 cm, but, occasionally, tall species can be found with a height of 3 m [10]. In the course of the increasing growth and dense shrub structures, other ground vegetation is displaced, since it cannot exist under the shade of blueberries. Other presumed effects of blueberry cultivation are reduced evaporation rates and the influence on nutrient cycles in wetlands, which, in turn, can have an impact on existing plant species. Therefore, human interventions are necessary to protect the sensitive rare structures and characteristic plants of wetlands [10].

2.3. Data Collection and Pre-Processing

Image collection was performed by using a DJI phantom 4 UAV in autumn 2018, because of the seasonal red colouring of blueberry leaves, which makes them easily detectable (Figure 2). For sites B1 to B4, 490 to 584 images were collected, while 1346 images were taken for B6. The flight height was 50 m and front and side overlaps of 80% were chosen. These images were then processed while using the Metashape software [40] to align images in order to produce one orthomosaic and DEM (Digital Elevation Model) for each site (Figure 1). It should be mentioned that an overlap between the sites B1 to B4 was chosen, so the east and west borders of the orthomosaics B1 to B4 are overlapping. All of the obtained orthomosaics were annotated while using the open source image manipulation software GIMP [41]. For three of them (B1, B2, and B3), the whole orthomosaic was annotated and each pixel was given one of the following six labels: blueberries, trees, yellow bushes, soil, water, and dead trees (Figure 1). The class trees contains pine trees (*Pinus sylvestris*), the class yellow bushes contains shrubby birches (predominantly *Betula pubescens*, secondarily *Betula pendula*). Binary layers for each of the six classes were created for each of the three orthomosaics while using the pixel-level labels. These annotations were based on colour, shape, and context information contained in the orthomosaics. In the last two orthomosaics (B4 and B6), only blueberry bushes were annotated.

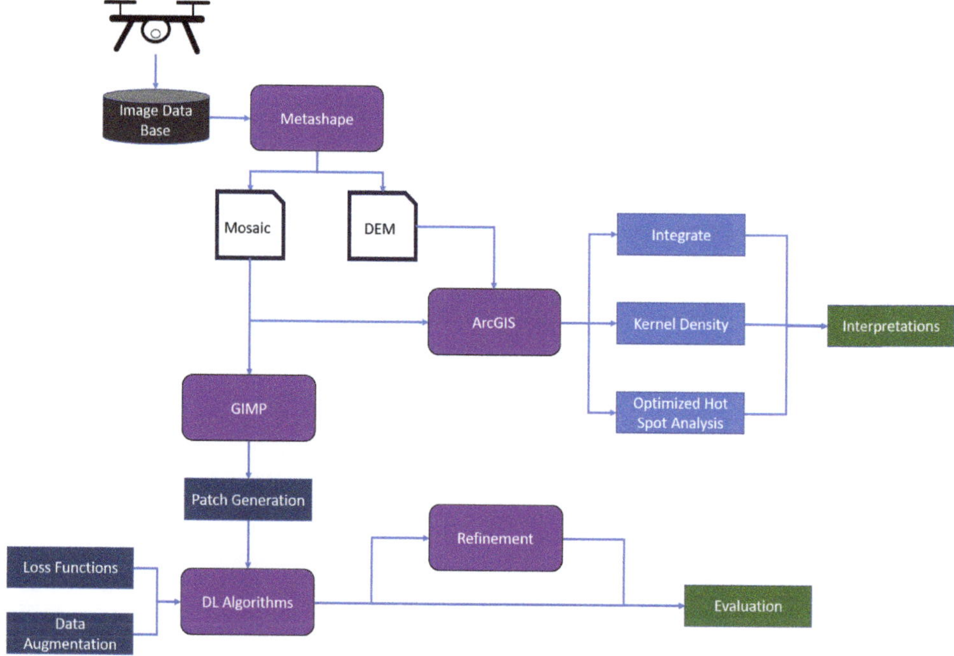

Figure 2. Workflow of the paper. Gray is our data base; purple boxes show softwares/programs used in this study; white boxes with a blue outline are generated files; dark blue are processes used for the deep learning (DL) classification and detection; light blue are tools in ArcGIS.

In order to train deep learning models to detect blueberry bushes orthomosaics annotated with all of the aforementioned labels were needed. Therefore, only B1 to B3 were used in the deep learning section of this study. These three orthomosaics, as well as the corresponding annotated binary layers were divided into axis-parallel patches of side length = 100 (hereafter "patch size"). This value was determined by taking the sizes of the blueberry bushes in the images, which ranged from 20 to 100 pixels in radius, into account. The classes in each patch were stored in a separate *label* list. In general, patches contained more than one class and, therefore, our problem can be defined as a multi-label patch classification problem.

2.3.1. Data Processing Using Arcgis

This section deals with data processing performed with ArcGIS pro 2.4.1 and python in order to identify parts of the orthomosaic containing blueberry bushes, in order to visualize them and analyse their characteristics. ArcGIS is a Geographic Information System software that visualises and comprehends geographic data. The software provides over 1000 tools to analyse real world data, including UAV-acquired images. The mapping options allow to visualise the gathered data within the correct location in an eligible base map. The primary purpose of this study was to provide information about the location and distribution of invasive blueberry species and map them for management.

In this context, the positions of the blueberry bushes in the five orthomosaics were digitised by hand as point data in an ArcGIS shapefile and several analytical tools were then used (Figure 2). The *kernel density* tool functions were used to calculate the magnitude-per-unit area from the blueberry points. Smaller search radiuses were used to show a detailed density raster. The tool *integrate* was used to group blueberry bushes that fall into a specified distance, as specified distance, 3 and 6 m were used. With the tool *collect event*, the number of points which were integrated before, were summarized in a new layer. Those steps were necessary to perform an *optimized hotspot analysis* of the blueberry abundance with the blueberry point shapefile. The hotspot analysis identifies the significant difference between the neighbourhood of a feature in comparison to the extent of the respective study area. Is the value of a feature significantly higher, it is considered to be a hotspot and the tool provides a feature map with three levels of confidence (90%, 95%, and 99%). As input for the hot spot analysis, the created layer of the tool *collect events* was used and analysis field counts were chosen. Furthermore, it also indicates the significantly lower features.

On the basis of the annotation made for all orthomosaics, several simple python codes and ArcGIS were used to perform image analysis. Pixels were counted for all orthomosaics, as well as all annotated layers. The number of black pixels in the annotated layers were specifically counted in order to obtain the percentage and area in m^2 per class. Additionally, the overall area presented in the orthomosaics was calculated in hectares. Because the focus of this study was on the blueberry bushes, several statistical values were generated: the number of blueberry bushes was counted and the number of blueberry bushes per ha was calculated for each orthomosaic. Additionally, the total area, as well as the area per blueberry bush, were computed in m^2. Furthermore, the proportion of blueberry bushes in relation to the vegetation was calculated in %. Finally, height values were computed on the basis of DEMs, annotations of the blueberry bushes, and annotations of ground points per site. Ground points were annotated close to blueberry bushes in order to increase the accuracy of the computed height. Maximum height values were estimated in a first analysis and the median height in a second analysis to evaluate the results.

2.4. Persistent Homology

Persistent homology provides topological information of complex datasets [42] at different spatial resolutions. This information deals with the connectivity of nearby points and it can be computed in different dimensions. For the purpose of this study, we worked with 0-D persistent homology (usually expressed in the form of H0 diagrams). H0 homology can be seen as growing disk at uniform radius-increase speed around pre-defined data points. In order to use this tool, we first discredited the manually annotated blueberry regions by uniformly sampling them. Subsequently, the radius around each sample point was increased to grow blueberry bush regions. When two blueberry regions were merged because of the expansion, one of the regions was considered to be dead because it was absorbed by the other region. As time passed, the number of connected blueberry regions decreased, and finally all of the regions were connected in one region.

The H0 diagram shows the change in the connectivity of the blueberry bushes by plotting the time/radius when blueberry regions get connected. False regions that were produced by our sampling of the annotated regions were discarded and only those parts of the diagrams that were obtained after all sampling points in each region had been joined together were considered. Based on this outcome, the radius was calculated until 1%, 10%, 50%, and 90% of the blueberry regions were connected and plotted. The diagram will vary greatly, depending on the number and position of blueberry bushes in the input image.

2.5. Deep Learning Techniques

Deep learning is a trending field of machine learning that focuses on fitting large models with millions of parameters for a variety of tasks, such as image classification and segmentation. These approaches have been rapidly gaining attention in computer vision tasks, due to their recently increased accuracy. Deep learning models commonly learn from examples in a supervised manner. First, an *architecture* or a graph of connected nodes is defined. These nodes are often grouped in *layers* that perform a specific operation, and the combination of a large number of layers is referred to as a *deep neural network* (**DNN**).The typology of the nodes, the number of nodes per layer and the connections between them determine the behaviour of the network. In general, two main types of nodes are used: linear nodes, expressed as matrix multiplications and nodes that introduce non-linear functions (such as the sigmoid function). The weights in linear nodes are usually initialised with random values following a specific distribution. Afterwards, the network is given samples of the data, known as *training samples*, which contain instances of the problem (i.e., image intensities) with their corresponding solutions (i.e., labels). These samples are iteratively run through the network in order to evaluate its current accuracy and the weights are updated following an optimization process.

In this study, DNNs were used to locate and identify the six classes that are defined in Section 2.3, with a focus set on the blueberry class. The basis for this deep learning approach is described in previous studies [20,43], which led to the use of the algorithms that are described in this section.

Our approach is based on a patch classification model that uses the patches of 100×100 pixels described in Section 2.3. A patch of orthomosaic B1 and B3 covered; therefore, 700 cm $\times$ 700 cm and a patch of orthomosaic B2 500 cm $\times$ 500 cm. For each patch, a list containing the class labels was created from the binary maps for each class. This classification is usually referred as multi-label, since each input patch might contain different labels (i.e., a part of the patch may contain soil, while other parts of the same patch could also contain bushes or blueberries).

Deep neural networks for classification have two major components: a feature extraction stage and a prediction stage. At the first stage, convolutional operators are trained in order to extract salient and meaningful features (such as texture) while at the second stage these features are used to predict the final labels for the given input patch or image. In order to train general and robust feature extractors, a large pool of heterogeneous images with different properties (lightning, colour, view, etc.) is needed to capture all of the possible image variabilities. However, as proven by our previous work [20,43], transfer learning is a useful tool for image analysis applications, where the training dataset is too small to properly train these feature extractors from scratch.

In our case, only three different orthomosaics are available; hence, we decided to use transfer learning by loading a pre-trained ResNet50 architecture with weights from ImageNet, due to its accuracy and reduced training time. ImageNet is one of the largest image databases for image classification research, with more than 80,000 labels and at least 1000 images for each label.

In order to perform the evaluation of the proposed model, two of the three orthomosaics were used for training and validation and the third one was used for testing. This cross-validation strategy is usually referred to as a leave-one-out strategy. All of the orthomosaics were used once for testing, training, and validation by rotating them for each experiment. Patches from the testing orthomosaic were not included for training or validation to avoid data leakage during training.

Two main approaches were used in order to obtain a higher detection rate for the blueberry class:

- Data augmentation is a commonly used strategy in deep learning and it can increase the size of the training datasets without the need to collect new data. In this case, data augmentation was used to generate new synthetic patches of the blueberry class, which was the less frequent class (see Section 3.2 for details). Six image transformations to augment the data were used: up/down and left/right flips; small central rotations with a random angle, in order to simulate different perspectives of the bushes; Gaussian blurring of the images, which simulates blurring due to the movement of the UAV; linear and small contrast changes, which can represent different light and shadow conditions and localised elastic deformations. In order to implement this transformations we used the "imgaug" library [44].
- Loss functions are used to compute the accuracy of the network and update their parameters. By giving different weights to different classes, their importance can be changed during training. In this study, two loss functions were used; the first function checks if a patch contains a blueberry or not, while the second one checks the fraction of blueberry pixels inside the patch. The optimal training settings followed the ones that were used in our previous study [43].

We considered labels for all the patches used and the relation between (1) predicted values resulting from our algorithm and (2) real values as stated in the manually-annotated ground truth in order to assess the predictive power of our algorithms.

We then broke all of the patches into the usual classification categories of:

- **True Positives** or TP, predicted to contain the blueberry class and also marked in the ground truth as containing them.
- **False Positives** or FP, predicted to contain the blueberry class but NOT marked as such in the ground truth. These patches correspond to over-prediction errors where the algorithm "sees" the blueberry class when it is not really there.
- **True Negatives** or TN, not containing the blueberry class in the prediction or in the ground truth.
- **False Negatives** or FN, not predicted to contain blueberries, but actually being marked as containing them in the ground truth. These patches correspond to under-prediction errors where existing blueberry instances are missed by the algorithm.

We used the two following metrics (True Positive Rate or Sensitivity and Accuracy) in order to summarize the occurrences of the four categories.

$$TPR = SENS = \frac{TP}{TP + FN} \qquad ACC = \frac{TP + TN}{TP + TN + FP + FN} \tag{1}$$

Finally, we modified the algorithm to present the results in a way that was more usable for end-users. The 100×100 patches used for prediction managed to capture most of the occurrences of the Blueberry class (see Section 3.2 for details). However, these rather large patches also included large areas that actually contained no blueberries. While expert users could easily use these results as a starting point in order to quickly identify the exact location of blueberry bushes, we felt that refining our prediction using smaller patches would make their work faster, while also providing clearer information for non-expert users. Consequently, we divided each of the predicted 100×100 patches into

Table 1. Area and counting measures of blueberry bushes detected in the orthomosaics.

	Orthomosaic Area in ha	Number of Blueberry Bushes	Blueberry Bushes per ha	Blueberry Bush Area in m^2	Area per Blueberry Bush in m^2
B1	11.64	375	32.21	1331.42	3.55
B2	10.64	687	64.55	1885.51	2.74
B3	12.47	566	45.40	1470.24	2.60
B4	12.44	405	32.54	870.33	2.15
B6	15.14	235	15.53	278.07	1.18

The number of blueberry bushes varied from 235 in orthomosaic B6 to 687 in B2 (Table 1). The site areas of orthomosaic B1, B3 and B4 were similar, while orthomosaic B6 is the largest site containing the least number of blueberry bushes and orthomosaic B2 contains the greatest number of blueberry bushes in the smallest area. The ratio could be confirmed by calculating the blueberry bushes per ha (Table 1). In another step, annotations were used in order to calculate the total area covered by blueberry bushes. In orthomosaic B6, an area of 278.07 m^2 was covered by blueberry bushes, which represents the smallest area and it resulted in an area of 1.18 m^2 per blueberry bush. The largest area was covered by blueberry bushes in B2 with 1885.51 m^2 . The average size of the bushes were similar for B2 and B3. In site B1 the average size of blueberry bushes was the highest with 3.55 m^2, while the covered area was third lowest with 1331.41 m^2.

Because the covered area and the average size of the bushes could be calculated, the next point of interest was the area and height per blueberry bush (Figure 5). Bushes were grouped into six to smaller than 10 m^2 and over 10 m^2 because the percentage of blueberries decreased towards larger cover areas. B1 and B2 had approximately 30 bushes between 6 and 8 m^2, which was the maximum of all sites. The mean areas were computed for all orthomosaics, indicating that B1 had a high number of large bushes with a mean area of 3.57 m^2. The smallest blueberry bushes could be found in orthomosaic B6 indicated a mean value of 1.18 m^2. In general, most of the blueberry bushes showed areas of up to 2 m^2, a lower amount distributed between 2 and 4 m^2 and the lowest numbers distributed in areas greater than 4 m^2. B1 was an exception, with around 10% per class over 4 m^2. The highest areas calculated range between 17 to 25 m^2, with B1 containing four bushes in that range and 27 with areas above 10 m^2. The lowest areas were found to be less than 10 cm^2 for B1/B4 and approximately 15 cm^2 for all other orthomosaics.

A similar distribution can be seen in Figure 5, where the number of blueberry bushes were plotted against the height. Classes were chosen for each 0.5 m starting with 0 m up to lower than 3.5 m and more than 3.5 m. This distribution was chosen due to the characteristics of shallow blueberry bushes that reach 60 cm and tall species that reach 3 m. Regarding the maximum height, no height was computed for 2.3% (B1) up to 15.5% (B6) while the numbers were higher when the median height was considered (11.0% for B1 up to 35.2% for B6). In general, the median height values were higher for the classes 0 m and 0.5 m in comparison to the maximum height, while the values are lower from 0.5 m.The lowest height values started from 0.01 m (B6), 0.03 m (B1), 0.07 m (B3/4), and 0.1 m (B2) for both max. and median height. In general, the maximum and the median height distribution of the orthomosaics was similar. Almost all of the blueberry bushes in orthomosaic B6 were within the class < 0.5 (83.3%). In B1 the number of blueberry bushes in this same class was 79% and 15.2% was between 0.5 and <1 m, which was similar to B6. Orthomosaic B2 to B4 showed a Poisson distribution, whereby B2 had the highest number in 1 < 1.5 m with 21.2 m, while 41.6 of the blueberry bushes in B4 had the highest value in <0.5 m. Furthermore, in B2 and B3, more than 50% of the blueberry bushes reached heights between 1 m and 3.5 m (57% and 57.7%).It has to be considered that the area was calculated on the shapefiles in ArcGIS, while our developed python code was used in order to calculate the height of the blueberry bushes. Because the input for the code was the annotations of the blueberry bushes that were stored as a PNG file, those bushes that were close together were grouped. Therefore, the height values were not always calculated for an individual bush, which resulted in a different number of blueberry bushes per site: 309 (B1), 519 (B2), 461 (B3), 394 (B4), and 219 (B6).

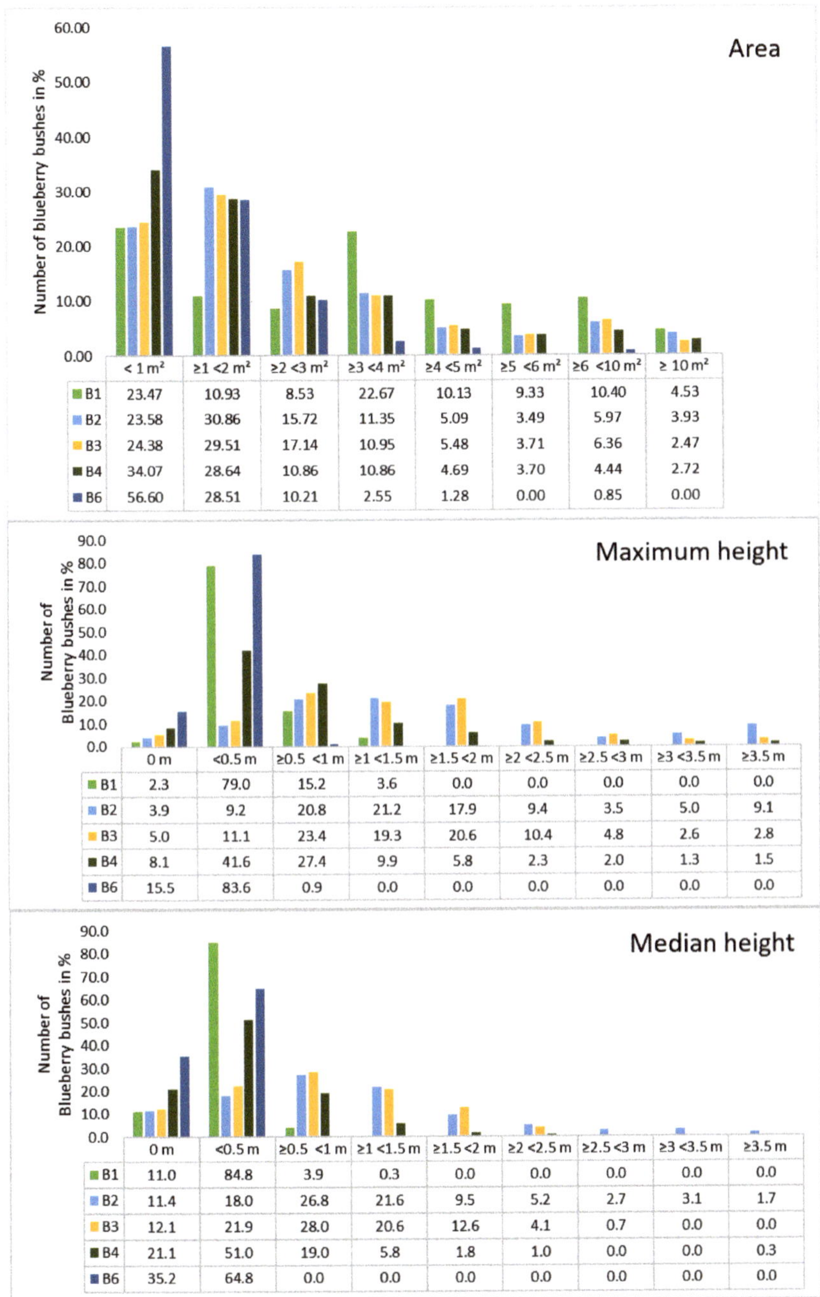

	< 1 m²	≥1 <2 m²	≥2 <3 m²	≥3 <4 m²	≥4 <5 m²	≥5 <6 m²	≥6 <10 m²	≥ 10 m²
B1	23.47	10.93	8.53	22.67	10.13	9.33	10.40	4.53
B2	23.58	30.86	15.72	11.35	5.09	3.49	5.97	3.93
B3	24.38	29.51	17.14	10.95	5.48	3.71	6.36	2.47
B4	34.07	28.64	10.86	10.86	4.69	3.70	4.44	2.72
B6	56.60	28.51	10.21	2.55	1.28	0.00	0.85	0.00

	0 m	<0.5 m	≥0.5 <1 m	≥1 <1.5 m	≥1.5 <2 m	≥2 <2.5 m	≥2.5 <3 m	≥3 <3.5 m	≥3.5 m
B1	2.3	79.0	15.2	3.6	0.0	0.0	0.0	0.0	0.0
B2	3.9	9.2	20.8	21.2	17.9	9.4	3.5	5.0	9.1
B3	5.0	11.1	23.4	19.3	20.6	10.4	4.8	2.6	2.8
B4	8.1	41.6	27.4	9.9	5.8	2.3	2.0	1.3	1.5
B6	15.5	83.6	0.9	0.0	0.0	0.0	0.0	0.0	0.0

	0 m	<0.5 m	≥0.5 <1 m	≥1 <1.5 m	≥1.5 <2 m	≥2 <2.5 m	≥2.5 <3 m	≥3 <3.5 m	≥3.5 m
B1	11.0	84.8	3.9	0.3	0.0	0.0	0.0	0.0	0.0
B2	11.4	18.0	26.8	21.6	9.5	5.2	2.7	3.1	1.7
B3	12.1	21.9	28.0	20.6	12.6	4.1	0.7	0.0	0.0
B4	21.1	51.0	19.0	5.8	1.8	1.0	0.0	0.0	0.3
B6	35.2	64.8	0.0	0.0	0.0	0.0	0.0	0.0	0.0

Figure 5. Distribution of the area and height values of blueberry bushes. From top to bottom: area, maximum height, and median height.

3.1.2. Analysis of Spread Patterns

In a second step of our quantitative analysis of the blueberry invasion, the concentration, density, and spread patterns were examined. GIS and persistent homology were used to assess these issues. Characterisations of concentrations and densities can indicate the number of blueberry bushes within a given region of the orthomosaic, which exceeds a simple location because the distribution of the bushes can be analysed precisely. Clustering bushes and mapping densities further increase the understanding of the distribution. Together with the persistent homology and hotspot analysis, the spread can be defined for all orthomosaics, which helps to characterise the invasion.

The first step was to cluster blueberry bushes by using specified distances, of which 3 and 6 m were chosen, due to the calculated area of the blueberry bushes. The average diameter was considered to be 2 m for the different sites, and therefore a diameter of 3 m was found to be appropriate to especially group blueberry bushes that were close to each other. The results of both distances were compared and they are listed in Table 2. When blueberry bushes were located in a range of 3 m, they were clustered with the following results. From orthomosaic B3 to B1, 35.51% to 39.87% were clustered. The highest number of clustered bushes were 33 in B3, followed by 25 in B2 and 10 in B1. In comparison to B1 to B3, B4 and B6 had around 28.3 % clusters with three and more blueberry bushes. The highest number within a cluster was nine for B4 and 13 for B6. After increasing the range to 6 m, the number of blueberry bushes clustered in the group 3 or more bushes increased to 69.57%, which is more than 30 percentage points. B1, B3, and B4 had a similar increase of around 15 percentage points and reached 56.63 % in B1, 50.43% in B3 and 44.35 in B6. With less than 10 percentage points, 37.68 % of the blueberry bushes were grouped together with more than three bushes.

Table 2. Clustering results that are based on point shapefiles of the sites.

	3 m Clustering			6 m Clustering		
	Grouped 3 or More (in %)	Number of Single Bushes	Highest Count in One Group	Grouped 3 or More (in %)	Number of Single Bushes	Highest Count in One Group
B1	39.87	89	10	56.63	18	25
B2	36.82	87	25	69.57	23	50
B3	35.51	98	33	50.43	36	50
B4	28.34	92	9	44.35	39	22
B6	28.28	51	13	37.68	28	21

Based on the point shapefiles of the blueberry bushes, density maps were generated to see how the bushes were distributed within the map. Figure 6 provides three examples of the orthomosaics B1 to B3. Areas with a high density are marked in red and low densities in dark green. Orthomosaic B1 has one large density spot in the northwestern part of the map, while the southeast direction the density decreases with only single or paired bushes. Orthomosaic B2 shows four density spots. Two smaller ones were located in the northwest, a larger spot is in the middle of the orthomosaic and a final one in the southeast. The space between the middle and southeast spot is covered by blueberry bushes, which was similar to the distribution of B3. In orthomosaic B4, nearly the whole area is covered with green to reddish colours. There are three dense spots in the northwest, two spots in the middle and one in the southeast. In comparison to B2 and B3, the spots are smaller. Orthomosaic B6 covers a larger area than all other orthomosaics, but only three density spots could be identified in the middle of the orthomosaic. There were smaller groups of blueberry bushes along the borders of the orthomosaic, and single ones are distributed close to the groups of bushes.

Another analysis focused on the point analysis to generate a map of hotspot areas in order to analyse the spread of the blueberry species. The point analysis used the manually marked blueberry bushes to identify where the proximity of the bushes was significantly different (hot and cold), and to quantify those that were not identified as significantly different. In B1, two 90% confidence hotspots were found in the north. 21 clusters were identified to be 90% significantly different from the study area. The hotspots in B2 were concentrated in the south-easternmost part of the orthomosaic. 26 clusters (out of 220) were significantly different to the study area with a confidence interval of 99%. These points contained all of the bushes located in the south-easternmost part of the orthomosaic. In B3 16 out of the 214 clusters fell into the 99% confidence interval, all located in the south-east of the orthomosaic. The same characteristic was found in B4. 23 clusters out of 248 were found to be significant with a 99% confidence and seven points with 90% to 95% confidence. B4 was the only orthomosaic containing two points considered to be cold spots with 90% confidence in the centre of the orthomosaic.

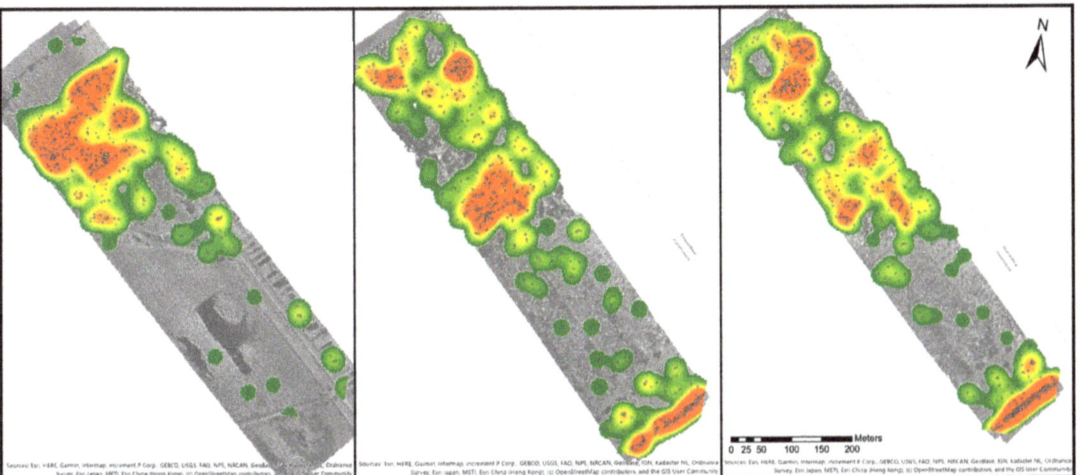

Figure 6. Density map for the blueberry species for the sites B1 to B3 (location see Figure 1). Areas of low densities are marked in green and high densities are red coloured. A gradient between green and red represents values of medium density. White points mark the location of the blueberry bushes.

Finally, the persistent homology was performed, as described in Section 2.3.1. The radius was plotted against the fused region, as can be seen in Figure 7. The orthomosaics B2, B3, and B4 show a similar trend, while B1 and B6 also follow a different, but similar, trend to each other. B2 is the first orthomosaic, where 1% of the blueberry bushes were fused with a radius of 386 and B1 needed the largest radius with a value of 497. In all orthomosaics the radius needed to fuse up to 10% of the blueberry bushes is similar with values between 415 and 557. There is a small gap of approximately 150 between B3 (945), B6 (998), B1 (824), B2 (768), and B4 (831), when 50% of the blueberry bushes were fused. The radius needed to fuse 90% for B1 and B6 are 3279 and 3611, while, for B2 to B4, it is 1610 to 1709.

3.2. Deep Learning Results

In this section, we describe the usefulness of deep learning techniques in order to automatically determine the location of blueberry bushes in our data. As stated in Section 2.5, a widely used network (ResNet50) was chosen and two main aspects were studied: how the data balance affects the final classification results and whether using transfer learning resulted in improved results. Several experiments were presented in a previous study [43]; however, we will only focus on the optimal results for this current study.

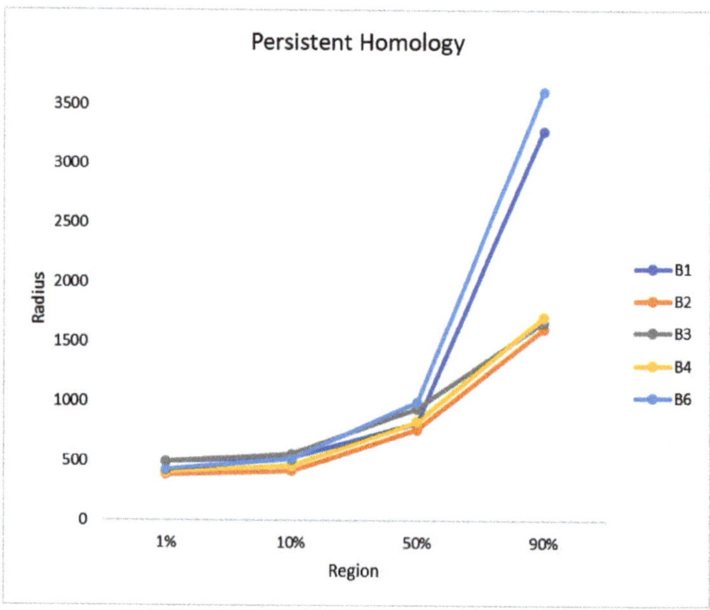

Figure 7. The persistent homology is plotted by radius against the region. Four fused regions were considered: 1%, 10%, 50%, and 90% and all sites are plotted.

Regarding transfer learning, *unfrozen* ImageNet [45] weights were considered to initialise the network. When the model weights are *unfrozen*, all of the layers are normally trained and, thus, all the weights are updated. Regarding the data balance, the blueberry patch loss was weighted eight times that of the soil class, and four times the amount of the other classes. We also performed upsampling of the blueberry class by creating 12 new samples for each patch, as detailed in Section 2.5. Finally, the soil class was downsampled to 50% of its original number of patches.

Because three orthomosaics were available, a leave-one-mosaic-out cross validation approach was applied in order to evaluate the results. One of the orthomosaics was used for training, another for validation and the last one for testing. In order to ensure that all orthomosaics were used at least once for training, validation, and testing, we rotated them accordingly. This section presents the averages results for the TPR and the accuracy results of the three testing stages that correspond to each orthomosaic.

By using the optimal settings that are presented in this section, the model improved from a low TPR value of 63.8% for the blueberry class when no data balancing was applied to a value of 93.39%. In both cases, the overall accuracy for all classes remained similar (98.83% and 98.10%, respectively). Furthermore, it could be observed that unfreezing the weights had a positive effect on the TPR. The best TPR value for the *frozen* weights was 37.12% without data augmentation while maintaining an overall high accuracy value of 98.01%. However, when using high data augmentation with *frozen* weights, the best TPR value of 87.99% was obtained at the expense of a lower overall accuracy value of 75.20%. These results suggest that ImageNet weights can be used for this problem, but they need to be updated (and *unfrozen*) when using data augmentation to focus on blueberry bushes due to the differences in images.

Regarding the refinement step of the algorithm, Figure 8 presents an example of the obtained results. For most of the predicted patches, the segmentation that was obtained with the refined version of the algorithm is much closer to the manual annotation. However, in a few cases, some of the bushes that had originally been detected are missed after the refinement. Table 3 presents the detailed results of these two issues.

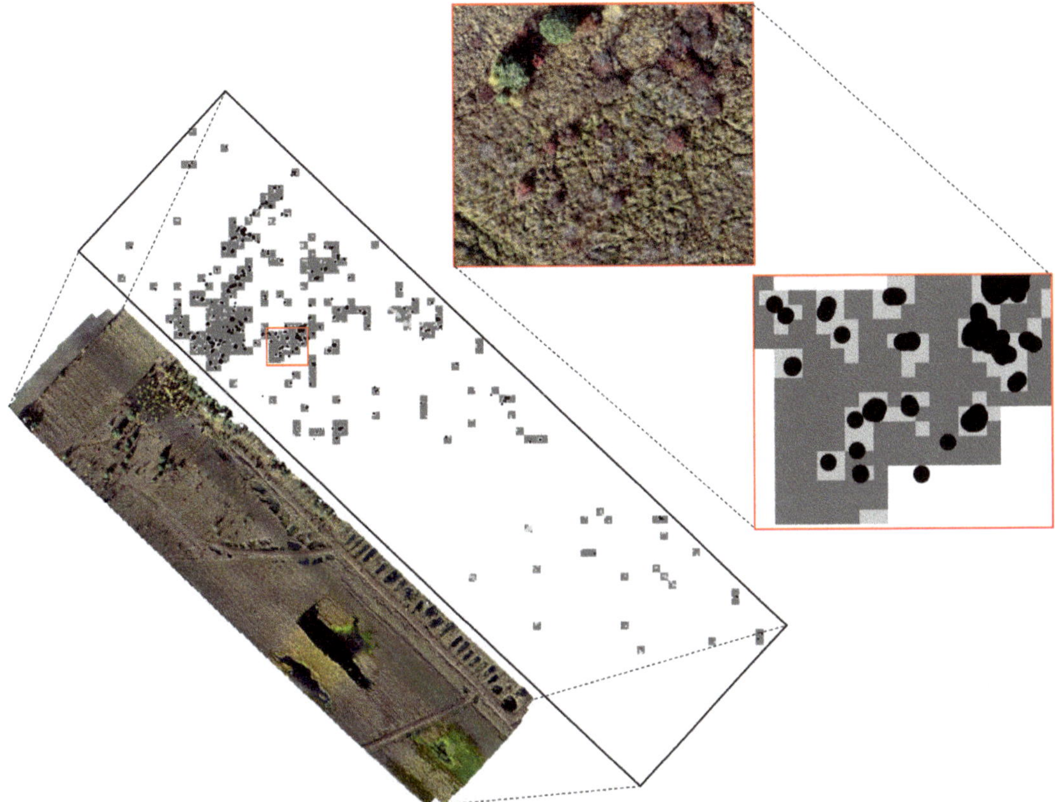

Figure 8. Orthomosaic B1 is displayed with a combination of the manual annotations (black marked spots), the coarse mask (dark grey) and the refined mask (light grey). The red box was zoomed in to show a detailed view on the image and masks.

Table 3. Numerical evaluation of the refinement step of the DNN. The rows marked "refined" stand for the algorithm after refinement, while the rows marked "Coarse" correspond to the algorithm without refinement for each of the three studied orthomosaics. The Dice coefficient, as well as the ratio of blueberry pixels in the Ground truth covered by each of the two masks, are presented.

Orthomosaic	Mask Type	Dice	GT Cover
1	Coarse	0.187	0.953
	Refined	0.526	0.860
2	Coarse	0.264	0.949
	Refined	0.624	0.874
3	Coarse	0.223	0.884
	Refined	0.587	0.789

On the one hand, the Dice coefficient, which evaluates the overlap and number of non-GT pixels that are contained in the predicted patches, improved significantly with the refinement algorithm from values around 0.2 to values in the 0.5–0.6 range. On the other hand, the ratio of GT pixels that are covered by the mask, which ranges from 0.88 to 0.95 for the non-refined mask was slightly inferior in the refined version (0.78 to 0.87).

4. Discussion

The applied methodology used UAVs to gather information in a restricted access area. Techniques from several research areas were then applied in order to gain knowledge regarding the distribution and properties of the bushes of the invasive blueberry species. In this section, the results that are presented in Section 3 are interpreted in order to assess the stage of the invasion in each of the mosaics.

4.1. Difficulties with Data

Blueberry plants show a characteristic red leaf colour in autumn, which make them easily recognisable and identifiable in comparison to other classes of vegetation. Both a simple identification and segmentation by colour were proposed and applied in one of the first segmentation approaches. However, partly visible soil with reddish tones constrained blueberry identification. This problem was especially critical for small blueberry bushes. In autumn, the leaf colours can vary between red, red with a yellowish tone, and partly black. This caused challenges for the annotations and for the deep learning algorithm, since the number of blueberry images was already low in comparison to the other classes and it made the colour approach not usable for this study. Further complications were given by light conditions during image taking. When the blueberry bushes had brighter red colours due to sun light, it was difficult to distinguish them from the ground. Bushes, which were located in the shadows, especially the ones that had a predominately black colour, were barely recognisable.

The analysis presented some difficulties in the calculations of the height and surface area of the blueberry bushes. The main problems to determine bush height are occlusions, due to nearby trees and difficulties due to dense floor covering vegetation. As the cluster analysis shows, the high density of blueberry bushes in some areas and their proximity increased the possibilities that the bushes were partly covered and the whole bush area was not visible, as already pointed out by [10]. Furthermore, bushes were often located close to trees that have canopies that can cover most of a blueberry bush. The areas calculated for the blueberry bushes, exceeding 4–5 m^2, indicated that there has to be more than one bush, which was difficult to identify in the images, as well as for calculating the height. Wetland regions, imaged in the orthomosaics are grassland and covered with dense hassocks. Therefore, the soil is often not visible and the ground annotations often represent the height of the hassocks, which resulted in values of 0 m maximum height and even more bushes showed a value of 0 m, when the median height was calculated for smaller bushes, especially in B4 and B6. Therefore, it is assumed that the hassocks can reduce the real height of the bushes by 30 cm. However, the calculated height values exceeded 3 m, which is unusual for the blueberry species that are studied here. There were errors produced when the annotations contained parts of an overlapping tree canopy, which increased the maximum height. The median height was resistant to outlier values. When ground areas were annotated, the median generally decreased the height values, especially for small bushes. Annotations of the ground need to be set carefully, since the wetland was uneven and depressions could increase the height values of the blueberry bushes. Furthermore, the differences between the max. height and median height can be influenced by the structure of the bush canopy. Because of these difficulties, a correlation between the height and area values of blueberry bushes was not considered, but a comparison of the distribution of these values showed that the distribution was similar and the bush area was larger than the height.

Even though these values are estimations, the applied methodology gave a good overview over a large study area, which cannot otherwise be done by extensive field measurements due to wetland protection regulations. Therefore, despite the difficulties, the achieved results emphasised the following discussion of applications and the qualitative use of the introduced methodology.

4.2. Application in Landscape Management

The collection of high resolution images and the gathered information can help to map and visualise the findings of this study. This information can be used in order to easily establish management measures against the further invasion of alien species into the wetlands, as pointed out in previous studies [31,35].

The area that was occupied by blueberry bushes was low in terms of the studied area (covering 1 to 1.5 %), which is lower than the identified 3 to 5% in [15]. However, when only the living vegetation was considered, the number of blueberry bushes was found to increase from 8.2 up to 21.1%. These percentages can be considered to be very high, due to the fact that the species is invasive and it does not belong to this sensitive ecosystem. B6 has the largest area and it contains the smallest number of blueberry bushes, with the smallest height and area values measured. Therefore, the invasion seems to be in an early stage and it should be easier to manage. Nevertheless, as shown by persistent homology and the high number of single bushes after clustering, bushes were wide spread, which increases the area where measures against the blueberry bushes need to be considered. The hotspot analysis and density map of B6 indicated that there are some bushes, which are concentrated in a dense spot in the middle of the area and distributed from there homogeneously. These findings allowed for determining that the progress of the blueberry bushes into this site is low.

B1 has a similar distribution, while the number of blueberry bushes per hectare is doubled in comparison to B6. The density map showed high densities in the northern part of the orthomosaic and a gradual decrease of blueberry bushes in the southern direction, which was confirmed by the hotspot analysis. The density of the bushes was higher than in B6, but, as indicated by the persistent homology, the spread was greater. This indicates that the blueberries invaded but did not reach every region of the site. Furthermore, the blueberry bushes in B1 have the largest bush area within all studied sites. The identified bushes maturity suggests that the blueberry species has invaded the area a long time ago. Because the density map provided the information that the blueberry bushes are mainly distributed in the north, there must be conditions in the south, which prevented further spread.

In comparison, B2, B3, and B4 showed a high spread in the southern direction, because blueberry bushes of various size were found everywhere and there were higher concentration areas and several spreading centres. These findings can be confirmed with the persistent homology and density map, indicating a high progress of the invasion. The three orthomosaics seem to have a homogeneous distribution and a gradual change to lower numbers in the southern direction. However, the significant differences that were found by the hotspot analysis indicate that there are conditions that influence the distribution of the blueberry bushes, as in site B1. In addition, all three orthomosaics show an area with a small density of bushes, which can probably be explained by the high water content in the soil correlated with unsuitable conditions regarding plant growth. The invasion of the blueberry species was characterised as such and far advanced for B2 to B4. Only B4 has smaller bushes that cover a smaller area, which suggests that the invasion is less advanced than for B2 and B3. The spread will probably increase there in the following years.

Furthermore, the distribution of the blueberry bushes can be connected to the proximity of trees, especially in B1 to B3, where bushes are mainly found around pine trees and shrubby birches, since birds use these as rest areas and mainly distribute seeds where they rest. An exception is B4, where the trees are located next to depressions that are filled with water, which confirms the unsuitable living conditions for blueberry species. In general, it seems that more blueberry bushes occur when the density of the living vegetation is higher, which can be explained by better living conditions and a better distribution through birds.

To sum up the results and interpretations, it was found that B6 showed an early stage of invasion. B1 shows an advanced stage of invasion with limitations in the south, while B2 to B4 show a critical, advanced stage of invasion, since the blueberry bushes can be found in the whole study site. The methodology used here helped to assess the stages of invasion.

Our study shows a method for helping preserve a sustainable and adaptive conservation of natural ecosystems. Together with further expert interpretations, a deeper understanding of wetland ecosystems can be achieved [30]. The calculated properties, height and area, are indices that can be used for plant growth monitoring [46], and, therefore, they provide useful information for the practical management of wetlands.

4.3. Contribution to Invasive Blueberry Studies in Wetlands

Previous studies were conducted to identify areas of blueberry bush invasions. The first studies were done by [9,14] covering up to 12.5 km^2 with estimations of blueberry bush coverage areas. Another study focusing on the invasive blueberry species was presented by [15], studying an area of 4.7 ha, which is nine times smaller than the area that was studied in our work. Reference [10] studied an area of 230 ha and characterised the blueberry invasion. The author conducted 11 days of fieldwork in order to capture the information regarding the cover area of blueberry bushes. The study area is less in our work with 63 ha, but, in comparison to the manual fieldwork, image capturing only took half a day. The pre-processing step needed the longest time, about 4 to 5 h per orthomosaic, since manual annotations were done in the beginning of the work. Therefore, the presented workflow of our study reduces the amount of time for gathering information significantly. The characteristics that were provided by [10] are mainly focused on the blueberry bush cover area, which were 16% of the 230 ha. Other characteristics were provided by general statements about average heights and clustering behaviours. Our study provided height and area measurements for each blueberry bush, which has not been done in previous studies. Based on the orthomosaics, further measures, like hot spot analysis, density calculations, and spread measures, could be performed in order to characterise the blueberry invasion in much more detail than previous studies have shown [10,15]. Furthermore, refs. [10,15] stated a high probability of loss of identification during fieldwork due to the density of the bushes in certain areas near former blueberry plantations, which were therefore predominantly investigated. In comparison to the mentioned studies, our work presents an objective grading, a precise cover area and each bush was detected in the captured high-resolution images. This information is essential in order to compare both the invasion and monitoring in different areas.

4.4. Automatic Masks Generation

Deep learning techniques were used in order to assess whether DNN can be used to automatically detect the presence of blueberry bushes. Since the manual annotation of the blueberry bushes is the most time-consuming step of this workflow, this has the potential to greatly extend the range of studies. The results in Section 3.2 show that the ResNet50 network succeeded at the classification tasks associated to our problem and that the best results were obtained by re-training the whole network. In this respect, relying on pre-trained weights from ImageNet to solve our problem after minor re-training of the last layer is suboptimal. A dataset must be large enough to re-train the full networks. In our case, this meant using images that were taken from three orthomosaics covering a total of 33 hectares.

At the same time, the results also quantify how a data imbalance may result in a network that classifies most of the patches correctly, wven if the blueberry bushes were mainly misclassified. To address this problem, data augmentation as well as a loss function that took into account the number of pixels for each class were used to influence the weights which helped to reduce bias of the training towards the correct classification of blueberry bushes.

Wetland image classification was performed by [36,37] identifying four and six classes, reaching an overall accuracy of 94.82% and 82.02%. Transfer learning which was only applied in [36] showed the effectiveness of this technique for natural environment studies, when the dataset is large enough. Interestingly, the best result of [37] was obtained when using multiview images. We will consider this type of images in our future work. A similar approach to detect invasive species was used by [38]. Their approach applied deep learning with data augmentation and transfer learning. Their highest accuracy was 97.6%, which is slightly lower than our overall accuracies of 98.83% and 98.10%. It is important to consider that, in [38], the dataset was mainly composed of images of the invasive species, while, in our case, only 2.64% of the generated patches contained the invasive species. This illustrates the effectiveness of the techniques used in our approach to overcome the imbalance in our dataset.

The results of the automatic classification were used in order to map the invasive species, which is an important first step, whereby a refined segmentation is essential to effectively determine the exact location of plant invasions [35]. The deep learning applications used classified images, and a refined mask provided the blueberry bush locations. The refinement step was necessary in this application, since most of the blueberry bushes are small. Hence, the refined mask offered a more precise localisation of the blueberry bushes. Even though not all bushes were found and some soil areas were misclassified as blueberry bushes, the refined mask could significantly reduce the time of manual annotations and provide maps of the studied area. Increasing the amount of blueberry data can help to optimize the classification accuracy and the localisation of blueberry bushes. This will be considered in our future research.

The generation of automatic annotation masks, as performed here, will allow for large scale studies with a minimum of disturbances in the studied environment. Therefore, UAVs and image analysis provide accurate and cost-effective surveys that are needed when studying invasive species [31]. The used techniques provided more information than that gathered by previous field surveys in wetland areas [10,14].

4.5. Limitation and Future Works

The presented work provided a methodology to analyse invasive species from UAV images. Nevertheless, we faced limitations regarding the data that are described in Section 4.1. The autumn season seemed to be a good timing for image taking, but the difficulties regarding the varying colour of the blueberry bushes should be taken into account more carefully. Image taking could be done when the weather is cloudy and not windy, as suggested in previous studies [47,48]. The chosen flight height of 50 m resulted in high-resolution images, which created heavy orthomosaics. The resolution was reduced to 5–7 cm/pixel to be able to use image processing software. This led to difficulties in identifying blueberry bushes and their properties. For the future, different flight settings can be tested, as presented in [49,50], especially smaller flight areas and a reduced flight height can help to increase the resolution and precision of the results of blueberry bush properties.

The use of deep learning proved to be effective for this problem, but also challenging. As in all artificial intelligence approaches, a sufficient number of images that represent the variability of the desired target is necessary in order to train a supervised model that is usable in practice (in the current studies, this amounted to images that corresponded to 33 hectares). Furthermore, with the size of the current blueberry bush data in these images, careful use of data augmentation as well as a dedicated balancing approach was necessary. In order to improve the detection of blueberry bushes, a larger training dataset would likely be needed. For instance, because blueberry bushes are often planted for blueberry production, those fields could offer a good opportunity to collect new images.

Therefore, the detailed mapping within this study can further improve the results of the current status of the blueberry species invasion. Repeated data collection in the same area, as planned, will provide a year-to-year comparison, which will allow for the monitoring and analysis of the ongoing spread in the wetlands. Changes will be easily detectable within the wetlands and the studied blueberry species without disturbances of vulnerable plant and animal species and habitats, as already mentioned by [30,35].

5. Conclusions

In this paper, we introduced a multi-disciplinary methodology to quantitatively evaluate the role of plant species in ecosystems, including invasive species. The use of UAVs makes the approach applicable, even in restricted access areas and it increases the total area that can be studied, greatly exceeding the range of existing field studies. We used this methodology to gather information regarding wetland vegetation. Simple and time-saving methods were applied to classify vegetation and provide information about the properties of the invasive blueberry species found in our study site. The distribution of blueberry bushes was analysed in terms of their density, clustering, and spread. The relative importance of blueberries in the wetland was analysed (number of bushes, bush area, and bush height). This information was transformed into location, density, and hotspot maps to provide advanced visualization tools. Deep learning techniques were used in order to automatically detect and segment blueberry bushes, opening the possibility to further extend the range of similar studies.

Author Contributions: S.K., M.C., L.T., J.G., B.B., K.W., M.L.L.C. and Y.D., conceived the conceptualization and methodology, supported the writing—review and editing. S.K., M.C. and Y.D. developed the software, performed the validation, investigation and writing—original draft preparation. S.K., M.C. and Y.D. carried out formal analysis. S.K. was in charge of the visualisations. S.K., and J.G. administrated the data. J.G, B.B., M.L.L.C. and Y.D. supervised the project and provided resources. J.G., B.B. and Y.D. directed the project administration. All authors have read and agreed to the published version of the manuscript.

Funding: This research received no external funding.

Institutional Review Board Statement: Not applicable.

Informed Consent Statement: Not applicable.

Data Availability Statement: The data presented in this study are available on request from the corresponding author.

Acknowledgments: The publication of this article was funded by the Open Access Fund of the Leibniz Universität Hannover. We thank Angie Faust for language corrections and Thomas Beuster (ÖSSM - Ecological protection station Steinhuder Meer) for the opportunity to take aerial photography in the study site and for his helpful contributions to the ecology of blueberries in the Lichtenmoor.

Conflicts of Interest: The authors declare no conflict of interest.

References

1. Prentis, P.J.; Wilson, J.R.; Dormontt, E.E.; Richardson, D.M.; Lowe, A.J. Adaptive evolution in invasive species. *Trends Plant Sci.* **2008**, *13*, 288–294. [CrossRef]
2. Pyšek, P.; Richardson, D.M. Invasive Species, Environmental Change and Management, and Health. *Annu. Rev. Environ. Resour.* **2010**, *35*, 25–55. [CrossRef]
3. Pimentel, D.; Zuniga, R.; Morrison, D. Update on the environmental and economic costs associated with alien-invasive species in the United States. *Ecol. Econ.* **2005**, *52*, 273–288. [CrossRef]
4. Caffrey, J.; Baars, J.R.; Barbour, J.; Boets, P.; Boon, P.; Davenport, K.; Dick, J.; Early, J.; Edsman, L.; Gallagher, C.; et al. Tackling Invasive Alien Species in Europe: The Top 20 Issues. *Manag. Biol. Invasions* **2014**, *5*, 1–20. [CrossRef]
5. Rabitsch, W.; Genovesi, P. *Invasive Alien Species Indicators in Europe*; EEA Technical Report; 2012. Available online: https://op.europa.eu/en/publication-detail/-/publication/0e70dca6-0213-4420-b04a-8951cb9a0df7/language-en (accessed on 8 January 2021).
6. Didham, R.K.; Tylianakis, J.M.; Hutchison, M.A.; Ewers, R.M.; Gemmell, N.J. Are invasive species the drivers of ecological change? *Trends Ecol. Evol.* **2005**, *20*, 470–474. [CrossRef]
7. Piria, M.; Copp, G.H.; Dick, J.T.; Duplić, A.; Groom, Q.; Jelić, D.; Lucy, F.E.; Roy, H.E.; Sarat, E.; Simonović, P.; et al. Tackling invasive alien species in Europe II: Threats and opportunities until 2020. *Manag. Biol. Invasions* **2017**, *8*, 273–286. [CrossRef]
8. Hollenbach, M. Verstärktes Vorgehen der Naturschutzbehörde Gegen Die Nordamerikanische Kulturheidelbeere. 2014. Available online: https://www.nlwkn.niedersachsen.de/naturschutz/fach_und_forderprogramme/life/hannoversche_moorgeest/aktuelles_termine/verstaerktes-vorgehen-der-naturschutzbehoerde-gegen-die-nordamerikanische-kulturheidelbeere-126107.html (accessed on 11 November 2020).
9. Schepker, H.; Kowarik, I. Invasive North American Blueberry Hybrids (*Vaccinium corymbosum* x *angustifolium*) in Northern Germany. In *Plant Invasions: Ecological Mechanisms and Human Responses*; Backhuys Publishers: Leiden, The Netherlands, 1998; pp. 253–260.
10. Stieper, L.C. Distribution of Wild Growing Cultivated Blueberries in Krähenmoor and Their Impact on Bog Vegetation and Bog Development. Bachelor's Thesis, Leibniz University of Hannover, Hanover, Germany, 2018.

11. Nehring, S.; Kowarik, I.; Rabitsch, W.; Essl, F.; Lippe, M.; Lauterbach, D.; Seitz, B.; Isermann, M.; Etling, K. Naturschutzfachliche Invasivitätsbewertungen für in Deutschland wild lebende gebietsfremde Gefäßpflanzen. *BfN-Skripten* **2013**, *352*, 1–202.

12. Deilmann, H.C.; Eichhorn, G.; Falkenberg, H.; Günther, J.; Hayen, H.; Kuntze, H.; Pollak, E.; Schmatzler, E.; Steffens, P.J.T. Moor und Torf in Niedersachsen. *Niedersächsische Akad. Geowiss.* **1990**, *5*. Available online: https://www.schweizerbart.de/publications/detail/artno/183010500/Moor-und-Torf-in-Niedersachsen (accessed on 11 January 2021).

13. Kowarik, U.S.I. *Vaccinium angustifolium* x *corymbosum*. 2003. Available onlne: https://neobiota.bfn.de/handbuch/gefaesspflanzen/vaccinium-angustifolim-x-corymbosum.html (accessed on 11 November 2020).

14. Schepker, H.; Kowarik, I.; Grave, E. Verwilderung nordamerikanischer Kultur-Heidelbeeren (Vaccinium subgen. Cyanococcus) in Niedersachsen und deren Einschätzung aus Naturschutzsicht. *Nat. Landsch.* **1997**, *72*, 346–351.

15. Essl, F. Erstfund eines verwilderten Vorkommens der Kultur-Heidelbeere (*Vaccinium angustifolium* x *corymbosum*) in Östereich. In *Linzer Biologische Beiträge*, Austria; 2004. Available online: https://www.zobodat.at/pdf/LBB_0036_2_0785-0796.pdf (accessed on 11 January 2021).

16. Grenzdorffer, G.; Teichert, B. The photogrammetric potential of low-cost UAVs in forestry and agriculture. *Int. Arch. Photogramm. Remote Sens. Spatial Inf. Sci.* **2008**, *31*, 1207–1214.

17. Raparelli, E.; Bajocco, S. A bibliometric analysis on the use of unmanned aerial vehicles in agricultural and forestry studies. *Int. J. Remote Sens.* **2019**, *40*, 9070–9083. [CrossRef]

18. Natesan, S.; Armenakis, C.; Vepakomma, U. Resnet-based tree species classification using UAV images. *ISPRS-Int. Arch. Photogramm. Remote Sens. Spat. Inf. Sci.* **2019**, *XLII-2/W13*, 475–481. [CrossRef]

19. Gambella, F.; Sistu, L.; Piccirilli, D.; Corposanto, S.; Caria, M.; Arcangeletti, E.; Proto, A.R.; Chessa, G.; Pazzona, A. Forest and UAV: A bibliometric review. *Contemp. Eng. Sci.* **2016**, *9*, 1359–1370. [CrossRef]

20. Kentsch, S.; Lopez Caceres, M.L.; Serrano, D.; Roure, F.; Diez, Y. Computer Vision and Deep Learning Techniques for the Analysis of Drone-Acquired Forest Images, a Transfer Learning Study. *Remote Sens.* **2020**, *12*, 1287. [CrossRef]

21. Heipke, H.; Pakzad, K.; Straub, B.M. Image Analysis for GIS Data Acquisition. *Photogramm. Rec.* **2000**, *16*, 963–985. [CrossRef]

22. Carlsson, G. Persistent Homology and Applied Homotopy Theory. *arXiv* **2020**, arXiv:math.AT/2004.00738.

23. Mahdavi, S.; Salehi, B.; Granger, J.; Amani, M.; Brisco, B.; Huang, W. Remote sensing for wetland classification: A comprehensive review. *GISci. Remote Sens.* **2018**, *55*, 623–658. [CrossRef]

24. Wu, Q. 2.07-GIS and Remote Sensing Applications in Wetland Mapping and Monitoring. In *Comprehensive Geographic Information Systems*; Huang, B., Ed.; Elsevier: Oxford, UK, 2018; pp. 140–157.

25. Twumasi, Y.A.; Merem, E.C. GIS and Remote Sensing Applications in the Assessment of Change within a Coastal Environment in the Niger Delta Region of Nigeria. *Int. J. Environ. Res. Public Health* **2006**, *3*, 98–106. [CrossRef]

26. Lang, S. Object-based image analysis for remote sensing applications: Modeling reality—Dealing with complexity. In *Object-Based Image Analysis*; Springer: Berlin/Heidelberg, Germany, 2008; pp. 3–27.

27. Joshi, C.; de Leeuw, J.; van Duren, I. Remote sensing and GIS applications for mapping and spatial modelling of invasive species. In Proceedings of the XXth ISPRS Congress: Geo-Imagery Bridging Continents, Istanbul, Turkey, 12–23 July 2004; pp. 669–677.

28. Rebelo, L.M.; Finlayson, C.; Nagabhatla, N. Remote sensing and GIS for wetland inventory, mapping and change analysis. *J. Environ. Manag.* **2009**, *90*, 2144–2153. [CrossRef]

29. Gandhi, G.M.; Parthiban, S.; Thummalu, N.; Christy, A. Ndvi: Vegetation Change Detection Using Remote Sensing and Gis—A Case Study of Vellore District. *Procedia Comput. Sci.* **2015**, *57*, 1199–1210. [CrossRef]

30. Dronova, I. Object-Based Image Analysis in Wetland Research: A Review. *Remote Sens.* **2015**, *7*, 6380–6413. [CrossRef]

31. Tóthmérész, B.; Wan, H.; Wang, Q.; Jiang, D.; Fu, J.; Yang, Y.; Liu, X. Monitoring the Invasion of Spartina alterniflora Using Very High Resolution Unmanned Aerial Vehicle Imagery in Beihai, Guangxi (China). *Sci. World J.* **2014**, *2014*, 638296.

32. Boon, M.; Greenfield, R.; Tesfamichael, S. Wetland assessment using unmanned aerial vehicle (UAV) photogrammetry. *ISPRS-Int. Arch. Photogramm. Remote Sens. Spat. Inf. Sci.* **2016**, *XLI-B1*, 781–788. [CrossRef]

33. Boon, M.; Greenfield, R.; Tesfamichael, S. Unmanned Aerial Vehicle (UAV) photogrammetry produces accurate high-resolution orthophotos, point clouds and surface models for mapping wetlands. *South Afr. J. Geomat.* **2016**, *5*, 186. [CrossRef]

34. Dvořák, P.; Müllerová, J.; Bartaloš, T.; Brůna, J. Unmanned aerial vehicles for alien plant species detection and monitoring. *Remote Sens. Spat. Inf. Sci.* **2015**, *XL-1/W42015*. [CrossRef]

35. Mafanya, M.; Tsele, P.; Botai, J.; Manyama, P.; Swart, B.; Monate, T. Evaluating pixel and object based image classification techniques for mapping plant invasions from UAV derived aerial imagery: Harrisia pomanensis as a case study. *ISPRS J. Photogramm. Remote Sens.* **2017**, *129*, 1–11. [CrossRef]

36. Rezaee, M.; Mahdianpari, M.; Zhang, Y.; Salehi, B. Deep Convolutional Neural Network for Complex Wetland Classification Using Optical Remote Sensing Imagery. *IEEE J. Sel. Top. Appl. Earth Obs. Remote Sens.* **2018**, *11*, 3030–3039. [CrossRef]

37. Liu, T.; Abd-Elrahman, A. Deep convolutional neural network training enrichment using multi-view object-based analysis of Unmanned Aerial systems imagery for wetlands classification. *ISPRS J. Photogramm. Remote Sens.* **2018**, *139*, 154–170. [CrossRef]

38. Ashqar, B.; Abu-Naser, S. Identifying Images of Invasive Hydrangea Using Pre-Trained Deep Convolutional Neural Networks. *Int. J. Acad. Dev.* **2019**, *3*, 28–36. [CrossRef]

39. Schneekloth, H.; Tuexen, J. Die Moore in Niedersachsen. GOTTINGEN Kommissionsverl. Goettinger Tageblatt, 1975, P. 1 A 198. 1975. Available online: http://pascal-francis.inist.fr/vibad/index.php?action=getRecordDetail&idt=PASCALGEODEBRGM7620141242 (accessed on 8 January 2021).

40. Agisoft. Agisoft Metashape 1.5.5, Professional Edition. Available online: http://www.agisoft.com/downloads/installer/ (accessed on 19 August 2019).
41. Team, T.G. GNU Image Manipulation Program. Available online: http://gimp.org (accessed on 19 August 2019).
42. Asaad, A. Persistent Homology for Image Analysis. Ph.D. Thesis, University of Buckingham, Buckingham, UK, 2020.
43. Cabezas, M.; Kentsch, S.; Tomhave, L.; Gross, J.; Caceres, M.L.L.; Diez, Y. Detection of Invasive Species in Wetlands: Practical DL with Heavily Imbalanced Data. *Remote Sens.* **2020**, *12*, 3431. [CrossRef]
44. Jung, A.B.; Wada, K.; Crall, J.; Tanaka, S.; Graving, J.; Reinders, C.; Yadav, S.; Banerjee, J.; Vecsei, G.; Kraft, A.; et al. Imgaug. 2020. Available online: https://github.com/aleju/imgaug (accessed on 1 July 2020).
45. Krizhevsky, A.; Sutskever, I.; Hinton, G.E. ImageNet Classification with Deep Convolutional Neural Networks. *Commun. ACM* **2017**, *60*, 84–90. [CrossRef]
46. Patrick, A.; Li, C. High Throughput Phenotyping of Blueberry Bush Morphological Traits Using Unmanned Aerial Systems. *Remote Sens.* **2017**, *9*, 1250. [CrossRef]
47. Zmarz, A. UAV—A useful tool for monitoring woodlands. *Misc. Geogr.–Reg. Stud. Dev.* **2013**, *18*, 46–52. [CrossRef]
48. Wierzbicki, D.; Kedzierski, M.; Fryskowska, A. Assesment of the influence of uav image quality on the orthophoto production. *ISPRS-Int. Arch. Photogramm. Remote Sens. Spat. Inf. Sci.* **2015**, *XL-1/W4*, 1–8. [CrossRef]
49. Dandois, J.P.; Olano, M.; Ellis, E.C. Optimal Altitude, Overlap, and Weather Conditions for Computer Vision UAV Estimates of Forest Structure. *Remote Sens.* **2015**, *7*, 13895–13920. [CrossRef]
50. Frey, J.; Kovach, K.; Stemmler, S.; Koch, B. UAV Photogrammetry of Forests as a Vulnerable Process. A Sensitivity Analysis for a Structure from Motion RGB-Image Pipeline. *Remote Sens.* **2018**, *10*, 912. [CrossRef]

Article

Hierarchical Mission Planning with a GA-Optimizer for Unmanned High Altitude Pseudo-Satellites

Jane Jean Kiam [1,*], Eva Besada-Portas [2] and Axel Schulte [1]

[1] Institute of Flight Systems, Bundeswehr University Munich, 85579 Neubiberg, Germany; axel.schulte@unibw.de

[2] Department of Computer Architecture and Automation, Universidad Complutense de Madrid, 28040 Madrid, Spain; ebesada@ucm.es

* Correspondence: jane.kiam@unibw.de

Abstract: Unmanned Aerial Vehicles (UAVs) are gaining preference for mapping and monitoring ground activities, partially due to the cost efficiency and availability of lightweight high-resolution imaging sensors. Recent advances in solar-powered High Altitude Pseudo-Satellites (HAPSs) widen the future use of multiple UAVs of this sort for long-endurance remote sensing, from the lower stratosphere of vast ground areas. However, to increase mission success and safety, the effect of the wind on the platform dynamics and of the cloud coverage on the quality of the images must be considered during mission planning. For this reason, this article presents a new planner that, considering the weather conditions, determines the temporal hierarchical decomposition of the tasks of several HAPSs. This planner is supported by a Multiple Objective Evolutionary Algorithm (MOEA) that determines the best Pareto front of feasible high-level plans according to different objectives carefully defined to consider the uncertainties imposed by the time-varying conditions of the environment. Meanwhile, the feasibility of the plans is assured by integrating constraints handling techniques in the MOEA. Leveraging historical weather data and realistic mission settings, we analyze the performance of the planner for different scenarios and conclude that it is capable of determining overall good solutions under different conditions.

Keywords: HAPS; UAV; monitoring; constrained multiple objective optimization; temporal hierarchical task planning

Citation: Kiam, J.J.; Besada-Portas, E.; Schulte, A. Hierarchical Mission Planning with a GA-Optimizer for Unmanned High Altitude Pseudo-Satellites. *Sensors* **2021**, *21*, 1630. https://doi.org/10.3390/s21051630

Academic Editor: Alfred Colpaert

Received: 4 January 2021
Accepted: 16 February 2021
Published: 26 February 2021

Publisher's Note: MDPI stays neutral with regard to jurisdictional claims in published maps and institutional affiliations.

1. Introduction

Regular monitoring of land development (e.g., agricultural activities, big construction sites, essential infrastructure, wildforest, etc.) can be done using either satellites or airplanes. Recently, Unmanned Aerial Vehicles (UAVs) are preferred for a more cost-efficient and flexible deployment. However, UAVs flying at low altitude may not always be a solution, as their missions depend on the possibility of obtaining a permit-to-fly, on weather conditions that can be quite challenging at low altitude, and on the required flight range. In the case of fixed-wing UAVs, the takeoff and landing can also be troublesome for regular deployments or may not even be an option from surroundings with unfavorable topologies.

Solar-powered unmanned High Altitude Pseudo-Satellites (HAPSs) are considered a viable alternative to overcome the challenges arising from using satellites with a fixed orbit, manned airplanes, or UAVs for regular monitoring. As [1] explains, HAPSs are a type of light-weight High Altitude Long Endurance (HALE) aerial platforms that fly at low speed (in order to be energy efficient), in the lower stratosphere (where the airspace is quite calm and little congested), with extremely long endurance (e.g., [2] reports a continuous HAPSs flight of almost 26 days). Moreover, although still at its infancy, the development of HAPSs is promising and is expected to provide multiple benefits. However, given their light-weight build, operating HAPSs can also be challenging. Table 1 summarizes some of the general benefits (+) and challenges (−) of these platforms, according to the relevant characteristics of the HAPSs that contribute to each of them.

Table 1. Relationships among characteristics of HAPSs and their benefits (+) and challenges (−).

Properties	Benefits and Challenges During Operation
Light-weight material	(+) Energy efficient (−) Fragile and vulnerable to adverse weather
Limited payload	(+) Energy efficient (−) Limited onboard computation power
Fixed-wing, large wingspan	(+) More surface for harvesting solar power (−) Limited maneuverability with respect to turn rate and mid-air still-stop
High flight altitude	(+) Calmer weather (−) Takeoff and landing are time consuming
Extreme long endurance	(+) Suitable for longer missions (+) No frequent takeoff and landing necessary (−) High operating cost
Low-power electro-motor, low airspeed	(+) Energy efficient (−) Wind effect cannot be neglected

HAPSs operations contain the typical space flight phases, such as planning, processing, departure and flight operations, return and landing, refurbishment, and turnaround [3]. However, these phases present some peculiarities, due to the HAPSs characteristics. For instance, and according to the analysis presented in [4] on the trajectories obtained from a test flight conducted using the Kelleher platform in Arizona in 2018, this HAPS takes around a day to ascend/descend to/from its operating altitude by flying within a safe vertical corridor allocated for takeoff and landing. Subsequently, the platform stays as long as possible in the air and at the operating altitude in the lower stratosphere.

Given these continuous and extremely long operations, increasing HAPSs autonomy is essential. Besides, it is also useful from a safety and pragmatic point of view, as well as to reduce manpower and human error. Finally, and according to [5], the deployment of HALE platforms can be "cost-efficient", since by increasing autonomy and decreasing piloting, operation cost and be further reduced (without compromising safety and efficiency).

Hence, automated mission planning is convenient for the deployment of HAPSs that have to perform monitoring missions. However, although the airspace at this flight level is often relatively calm with mild winds, some rare weather conditions can pose serious safety-critical risks to the HAPSs. Moreover, since these platforms have limited maneuverability, reactive avoidance of risk zones may not always be possible. Therefore, weather risks must be addressed already in the mission planner on the Ground Control Station (GCS) to minimize the need of an onboard replanning or emergency landing. In particular, the following weather conditions must be considered:

- Cumulonimbus clouds: Although clouds are rare in the stratosphere where HAPSs operate, the anvil of Cumulonimbus clouds can reach high altitudes and is extremely dangerous. Hence, it must be avoided with substantial distance (∼37 km laterally and 1.5 km vertically) to prevent structural impairment to the platforms [6].
- Turbulences and Precipitation: These weather phenomena can be caused by strong winds and wind shear [7]. Although rare and harmless to bigger aircraft (e.g., airliners), turbulences and precipitations can cause extreme difficulties to HAPSs navigation and damage their structures.
- Wind field: Given the low airspeed of the HAPS, even mild wind (with wind speed up to 5 m/s) must be considered in planning for wind drift correction.

Besides, and although HAPSs airspace is little congested (since the airliners fly below), High-level Flight Rules (HFR) also apply to unmanned flights above Flight Level (FL) 600 [8]. This implies that the airspace regulations must also be considered in the mission planner (in order to avoid collisions with other stratospheric aircrafts), that it is recommendable to systematically organize and dynamically allocate the airspace [9], and that the flight plans must be communicated to the authorities before the execution of each mission. However, since HAPSs are long-endurance platforms intended to remain airborne, planning must also be performed during flight, before each monitoring mission starts.

Taking into account the previous considerations, this work focuses on increasing the autonomy and efficiency in mission planning that takes place on the GCS during flight operations but before the execution of the mission-related tasks. Our main goal is to optimize the mission success rate of monitoring the requested sites (i.e., to improve the chances of providing images of the sites with sufficient coverage and at the requested time windows), while reducing the risk of replanning by considering, at the planning phase, the predicted time-varying environment and the platform constraints. Furthermore, we assume the presence of one or several human operators in the mission-planning loop. Although their decision-making process is not considered in this work, our mission planner is developed to be part of a decision-support system that is responsible for automatically generating a group of feasible optimal plans and for presenting them as "suggestions" to the operators, who have to perform the selection of the final plan.

The work presented in this paper is closely related with the approach described in [10], which presented the preliminary version of our planner. The current version is improved, by (1) adopting a Multi-Objective Evolutionary Algorithm (MOEA) for constrained problems to optimize the mission plans and (2) by considering the uncertainty associated to the wind variability in the constraints. Besides, this paper presents new scenarios and analyzes the results of the new planner over a wider set of circumstances. Finally, it is worth noting that the relationships of other works with ours will be discussed later in Section 6, after the readers are acquainted with the main characteristics of our planner described through Sections 3 and 4.

The organization of this work is the following. Section 2 presents the problem at an abstract level and describes its main elements. Section 3 provides a more formal description of the problem, including the objective functions as well as the different components that conform with the constraint criteria. Subsequently, the implementation of the MOEA that supports the optimization process of the planner is described in Section 4, providing details on the encoding of the plan, on the hierarchical task decomposition process and on the particularities of our MOEA. Finally, results are illustrated and analyzed in Section 5, while a discussion on related work is provided in Section 6, followed by the conclusion and future work drawn in Section 7.

2. Problem Description

This work focuses on the task planning for multiple HAPSs, equipped with electro-optical (EO) mission cameras and contracted to monitor repeatedly areas on the ground at specific time windows.

This section presents the main elements of the problem, describing the monitoring scenario that will be considered in this paper, introducing how the mission plan is defined, characterizing the payload of the HAPSs, reporting the mission requirements and constraints, and finally, explaining how the weather conditions to be considered at the planning phase are extracted.

2.1. Monitoring Scenarios

The operation is assumed to take place in an organized airspace consisting of different types of dynamically allocated operation areas, in order to reduce congestion in the lower stratosphere. In particular, and as shown in Figure 1, the HAPSs will be able to operate in Mission Areas (MAs, represented in blue), Corridors (Cs, in gray), and Waiting Areas (WAs, in yellow).

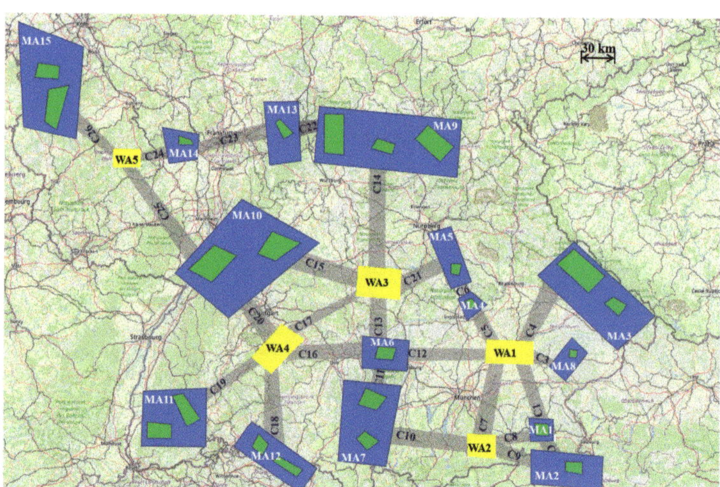

Figure 1. Mission scenario (plotted on © OpenStreetMap) for monitoring multiple Locations Of Interest (LOI#) on the ground. The operation airspace is organized using dynamically allocated mission elements of Corridors (C#), Waiting Areas (WA#), and Mission Areas (MA#) that encompass LOIs.

Besides, the Locations of Interest (LOIs, in green) are the projection of the ground areas to be monitored within the time windows and at the frequency requested by the clients. LOIs of the same client with the same set of mission requirements are grouped in a MA, which defines the airspace (at the operating altitude for the HAPSs), allocated to allow a HAPS to monitor the encompassed group of LOIs. Additionally, the WAs are airspace made available for the HAPSs to loiter freely (for example upon sunset) or to exploit as a "corridor" to reach another connected MA. HAPSs are allowed to move between MAs only through the designated Cs or through WAs. This also implies that MAs are not to be used as "corridors", i.e., a HAPS entering a MA has to monitor its corresponding LOIs before departing through a connected corridor.

Appendix A includes further details of the HAPSs considered in this work and of the mission scenario represented in Figure 1. In particular, the numerical information on the model of the HAPSs is adapted from [11] and summarized in Table A1, while the dimensions of the mission elements are presented in Table A2.

2.2. Hierarchical Task Plan

Execution of tasks for multiple HAPSs can be structured conveniently in a hierarchical manner, since the order of task execution depends substantially on the organisation of the airspace and on airspace-related constraints and requirements, which can be expressed at different levels of spatial resolution. In particular, we consider the following levels, ordered from the highest to the lowest level, according to the spatial resolution:

1. MA level, where the plans are the sequences of mission areas (MA#) and waiting areas (WA#) that each HAPS operates.
2. LOI level, where the plans are sequences of tasks to be performed in the mission elements expressed at one higher abstraction level (i.e., MA# and WA#). Examples of these tasks are flying through a WA ($\text{fly}_{WA\#}$) and monitoring a LOI in a given MA ($\text{monitor}_{LOI\#\in MA\#}$).
3. Waypoint (WP) level, where the plans consist of either executing a scan pattern (scan) over a LOI or flying to sequences of waypoint, which are: fly to the closest entrance of a given corridor ($\text{to}_{C\#}$), cross and fly to the end location of the given corridor ($\text{cross}_{C\#}$), and fly to the closest vertex of the LOI that has to be monitored (NPL).

Figure 2 illustrates the representation of a hierarchical task plan considered in this work. In particular, on the left side, Figure 2a shows, over a portion of the mission scenario, the execution of the task plan for a HAPS that, after the first task in WA1, continues monitoring first the unique LOI in MA6 and afterwards the two LOIs in MA7. This figure also shows the waypoints followed by the HAPS to move within MA1 and MA7. On the right side, Figure 2b shows the hierarchical structure of the task plan of the HAPS represented in Figure 2a and of a second HAPS (not depicted in Figure 2a). That is, it shows how the plans that govern the two HAPS to monitor the LOIs within the horizon $[T_{start}, T_{end}]$ are initially decomposed into the tasks expressed at the MA-level (represented at the two top timelines, one for each HAPS), followed by the tasks expressed at the LOI-level (represented at the two intermediate timelines) and finally by the actions presented at the WP-level (shown at the two bottom-most timelines). At the lowest level, vertical color bars without text annotation represent instantaneous tasks, for example, to turn on or off the mission camera.

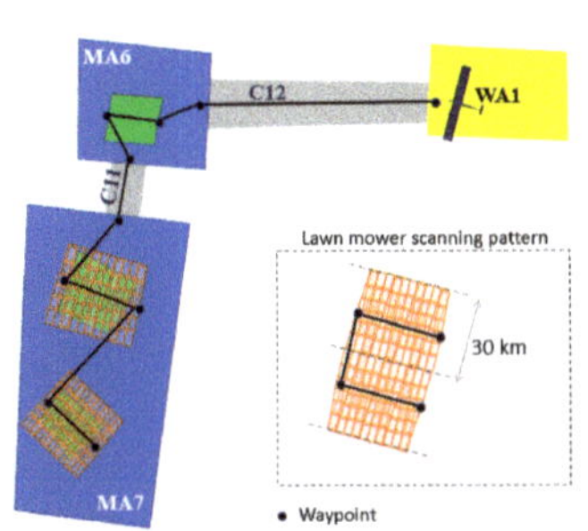

(**a**) Tasks execution viewed from different spatial resolutions of the airspace.

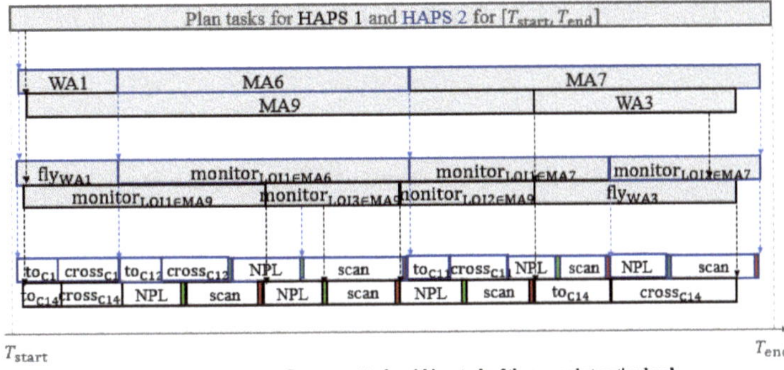

(**b**) Task execution organized in a hierarchical structure.

Figure 2. Hierarchical task execution for HAPS.

2.3. Mission Payload

The HAPSs are equipped with light-weight electro-optical mission cameras. The example camera considered in this work is inspired by the one described by Delauré et al. in [12], specially designed for unmanned HALE platforms. In particular, it is a light-weight ($\sim$2.6 kg) and energy-efficient (<50 W) camera with two custom CMOS image sensors and with resistance to low pressure (down to 60 mbar) and to a wide range of temperature (from $-70\,°C$ to $60\,°C$). Its pixel counts for the width w_I and height h_I of the image are 1200 px × 10,000 px. With a ground sampling distance of 30 cm, an image taken from an altitude of 18 km at Nadir position records an area of 360 m × 3000 m of the ground.

With this mission camera and a gimbal that performs a cross-track sweep scan within 10 s from $-45°$ to $45°$, a HAPS can record images covering a total width of more than 30 km, while advancing forward. Figure 2a illustrates the scanned footprint, which is a superposition of images taken during the scan. Even in the presence of a tailwind of 5 m/s, the HAPS, flying at the airspeed of 30 m/s (that is considered in Table A1) will not advance more than 360 m within a cross-track sweep, assuring some overlapping of the images between two periodic sweeps. Therefore, and as Figure 2a shows, we adopt a lawn mower scan pattern to monitor each LOI, with the distance between two consecutive tracks set at 30 km.

Finally, the cloud layers between the HAPS and the ground must be considered during the monitoring scans, as they reduce the coverage of the area recorded with the EO mission camera. For this reason, if the stitched image of any one of the LOIs of a MA has a coverage of the ground lower than requested, the monitoring of that MA will not be rewarded by the client.

2.4. Mission Requirements

The HAPS team is rewarded by the contracting client if the ensemble of all the monitoring tasks performed on the LOIs within a MA is considered "successful". Therefore, we consider this ensemble of tasks a "mission" unit, which is rewarded according to the amount agreed upon by each client.

In particular, monitoring a mission unit is successful if the following mission requirements (MRs) are fulfilled:

- **MR1**: The recorded image of each LOI has a coverage of the LOI that is bigger than the minimum required coverage for its corresponding MA.
- **MR2**: The captured images of each LOI are within the time windows requested by the client for each MA.
- **MR3**: The time-lapse between two consecutive successful visits to the MA is larger than the imposed minimum inter-visit time-lapse for the MA.
- **MR4**: The MA has not been visited more frequently per day than required by the client.

The coverage percentage and reward obtained for monitoring successfully each MA of the scenario presented in Figure 1 is presented in Table A3 of Appendix A. Besides, the rewarding time windows for each MA are directly depicted together with the mission plans obtained by our planner, which are presented in Figures 10–14 of Section 5, since they are required to observe if the MA can be successfully or unsuccessfully monitored. Finally, in the scenarios analyzed in this paper, the time-lapse between the start times of two consecutive successful visits is set to one hour and each MA must not be visited more than three times a day.

2.5. Mission Constraints

While mission requirements decide if a mission is successful, mission constraints (MCs) dictate the "feasibility" of a plan and are defined in the interest of operational safety by enforcing airspace regulation and measures for risk avoidance.

In particular, a plan is infeasible (i.e., it cannot be executed) if any of the following constraints is violated:

- **MC1**: any mission element that the HAPS is operating in has a wind field with a wind magnitude smaller than 5 m/s.
- **MC2**: the MA or WA that the HAPS is operating in (e.g., a MA, a WA, or a C) has an obstacle occlusion (related with zones of adverse weather) smaller than 30%.
- **MC3**: Only one HAPS can operate in a MA (i.e., the simultaneous coexisting of HAPSs in a MA is forbidden).
- **MC4**: Consecutive MAs or WAs have to be connected according to the mission scenario.
- **MC5**: LOIs are monitored exactly once at each visit to the MA.
- **MC6**: A MA cannot be used as a corridor, i.e., HAPS cannot pass the MA without monitoring all its encompassed LOIs.

2.6. Weather Conditions

Weather conditions also affect the HAPSs and can make a given mission plan unsuccessful and/or infeasible. To take them into account, high-resolution global weather forecast based on numerical weather prediction models can be used, because this approach is beneficial compared to wide area weather forecast to foresee risk zones and to consider wind effects.

In particular, for this study we use the COSMO-D2 (COnsortium for Small-scale MOdeling) numerical weather data from the German National Meteorological Service (Deutscher Wetterdienst, DWD), which are updated every couple of hours to provide information on the cloud coverage and on the wind vector field with a horizontal resolution of 2.2 km and a temporal resolution of one hour [13].

In order to argue for availability of weather data that fit the underlying framework, we also list here a set of alternative meteorological services that can be used in the mission planner described in this paper, which also provide numerical global weather data such as the Global Forecast System (GFS, Ref. [14]) and the European Center for Medium-Range Weather Forecast (ECMWF, Ref. [15]).

3. Formal Problem Statement

This section defines the problem formally, detailing the variables used to mathematically define a hierarchical plan, as well as the objective and the constraint functions used to evaluate them.

To help the reader understand the relationship of the elements presented in this section and the previous sections, Figure 3 shows how the MRs and MCs described in Sections 2.4 and 2.5 are mapped into the three Objective Functions (OF_{rew}, OF_{eff}, OF_{div}) and the three constraint criteria (φ_{saf}, φ_{coex}, φ_{con}) that are formally stated in this section. Furthermore, Figure 3 also illustrates the role of the operators as human decision makers, i.e., how they select a plan (π_k^{MA}, π_k^{LOI}, π_k^{WP}) among the feasible solution plans that form part of the first Pareto front determined by the planner that will be presented in Section 4.

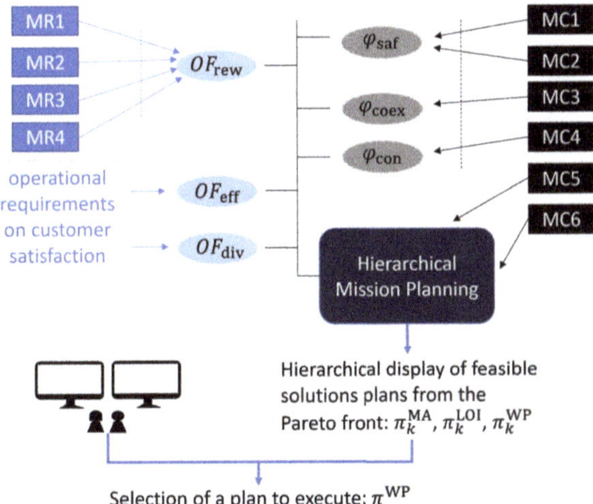

Figure 3. Relationships among the mission requirements (MRs) and mission constraints (MCs) described in Section 2 and the objective functions (OF) and constraints (φ) presented in Section 3.

3.1. Formal Definition of the Hierarchical Plan

The goal of the planning problem is to find a hierarchically structured plan such as the one depicted in Figure 2b that entails the sequence of tasks to be performed by each HAPS, as well as their expected initial time instant and duration.

Formally, the solution is a set of sequences of time-stamped tasks for each hierarchical level and HAPS. More in detail:

- At the MA level, the plan can be represented as the ordered list of tasks π_h^{MA} displayed in Equation (1), where $o_{h,i}^{\text{MA}}$ is the i-th mission task (i.e., a MA# or WA# of the mission scenario) on the list that will be performed by HAPS h, and $t_{h,i}^{\text{MA,start}}$ and $\delta_{h,i}^{\text{MA}}$ are the start time and the duration of the $i-th$ task of HAPS h.

$$\pi_h^{\text{MA}} =< o_{h,1}^{\text{MA}}(t_{h,1}^{\text{MA,start}}, \delta_{h,1}^{\text{MA}}), o_{h,2}^{\text{MA}}(t_{h,2}^{\text{MA,start}}, \delta_{h,2}^{\text{MA}}), o_{h,3}^{\text{MA}}(t_{h,3}^{\text{MA,start}}, \delta_{h,3}^{\text{MA}}), \cdots > \quad (1)$$

Under this formulation, the high level mission plan of the first HAPS displayed in Figure 2b will be represented as $\pi_1^{\text{MA}} =<$ WA1$(t_{1,1}^{\text{MA,start}}, \delta_{1,1}^{\text{MA}})$, MA6$(t_{1,2}^{\text{MA,start}}, \delta_{1,2}^{\text{MA}})$, MA7$(t_{1,3}^{\text{MA,start}}, \delta_{1,3}^{\text{MA}})>$.

- At the LOI-level, the plan π_h^{LOI} can be represented with Equation (2), where $o_{h,i}^{\text{LOI}}$ is the i-th mission task (i.e., fly$_{\text{WA\#}}$ or monitor$_{\text{LOI\#}\in\text{MA\#}}$) that will be performed by HAPS h, and $t_{h,i}^{\text{LOI,start}}$ and $\delta_{h,i}^{\text{LOI}}$ are the start time and duration of the $i-th$ task of HAPS h.

$$\pi_h^{\text{LOI}} =< o_{h,1}^{\text{LOI}}(t_{h,1}^{\text{LOI,start}}, \delta_{h,1}^{\text{LOI}}), o_{h,2}^{\text{LOI}}(t_{h,2}^{\text{LOI,start}}, \delta_{h,2}^{\text{LOI}}), o_{h,3}^{\text{LOI}}(t_{h,3}^{\text{LOI,start}}, \delta_{h,3}^{\text{LOI}}), \cdots > \quad (2)$$

Under this formulation, the middle level mission plan of the first HAPS displayed in Figure 2b will be represented as $\pi_1^{\text{LOI}} =<$fly$_{\text{WA1}}(t_{1,1}^{\text{LOI,start}}, \delta_{1,1}^{\text{LOI}})$, monitor$_{\text{LOI1}\in\text{MA6}}$ $(t_{1,2}^{\text{LOI,start}}, \delta_{1,2}^{\text{LOI}})$, monitor$_{\text{LOI1}\in\text{MA7}}(t_{1,3}^{\text{LOI,start}}, \delta_{1,3}^{\text{LOI}})$, monitor$_{\text{LOI2}\in\text{MA7}}(t_{1,4}^{\text{LOI,start}}, \delta_{1,4}^{\text{LOI}})>$. Moreover, we can relate the time variables of the MA and LOI level (e.g., $t_{1,1}^{\text{LOI,start}} = t_{1,1}^{\text{MA,start}}$, $t_{1,2}^{\text{LOI,start}} = t_{1,2}^{\text{MA,start}}$, $t_{1,3}^{\text{LOI,start}} = t_{1,3}^{\text{MA,start}}$, or $t_{1,4}^{\text{LOI,start}} = t_{1,3}^{\text{MA,start}} + \delta_{1,3}^{\text{LOI}}$) to signify the decomposition of the higher level task into lower-level tasks.

- A similar representation, where $o_{h,i}^{\text{WP}}$ are the actions that can be performed at the lower mission level, and $t_{h,i}^{\text{WP,start}}$ and $\delta_{h,i}^{\text{WP}}$ are its corresponding start time and duration, applies to the WP-level.

- Finally, we extend the previous notations as follows:

  - π^{MA}, π^{LOI} and π^{WP} represent the plans of the set of H HAPSs (i.e., $\pi^* =< \pi_1^*, \pi_2^*, \ldots, \pi_H^* >$, where * stands either for MA, LOI, or WP).

  - $\tilde{\pi}_{h,i:j}^{\text{MA}}$, $\tilde{\pi}_{h,i:j}^{\text{LOI}}$ and $\tilde{\pi}_{h,i:j}^{\text{WP}}$ represent the partial plans between the i-th and j-th task (i.e., $\tilde{\pi}_{h,i:j}^* =< o_{h,i}^*(t_{h,i}^{*,\text{start}}, \delta_{h,i}^*), o_{h,i+1}^*(t_{h,i+1}^{*,\text{start}}, \delta_{h,i+1}^*), \ldots, o_{h,j}^*(t_{h,j}^{*,\text{start}}, \delta_{h,j}^*) >$, where * stands either for MA, LOI, or WP.

At this point, it is necessary to highlight that the time-dependent variables ($t_{h,i}^{\text{MA,start}}, \delta_{h,i}^{\text{MA}}, t_{h,i}^{\text{LOI,start}}, \delta_{h,i}^{\text{LOI}}, t_{h,i}^{\text{WP,start}}, \delta_{h,i}^{\text{WP}}$) are probabilistic in our problem, except for the start time of the mission $t_0 = t_{h,1}^{\text{MA,start}} = t_{h,1}^{\text{LOI,start}} = t_{h,1}^{\text{WP,start}}$. The underlying reason of the probabilistic nature of these variables is that, at the lowest spatial resolution, the duration of each task can only be estimated, since neither the trajectory of the HAPS nor the exact wind vector are computed or considered yet.

For this reason, we model the duration $\delta_{h,i}^{\text{WP}}$ of a task at the WP level as a random variable *uniformly* distributed over:

$$[\delta_{h,i}^{\text{WP,min}}, \delta_{h,i}^{\text{WP,max}}] = [l(o_{h,i}^{\text{WP}})/(|v_a| + \max(|v_w|)), l(o_{h,i}^{\text{WP}})/(|v_a| - \max(|v_w|))], \quad (3)$$

where $l(o_{h,i}^{\text{WP}})$ is the total linear distance to travel between the waypoints associated to the task $o_{h,i}^{\text{WP}}$, $|v_a|$ is the cruising airspeed of the HAPS (see Table A1), and $\max(|v_w|)$ is the maximum wind magnitude (which is assumed to be 5 m/s in this study to ensure that MC1 is not violated). Hence, when we consider $|v_a| + \max(|v_w|)$ we assume that the HAPS is flying with tailwind, while by considering $|v_a| - \max(|v_w|)$, we assume that the HAPS is flying with headwind. Note that since $l(o_{h,i}^{\text{WP}})$ can only be estimated upon the decomposition down to the WP level, our hierarchical planning approach searches for a plan by adopting a downward-forward decomposition approach, which will be explained in a later section in Algorithm 1.

Assuming that lingering between the tasks at any level is forbidden, task $o_{h,i}^{\text{WP}}$ terminates at $t_{h,i}^{\text{WP,end}} = t_0 + \Sigma_{j=1}^{i}\delta_{h,j}^{\text{WP}}$, where t_0 is the deterministic start time of the plan, while $\Sigma_{j=1}^{i}\delta_{h,j}^{\mathrm{WP}}$ follows the distribution of the sum of i nonidentically distributed uniform random variables $\delta_{h,j}^{\mathrm{WP}}$. Therefore, the probability density function $f(t_{h,i}^{\mathrm{WP,end}})$ of completing the $i-th$ WP-level task of HAPS h at a given time can be calculated with the following expression, as derived in [16]:

$$f(t_{h,i}^{\mathrm{WP,end}}) = f(t_{h,i}^{\mathrm{WP,end}} - t_0) = f\left(\sum_{j=1}^{i}\delta_{h,j}^{\mathrm{WP}}\right) = \frac{\sum_{\vec{\epsilon}^k \in \mathcal{V}^i} (g(\vec{\epsilon}^k,\delta_{h,1:i}^{\mathrm{WP}}))^{i-1}\cdot\mathrm{sign}(g(\vec{\epsilon}^k,\delta_{h,1:i}^{\mathrm{WP}}))\prod_{j=1}^{i}\epsilon_j}{(i-1)!\,2^{i+1}\prod_{j=1}^{i}u_{\delta_{h,j}^{\mathrm{WP}}}} \tag{4}$$

where $\mathcal{V}^i$ comprises the set with all 2^i vectors of signs $\vec{\epsilon}^k = (\epsilon_1^k,\cdots,\epsilon_i^k) \in \{-1,1\}^i$, $u_{\delta_{h,i}^{\mathrm{WP}}} = (\delta_{h,i}^{\mathrm{WP,max}} - \delta_{h,i}^{\mathrm{WP,min}})/2$, $i!$ is the factorial of i, and $g(\vec{\epsilon}^k,\delta_{h,1:i}^{\mathrm{WP}}) = \sum_{j=1}^{i}\delta_{h,j}^{\mathrm{WP}} + \sum_{j=1}^{i}(\epsilon_j u_{\delta_{h,j}^{\mathrm{WP}}} - m_{\delta_{h,j}^{\mathrm{WP}}})$, with $m_{\delta_{h,i}^{\mathrm{WP}}}$ being the median value of $[\delta_{h,i}^{\mathrm{WP,min}},\delta_{h,i}^{\mathrm{WP,max}}]$.

The distribution of the higher levels (LOI and MA) time-dependent random variables can be modelled, given the hierarchical decomposition of the plan and the lack of lingering between tasks, by considering the distribution of the lowest level (WP-level) time-dependent random variables, i.e.,

$$f(t_{h,i}^{\mathrm{MA,end}}) = f(t_{h,j}^{\mathrm{LOI,end}}), \tag{5}$$

$$f(t_{h,j}^{\mathrm{LOI,end}}) = f(t_{h,k}^{\mathrm{WP,end}}), \tag{6}$$

where $t_{h,i}^{\mathrm{MA,end}}$ is the end time of the $i-th$ task of the MA level, $t_{h,j}^{\mathrm{LOI,end}}$ is the ending time of the $j-th$ task of the LOI level that terminates when the $i-th$ task of the MA level ends, and $t_{h,k}^{\mathrm{WP,end}}$ is the ending time of the $k-th$ task of the WP level that terminates when the $i-th$ task of the MA level and the $j-th$ task of the LOI level end. In other words, the distributions of the higher levels are associated to some of the distributions of the lower ones. Finally, it is worth noting that the estimated end time $t_{h,i}^{*,\mathrm{end}}$ of a task $(o_{h,i}^{*})$ at any level $*$ is the estimated start time $t_{h,i+1}^{*,\mathrm{start}}$ of the following task $(o_{h,i+1}^{*})$ of the same level $*$.

To understand better the implications of the previous distributions, we represent in Figure 4 the results of Equation (4) when considering up to five $\delta_{h,i}^{\mathrm{WP}}$ random variables with the median $m_{\delta_{h,i}^{\mathrm{WP}}}$ and half length $u_{\delta_{h,i}^{\mathrm{WP}}}$ provided at the figure caption. We can observe how the sum of more than two uniform distributed random variables assimilates towards a Gaussian distribution, while the variance grows with the number of random variables involved in the sum. This implies that the distribution becomes more wide-spread, and in our case, that the knowledge on the start or end time of a task further in the future is more "uncertain" than the knowledge on the start or end time of a task in the near future. Besides, it is possible to calculate the minimum and maximum values of $t_{h,i}^{*,\mathrm{end}}$, as the density functions calculated with Equation (4) have a limited support. Finally, it is worth noting that a correct estimation of the probability distribution of the sum of tasks durations implies the correct estimation of the start or end time of the tasks at each level. This is essential, especially at the MA-level, since some of the mission requirements and mission constraints stated in Sections 2.4 and 2.5 depend on the time-varying weather conditions and on the time-dependent requirements, associated to the mission time windows used to decide if the high-level tasks can be rewarded.

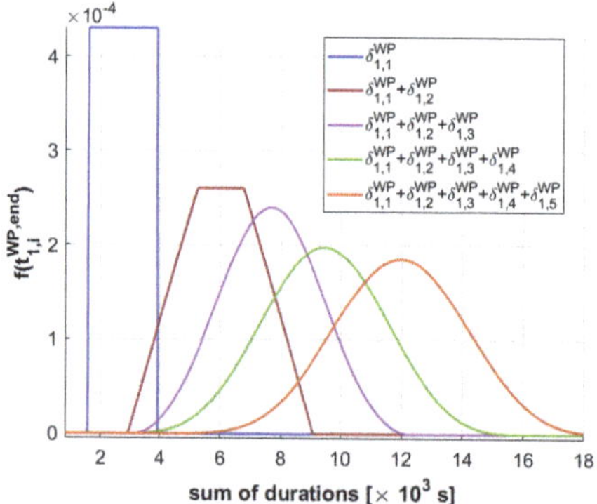

Figure 4. Probability distribution of the sum of durations of tasks, which are uniform distributed random variables of $(m_{\delta_{h,i}^{WP}}, u_{\delta_{h,i}^{WP}})$: (3132, 791), (4368, 1012), (2876, 698), (3856, 971), (4112, 1263).

3.2. Objectives

The aim of this work is to present a multi-HAPS planner that optimizes the HAPSs tasks plans, whose joint quality is evaluated by the three objective functions defined in the following subsections, each contributing to a different aspect of the overall operational performance.

3.2.1. Expected Cumulative Rewards per Hour

This objective focuses on the reward the team of HAPS can gain with the generated plan. Since the success of a task depends on its execution time (i.e., on its start and end time), which can only be probabilistically estimated using Equation (4), the reward can only be estimated with an expected cumulative reward function.

To do it, we exploit the Time-Dependent Markov Decision Process (TiMDP) of Boyan and Littman [17] to calculate the expected cumulative reward, obtained when applying at state $s_{h,i}$ (which in our case contains, among others, the current location of the HAPS) and at time t_i the remaining plan $\tilde{\pi}_{h,i:n}^{MA}$ under the weather w_{t_i} forecasted for t_i:

$$E(R|s_{h,i}, t_i, \tilde{\pi}_{h,i:n}^{MA}, w_{t_i}) = \Sigma_{\mu \in \{succ,fail\}} L(\mu|s_{h,i}, t_i, o_{h,i}^{MA}, w_{t_i}) \cdot$$
$$\cdot \int_{\mathbb{R}} f(t_{h,i}^{MA,end} = t_{i+1}) \cdot [R(\mu, o_{h,i}^{MA}, t_{i+1}) + E(R|s_{h,i+1}, t_{i+1}, \tilde{\pi}_{h,i+1:n}^{MA}, w_{t_{i+1}})]dt_{i+1}, \tag{7}$$

where $L(\mu|s_{h,i}, t_i, o_{h,i}^{MA}, w_{t_i})$ is the likelihood that action $o_{h,i}^{MA}$, performed at time t_i at state $s_{h,i}$ is successful (μ = succ) or not (μ = fail) under the weather conditions w_{t_i} at t_i; $f(t_{h,i}^{MA,end} = t_{i+1})$ is the probability density function of ending $o_{h,i}^{MA}$ at t_{i+1}, and $R(\mu, o_{h,i}^{MA}, t_{i+1})$ is the immediate reward obtained when performing $o_{h,i}^{MA}$ at time t_{i+1} successfully (μ = succ) or unsuccessfully (μ = fail). In the latter case, $R(\mu = fail, o_{h,i}^{MA}, t_{i+1}) = 0$.

Although Equation (7) was originally designed to devise a strategy aiming at optimizing the success rate of arriving in time at a destination using different combinations of means of transport, we do not seek to use TiMDP this way. Rather, we exploit the equation as a model for computing the expected cumulative reward of a complete plan $E(R|s_{h,0}, t_0, \pi_h^{MA}, w_{t_0})$, which can be done using a backward iteration, since the immediate reward $R(\mu, o_{h,i}^{MA}, t_{i+1})$ is piecewise constant with respect to t_{i+1}. Moreover, due to the piecewise constant weather data, $E(R|s_{h,i}, t_i, \tilde{\pi}_{h,i:n}^{MA}, w_{t_i})$ is piecewise constant too. Therefore, the integration can be performed in piecewise time intervals that are generated using the minimum and maximum of the start time of a task (according to $f(t_{h,i}^{MA,end})$), as well as the minimum and maximum bounding times of the piecewise constant coverage.

Exploiting the formulation in Equation (7) for computing the expected cumulative reward also leverages the following:

- It takes into account the immediate reward $R(\mu, o_{h,i}^{\mathrm{MA}}, t_{i+1})$ obtained after monitoring the selected mission area at end time t_{i+1}, as well as the reward of the remaining action plan $\tilde{\pi}_{h,i+1:n}^{\mathrm{MA}}$.
- It considers the likelihood of performing the task successfully and unsuccessfully, depending on the weather conditions, or more specifically, on the cloud coverage, which is related to the mission requirements (i.e., MR1) listed in Section 2.4.
- It exploits the weighting imposed by $f(t_{h,i}^{\mathrm{MA,end}} = t_{i+1})$ at the given times t_{i+1}. This is helpful since the weather forecast is constantly updated and a replanning can occur in the future. Therefore, while it is important to "look forward" in the plan to optimize it for a longer time horizon, we allocate more weighting according to immediacy, since a replanning could be triggered to improve the plan quality in the future.

Finally, since multiple HAPS can be involved and the start time of the plan for each HAPS can be different, we accumulate the expected reward of each HAPS to obtain the Objective Function (OF) of the expected cumulative reward OF_{rew}:

$$OF_{\mathrm{rew}}(\boldsymbol{\pi}^{\mathrm{MA}}) = \Sigma_h E(R | s_{h,0}, t_{h,0}, \pi_h^{\mathrm{MA}}, w_{t_0}). \tag{8}$$

3.2.2. Effort

Although the mission rewards are important, they are not the only objective to consider. Global client satisfaction must be taken into account too. That is, to keep the clientele satisfied, the HAPS team is required to perform monitoring tasks for as much of their time in the air as possible. Therefore, we consider the objective function of effort, which is the percentage of time spent on monitoring the LOIs:

$$OF_{\mathrm{eff}}(\boldsymbol{\pi}^{\mathrm{LOI}}) = \frac{\Sigma_h \Sigma_l E(\delta_{h,l}^{\mathrm{LOI}}) * \mathrm{isMonitor}(o_{h,l}^{\mathrm{LOI}})}{T_h^{\mathrm{max}} - t_{h,0}} = \frac{\Sigma_h \Sigma_l m_{\delta_{h,l}^{\mathrm{LOI}}} * \mathrm{isMonitor}(o_{h,l}^{\mathrm{LOI}})}{T_h^{\mathrm{max}} - t_{h,0}}, \tag{9}$$

where T_h^{max} is the end time of the plan horizon set for HAPS h, $E(\delta_{h,l}^{\mathrm{LOI}})$ is the expected duration of the monitoring task for $o_{h,l}^{\mathrm{LOI}}$ which, given the symmetric distribution of the random variable, is the median duration $m_{\delta_{h,l}^{\mathrm{LOI}}}$, and $\mathrm{isMonitor}(o_{h,l}^{\mathrm{LOI}})$ returns 1 if the action $o_{h,l}^{\mathrm{LOI}}$ consists of monitoring a LOI (i.e., if $o_{h,l}^{\mathrm{LOI}}$ equals $\mathrm{monitor}_{\mathrm{LOI\#} \in \mathrm{MA\#}}$) and 0 otherwise.

This objective function contributes to preventing the HAPS from trying too hard to reach more rewarding MAs by crossing multiple corridors and WAs.

3.2.3. Diversity

In the presence of missions that are much more rewarding than others, the plan computation can be extremely unfavorable for less rewarding missions. This has a long-term negative effect to the HAPS team in regard of "customer service". In order to satisfy a more diverse clientele pool, the diversity objective function OF_{div} is devised using the Simpson index [18]:

$$OF_{\mathrm{div}}(\boldsymbol{\pi}^{\mathrm{MA}}) = 1 - \frac{\Sigma_{c=1}^{N_{\mathrm{MA}}} n_c(n_c - 1)}{N(N-1)}, \tag{10}$$

where N_{MA} is the total number of MAs (or clients) considered in the mission scenario, n_c is the number of occurrences of MA_c in the task plan, and N is the total number of MAs within the task plan. Note that the function only considers what happens with the mission areas, ignoring what is occurring in the waiting areas.

Optimizing this objective reduces the probability of drawing the same MA when two of them are drawn without replacement from a given plan, preventing therefore the bias towards rewarding missions.

3.3. Constraints

While the missions' requirements presented in Section 2.4 are considered in the evaluation of the objective function OF_{rew}, the mission constraints presented in Section 2.5 are evaluated with different constraint criteria.

Besides, while **MC5** and **MC6** are directly encoded in the solutions manipulated by the EA-based planner described in Section 4.1 (and hence, they are never violated), the remaining criteria (**MC1**–**MC4**) are evaluated with the functions described in Sections 3.3.1–3.3.3.

Finally, it is worth noting that our constraint functions consider the number of times that each criterion is violated. Detailed information of this way of proceeding is presented in Section 3.3.4.

3.3.1. Safety

The safety constraint criterion comprises **MC1** and **MC2** and is violated if the MA# is a risk zone (due either to substantial obstacle occlusion or strong wind) while HAPS h is operating in it. Since the position of a HAPS is probabilistic due to the uncertainty in the task durations (as Equation (4) states), the constraint function associated to the safety violation $\varphi_{\text{saf}}(\pi^{\text{MA}})$ is incremented if the probability of operating any HAPS h in a MA representing a risk zone is greater than a predefined threshold p_{saf}:

$$P\left(\left[t_{h,i}^{\text{MA,start}}, t_{h,i}^{\text{MA,end}} \right] \cap T_{\text{risk}}(o_{h,i}^{\text{MA}}) \neq \varnothing \right) > p_{\text{saf}}, \tag{11}$$

where $t_{h,i}^{\text{MA,start}}$ and $t_{h,i}^{\text{MA,end}}$ are the start and end time of HAPS h performing the monitoring task $o_{h,i}^{\text{MA}}$, while $T_{\text{risk}}(o_{h,i}^{\text{MA}} = \text{MA\#})$ is the set of time windows where the MA# associated to $o_{h,i}^{\text{MA}}$ represents a risk zone. Alternatively, we can compute the same $\varphi_{\text{saf}}(\pi^{\text{MA}})$ by incrementing its value if

$$\exists t \in [\min(t_{h,i}^{\text{MA,start}}), \max(t_{h,i}^{\text{MA,end}})] \cap T_{\text{risk}}(o_{h,i}^{\text{MA}}), \ P(\text{pos}_h(t) \in o_{h,i}^{\text{MA}}) > p_{\text{saf}}, \tag{12}$$

where t belongs to the intersection of $T_{\text{risk}}(o_{h,i}^{\text{MA}})$ with the maximum time span that the HAPS could be performing task $o_{h,i}^{\text{MA}}$, $\text{pos}_h(t)$ is the position of HAPS h at time t, and $\text{pos}_h(t) \in o_{h,i}^{\text{MA}}$ indicates that HAPS h is positioned within the MA in which the monitoring task $o_{h,i}^{\text{MA}}$ takes place. The probability $P(\text{pos}_h(t) \in o_{h,m}^{\text{MA}})$ in Equation (12) can be further simplified as Equation (13) states, by taking advantage, in the second last step, of the fact that $P(t < t_{h,i}^{\text{MA,start}} \cap t > t_{h,i}^{\text{MA,end}}) = P(t < t_{h,i}^{\text{MA,start}} \cap t > t_{h,i}^{\text{MA,start}} + \delta_{h,i}^{\text{MA}}) \overset{\delta_{h,i}^{\text{MA}} > 0}{=} 0$.

$$
\begin{aligned}
P(\text{pos}_h(t) \in o_{h,i}^{\text{MA}}) &= \\
&= P(t_{h,i}^{\text{MA,start}} < t < t_{h,i}^{\text{MA,end}}) \\
&= P(t > t_{h,i}^{\text{MA,start}}) + P(t < t_{h,i}^{\text{MA,end}}) - P(t > t_{h,i}^{\text{MA,start}} \cup t < t_{h,i}^{\text{MA,end}}) \\
&= P(t > t_{h,i}^{\text{MA,start}}) + [1 - P(t > t_{h,i}^{\text{MA,end}})] - [1 - P(t < t_{h,i}^{\text{MA,start}} \cap t > t_{h,i}^{\text{MA,end}})] \\
&= P(t > t_{h,i}^{\text{MA,start}}) - P(t > t_{h,i}^{\text{MA,end}}) \\
&= \begin{cases}
0, & \text{if } t < \min(t_{h,i}^{\text{MA,start}}) \\
\int_{\min(t_{h,i}^{\text{MA,start}})}^{t} f(t_{h,i}^{\text{MA,start}} = t)dt, & \text{if } t < \min(t_{h,i}^{\text{MA,end}}) \\
\int_{\min(t_{h,i}^{\text{MA,start}})}^{t} f(t_{h,i}^{\text{MA,start}} = t)dt - \int_{\min(t_{h,i}^{\text{MA,end}})}^{t} f(t_{h,i}^{\text{MA,end}} = t)dt, & \text{otherwise.}
\end{cases}
\end{aligned} \tag{13}
$$

3.3.2. Coexistence

MC3 is violated and its corresponding violation index $\varphi_{\text{coex}}(\pi^{\text{MA}})$ is incremented by 1, if the probability of two HAPSs (h and h', with $h \neq h'$) operating at the same time t in the same MA# (i.e., $o_{h,i}^{\text{MA}} = o_{h',j}^{\text{MA}} = \text{MA\#}$) is greater than an imposed threshold p_{coex}. That is, if

$$\exists t \in [\min(t_{h,i}^{\text{MA,start}}), \max(t_{h,i}^{\text{MA,end}})], \ P(\text{pos}_h(t) \in o_{h,i}^{\text{MA}} \cap \text{pos}_{h'}(t) \in o_{h',j}^{\text{MA}}) > p_{\text{coex}}. \tag{14}$$

The probability in the previous expression can be expressed as the product of two probabilities, each of them computable using Equation (13):

$$P(\text{pos}_h(t) \in o_{h,i}^{\text{MA}} \cap \text{pos}_{h'}(t) \in o_{h',j}^{\text{MA}}) = P(\text{pos}_h(t) \in o_{h,i}^{\text{MA}}) \cdot P(\text{pos}_{h'}(t) \in o_{h',j}^{\text{MA}}). \quad (15)$$

To illustrate how the coexistence constraint is evaluated, Figure 5 shows the probabilistic evaluation of the existence of two HAPSs in MA6 for a particular π^{MA}. Given the probability density functions of the start and end time of each HAPS in MA6, obtained with Equations (4) and (5), and represented in Figure 5a,b, the probabilities of the presence of each HAPS in MA6 (i.e., $P(\text{pos}_h(t) \in o_{h,i}^{\text{MA}})$ and $P(\text{pos}_{h'}(t) \in o_{h',j}^{\text{MA}})$ with $o_{h,i}^{\text{MA}} = o_{h',j}^{\text{MA}} = \text{MA6}$) are estimated with Equation (13) and displayed in Figure 5c. Besides, for clarity of the representation, the time limits of Figure 5c are marked with vertical dashed lines in Figure 5a,b. The constraint function $\varphi_{\text{coex}}(\pi^{\text{MA}})$ associated to the coexistence of both HAPSs in the MA will be incremented if the product of the two probabilities represented in Figure 5c (i.e., $P(\text{pos}_h(t) \in o_{h,i}^{\text{MA}}) \cdot P(\text{pos}_{h'}(t) \in o_{h',j}^{\text{MA}})$) exceeds the threshold p_{coex}.

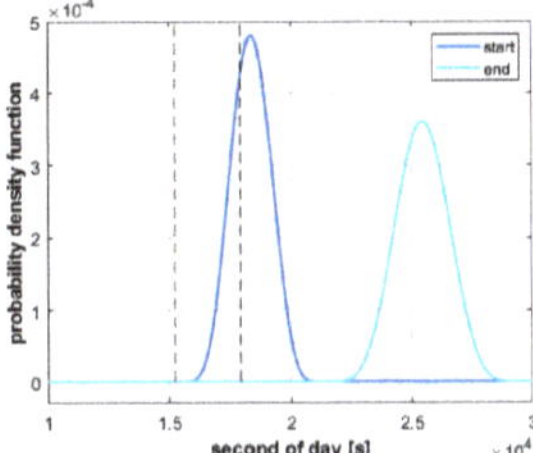

(**a**) Probability density function of start (dark blue) and end (light blue) time of HAPS1 in MA6 (i.e., $o_{1,i}^{\text{MA}} = \text{MA6}$)

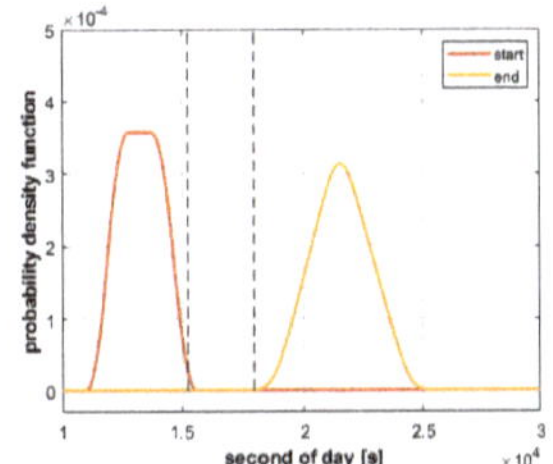

(**b**) Probability density function of start (dark orange) and end (light orange) time of HAPS2 in MA6 (i.e., $o_{2,i}^{\text{MA}} = \text{MA6}$)

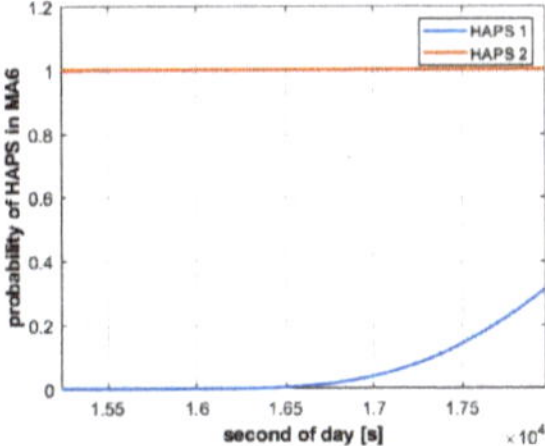

(**c**) Probability of HAPS1 (blue) and of HAPS2 (orange) operating in MA6 (i.e., $o_{1,i}^{\text{MA}} = o_{2,i}^{\text{MA}} = \text{MA6}$) between 15,200 s and 17,900 s

Figure 5. Probabilistic evaluation of the start and end time of the operation of HAPS in a MA and probability of their operation in the MA within the duration marked by the vertical dash lines.

3.3.3. Connection

This constraint considers the connectivity of mission elements of a plan represented at the MA-level. The mission elements (i.e., either MA# or WA#) are connected if and only if there is a corridor connecting two consecutive elements in the MA-level plan. Each lack of connection increments the constraint criterion $\varphi_{\text{con}}(\pi^{\text{MA}})$ by 1.

3.3.4. Overall Constraint Violation

As each constraint violation increments $\varphi_{\text{criteria}}(\pi^{\text{MA}})$ of its corresponding criteria; a non-null $\varphi_{\text{criteria}}(\pi^{\text{MA}})$ implies the infeasibility of the plan. For that reason, the overall constraint function of a given plan π^{MA} is simply the sum of all the constraint criteria:

$$\varphi(\pi^{\text{MA}}) = \varphi_{\text{saf}}(\pi^{\text{MA}}) + \varphi_{\text{coex}}(\pi^{\text{MA}}) + \varphi_{\text{con}}(\pi^{\text{MA}}). \tag{16}$$

Finally, note that in order to determine during the evaluation of $\varphi_{\text{saf}}(\pi^{\text{MA}})$ and $\varphi_{\text{coex}}(\pi^{\text{MA}})$ if there is a t where Equations (12) and (14) hold, the time variable t is discretized, within the corresponding intervals given in those equations, into equally spaced time instances.

4. Implementation of a GA-Guided Hierarchical Task Planner

The purpose of the planner presented in this section is to perform the task planning for a group of HAPSs that maximize the objective functions presented in Section 3.2 (i.e., reward, effort, and diversity), while ensuring that it is feasible according to the constraint criteria introduced in Section 3.3 (i.e., safety, coexistence, and connection).

To achieve it, we use the Genetic Algorithm (GA) based planner described in Section 4.3 that manipulates the codification of the solutions presented in Section 4.1, which encodes the sequence of MA-level tasks that determines the (sub)optimal temporal hierarchical decomposition of tasks governed by the approach presented in Section 4.2.

4.1. Plan Codification

The solutions that the planner must provide are hierarchical plans $(\pi^{\text{MA}}, \pi^{\text{LOI}}, \pi^{\text{WP}})$ to be presented as suggestions to the HAPS operator during the monitoring mission. Each plan π^* in the hierarchy, as its formal description in Section 3.1 shows, consists of a list of tasks and their start times and durations. However, as the latter are affected by the weather conditions, we decide to code only the tasks in the optimizer and estimate their timing, when required, in the evaluation of the objective and constraint functions.

Besides, a hierarchical plan decomposes the tasks at a given level into a set of tasks of a lower level, until the set of primitive tasks, or rather "actions", are obtained. In our case, the decomposition into tasks at the intermediate (LOI) level and at the lower (WP) level are given by a fixed set of rules. In particular, a $o_{h,i}^{\text{MA}} = $ WA# task is directly converted into $o_{h,j}^{\text{LOI}} = \text{fly}_{\text{WA\#}}$, while a $o_{h,i}^{\text{MA}} = $ MA# is decomposed into the sequence of $o_{h,j}^{\text{LOI}} = $ $\text{monitor}_{\text{LOI\#} \in \text{MA\#}}$ tasks that implies the sequential monitoring of all the LOIs (without revisit) in the MA before departing. As the number of possible sequences of LOIs in a MA is the number of their permutations, we fix the order in which the LOIs are visited, starting with the LOI closest to the HAPS entry point in the MA and following the order that minimizes the distance of the HAPS within the MA. This way of proceeding ensures the shortest travel distance within a MA, simplifies the optimization problem and accelerates the computation of the plans, as we can precalculate all the orders for a given MA, since we know beforehand all its possible entry and exit points. Besides, it is justified by the fact that the weather conditions do not vary much within a MA. Finally, the decomposition of LOI actions in waypoint actions is usually straightforward and the only possible choices are also fixed. As Figure 2a depicts, this can be done by connecting the entry and exit points of a corridor, the entry point at a MA to the start point of the scan, followed by the points that mark the start and end of a scan track and finally, the exit point of the MA.

Taking into account the previous ideas, the remaining effort to determine the (sub)optimal solution lies in the search for the optimal lists of high level tasks (i.e., MA# and WA#) that each HAPS must perform. As the number of possible WA and MA is finite, the elements of the lists can also be encoded with a finite alphabet of labels. Hence, for the GA-based planner, the solution will be encoded as an array of as many elements as HAPSs, where each element contains the list of the high level tasks (MA# and WA#) of each HAPS. Finally, to distinguish this encoding from the corresponding hierarchical plan $(\pi^{\text{MA}}, \pi^{\text{LOI}}, \pi^{\text{WP}})$, we represent sol_k as the k-th possible solution of the planner, $sol_k[h]$ as the part of the solutions for HAPS h, and $sol_k[h][i]$ as the $i-th$ task at the MA-level (i.e., MA# or WA#) to be performed by HAPS h of the k-th solution of the planner.

The next section explains how to obtain a hierarchical plan $(\pi_h^{MA}, \pi_h^{LOI}, \pi_h^{WP})$ from a given $sol_k[h]$.

4.2. Temporal Hierarchical Task Decomposition

Since the coding of the solutions manipulated by the GA that supports the search of (sub)optimal solutions in our planner is only a sequence of MA# and WA# actions, and the objective and constraint functions used to evaluate them require a hierarchical plan and the estimated end time of the tasks at different levels, in this section we detail, with the help of the pseudo-code presented in Algorithm 1, how the conversion from $sol_k[h]$ to π_h is carried out.

The algorithm inputs are the solution plan $sol_k[h]$ that encodes only the tasks at the MA level and the start mission time t_h^0 for HAPS h, and its output is the hierarchical plan $\pi_h = (\pi_h^{MA}, \pi_h^{LOI}, \pi_h^{WP})$. To start with, Line 1 initializes the hierarchical plan as empty sequences, while Line 2 initializes an empty list for the limits of the duration of each WP task (which will be used later to estimate the density functions of the end time of the tasks at WP level) and Line 3 initializes the Boolean flag b_{finish} (which is meant to keep track of the temporal plan length and ignore the tasks that start after the maximum plan horizon $T_h^{\max}$ has been reached).

After the initialization steps, three nested loops are implemented, in order to be able to decompose tasks at MA-level into primitive tasks at WP-level and to determine the probability distributions of the end time of the tasks of the highest levels from the primitive tasks of the lowest. As the number of nested loops depends on the depth of the decomposition, in our case, three loops are necessary, since the primitive tasks (at WP-level) lie two levels below the MA-level at which the initial $sol_k[h]$ is given.

The particular behavior implemented in the three loops is the following. At Line 7 the current MA task in $sol_k[h][i]$ is selected to be decomposed into an ordered list of LOI tasks at Line 8. Next, at Line 9, temporary partial plans of the lower level tasks ($\tilde{\pi}_h^{LOI}$ and $\tilde{\pi}_h^{WP}$) are initialized as empty lists to be able to temporarily store the sequences of tasks obtained after the decomposition of the selected task o_{current}^{MA} at the MA-level into the lists of tasks at the LOI-level or WP-level. This lowest level decomposition into primitive tasks happens at Lines 13 and 14, where the current LOI task o_{current}^{LOI} is selected and decomposed into the corresponding list of WP tasks. Next, we start processing sequentially each of the primitive tasks o_{current}^{WP} of our hierarchical task plan. For this, at Line 19 we determine (using Equation (3)) the limits (minimum and maximum) of the duration needed for the task and append them in Line 20 to the list of limits $list\_limits$. Next, at Line 21 the minimum temporal plan length up to the current o_{current}^{WP} is checked to see if the plan horizon $T_h^{\max}$ is exceeded. If that is the case, the decomposition must stop and all lower-level partial plans should not be accounted for. If $T_h^{\max}$ is not exceeded, at Line 26 the probability distribution on the end time of task o_{current}^{WP} is determined with Equation (4), and at Line 27 the current o_{current}^{WP} task and the determined probability distribution is appended to the temporal plan $\tilde{\pi}_h^{WP}$. Next and after looping over all the tasks at the WP-level (if the finishing time has not been reached) the probability distribution $f(t_{\text{current}}^{\text{end,LOI}})$ of the end time of the current LOI-level task is assigned the probability distribution $f(t_{\text{current}}^{\text{end,WP}})$ of the end time of the last WP-level task, and o_{current}^{LOI} and $f(t_{\text{current}}^{\text{end,WP}})$ are appended to the temporal plan $\tilde{\pi}_h^{LOI}$. Next, a similar process is repeated to obtain at Line 38 the probability of the end time of the current action of the MA level plan from the probability distribution of the end time of the last LOI level action and to update at Line 39 the MA plan π_h^{MA}. Finally, at Lines 40 and 41 the partial plans, $\tilde{\pi}_h^{LOI}$ and $\tilde{\pi}_h^{WP}$ are appended to their corresponding plans π_h^{LOI} and π_h^{WP}.

Algorithm 1: Temporal hierarchical task decomposition

Input: $sol_k[h]$, the encoded MA-level task sequence for HAPS h
Input: t_h^0, the initial mission time for HAPS h
Result: $\{\pi_h^{MA}, \pi_h^{LOI}, \pi_h^{WP}\}$, the hierarchical task plan for HAPS h

1   $\pi_h^{MA} =<>; \pi_h^{LOI} =<>; \pi_h^{WP} =<>;$ // Initialize the hierarchical plan as an empty plan
2   $list\_limits =<>;$ // Initialize an empty list for the limits of the duration of each WP task
3   $b_{finish} = false;$ // Boolean variable that finishes the algorithm because the plan has reached the maximum time
4   $i = 1;$ // Index to iterate that actions of the MA level
5   **while** $(i<=length(sol_k[h])$ & $b_{finish} = false)$ // Loop over the MA level
6   **do**
7      $o_{current}^{MA} = sol_k[h][i];$ // Current action at the MA level
8      $list\_LOI = decompose(o_{current}^{MA});$ // Decompose $o_{current}^{MA}$ in its corresponding actions list at LOI level
9      $\tilde{\pi}_h^{LOI} =<>; \tilde{\pi}_h^{WP} =<>;$ // Initialize temporary lists of lower-level tasks for $o_{current}^{MA}$
10      $j = 1;$ // Index to iterate the list of LOI
11   **while** $(j<=length(list\_LOI)$ & $b_{finish} = false)$ // Loop over the decomposition at the LOI level
12   **do**
13          $o_{current}^{LOI} = list\_LOI[j];$ // Current action at the LOI level
14          $list\_WP = decompose(o_{current}^{LOI});$ // Decompose $o_{current}^{LOI}$ in its corresponding actions list at WP level
15          $k = 1;$ // Index to iterate the list of WP
16   **while** $(k<=length(list\_WP)$ & $b_{finish} = false)$ // Loop over the decomposition at the WP level
17   **do**
18              $o_{current}^{WP} = list\_WP[k]$ // Current action at the WP level
19              $limit = [l(o_{current}^{WP})/(|v_a| + \max(|v_w|)), l(o_{current}^{WP})/(|v_a| - \max(|v_w|))]$
20              $list\_limits.add(limit);$ // Add the duration limits of $o_{current}^{WP}$ to the list of WP durations
21   **if** $(t_{h,0} + list\_limits.sum\_min() > T_h^{max})$ // Does the task ends after the maximum allowed time?
22   **then**
23                 $b_{finish} = true;$
24   **end**
25   **else**
26                 $f(t_{current}^{end,WP}) = compute(t_{h,0}, list\_limits)$ // Obtain, with Equation (4), the distribution for $t_{current}^{end,WP}$
27                 $\tilde{\pi}_h^{WP}.add(o_{current}^{WP}, f(t_{current}^{end,WP}))$ // Add current task and its distribution to the WP plan
28                 $k = k + 1;$
29   **end**
30   **end**
31   **if** $b_{finish} = false$ **then**
32              $f(t_{current}^{end,LOI}) = f(t_{current}^{end,WP})$ // Use the same distribution, as $t_{current}^{end,WP}$ for last WP equals $t_{current}^{end,LOI}$
33              $\tilde{\pi}_h^{LOI}.add(o_{current}^{LOI}, f(t_{current}^{end,LOI}))$ // Add current task and its estimated distribution to the LOI plan
34              $j = j + 1;$
35   **end**
36   **end**
37   **if** $b_{finish} = false$ **then**
38          $f(t_{current}^{end,MA}) = f(t_{current}^{end,LOI})$ // Use the same distribution, as $t_{current}^{end,LOI}$ for last LOI equals $t_{current}^{end,MA}$
39          $\pi_h^{MA}.add(o_{current}^{MA}, f(t_{current}^{end,MA}))$ // Add current task and its estimated distribution to the MA plan
40          $\pi_h^{LOI}.add(\tilde{\pi}_h^{LOI})$ // Add the partial hierarhical plan at the LOI-level
41          $\pi_h^{WP}.add(\tilde{\pi}_h^{WP})$ // Add the partial hierarhical plan at the MP-level
42          $i = i + 1;$
43   **end**
44 **end**

4.3. GA-Guided Search of the Best Plans

Algorithm 1 determines the hierarchical plan associated to a given HAPS h and list of MA-level tasks $sol_k[h]$. However, determining the best list of MA-level tasks $\boldsymbol{sol}_k$ for all the HAPSs in the mission is extremely complex, as we are facing a probabilistic time-dependent multiple-vehicle routing problem, where multiple objective functions (OF_{rew},

OF_{div}, and OF_{div}) and constraint criteria (φ_{saf}, φ_{coex} and φ_{con}) have to be considered. To tackle it, we develop a mission planner that exploits the Non-dominated Sorting Genetic Algorithm (NSGA-II, Ref. [19]) to look for the optimal sol_k. For the clarity and completeness of the paper, our implementation of NSGA-II is recapitulated in Algorithm 2, along with the specifics relevant to this work.

Algorithm 2: NSGA-II-guided search of nondominated solutions of hierarchical plans

---

Input: N_{P}, Population size
Input: I_{max}, Number of Iterations
Input: $p_{crossover}$, Probability of crossover
Input: p_{mut}, Probability of mutation
Input: $k_{tournament}$, Tournament arity
Result: *FirstFront*, information (MA-level plan, task decomposition, and evaluation criteria) of all the solutions in the first pareto front.

1   $Population = \varnothing$ `// Start an empty population set`
2   $k = 1$
3   **while** $k \leq N_{\mathrm{P}}$ `// Population Initialization and evaluation loop`
4   **do**
5     $sol_k = InitializeSolution()$ `// Initialize the lists of MA-level actions for all the HAPSs by ensuring connectivity among mission elements`
6     $[\pi_k, eval_k] = Decomponse\&Evaluate(sol_k)$ `// Obtain the hierarchical plan and evaluate it`
7     $Population.add(\{sol_k, \pi_k, eval_k\})$ `// Add solution to population`
8     $k = k + 1$
9   **end**
10   $i = 1;$ `// Initialize the iteration/generation counter`
11   **while** $i < I_{max}$ `// While the stop condition is not met`
12   **do**
13     $Children = \varnothing$ `// Start an empty children set`
14     $k = 1$
15   **while** $Children.size() < N_P$ `// While not enough children have been created`
16   **do**
17       $[par_k, par_{k+1}] = TournamentSelection(Population, k_{\mathrm{tournament}})$ `// Select pair of parents`
18       $[sol_k, sol_{k+1}] = Crossover(par_k, par_{k+1}, p_{\mathrm{crossover}})$ `// Create children` sol_k `and` sol_{k+1}
19   **for** $l=k{:}k{+}1$ **do**
20         $[sol_l] = Mutate(sol_l, p_{\mathrm{mut}})$ `// Mutate child` sol_l
21         $[\pi_l, eval_l] = Decompose\&Evaluate(sol_l)$ `// Decompose and evaluate child` sol_l
22         $Children.add(\{sol_l, \pi_l, eval_l\})$ `// Add child` sol_l `to children set`
23   **end**
24       $k = k + 2$
25   **end**
26     $Population = Recombine(Population, Children, N_P)$ `// Determine the new population based on the old population and on the children, using the nondominated sorting of NSGA-II. The solutions of the new population will be sorted in Pareto fronts. Duplicates will be discarded.`
27     $i = i + 1;$
28   **end**
29   $FirstFront = Population.FirstFront()$ `// Get information of the first Pareto front of the final population`

Between Lines 1 and 9, Algorithm 2 performs the initialization steps, consisting of generating N_p solutions of high-level lists of actions (sol_k), performing their decompositions into hierarchical plans (π), and evaluating their objective functions and constraint criteria ($eval_k = [OF_{\mathrm{rew}}, OF_{\mathrm{div}}, OF_{\mathrm{div}}, \varphi_{\mathrm{saf}}, \varphi_{\mathrm{coex}}, \varphi_{\mathrm{con}}]$). To do it, on the one hand, *InitializeSolution()* generates a population of solution plans that fulfill the connectivity constraint, by appending to $sol_k[h][i+1]$ a mission element (MA# or WA#) randomly selected (according to a uniform distribution) among the mission elements connected to the last one $sol_k[h][i]$. Besides, a predetermined minimum duration, used to decide how long the initialization of $\boldsymbol{sol}_k$ should take for each mission, is calculated based on the maximum ground speed ($v_a + |\max(v_w)|$) and on a travel distance that lower-bounds all realistic travel distances for the mission element derived from the tasks at the WP-level (i.e., the shortest diagonal distance of the mission element). On the other hand, $Decompose\&Evaluate(\boldsymbol{sol}_k)$ performs the decomposition into hierarchical plans of all the lists of high level actions $sol_k[h]$ using Algorithm 1 and evaluates the obtained plans $\boldsymbol{\pi}_k$ using the objective functions and constraint criteria. Finally, the *Population* is formed by the high-level list of actions $\boldsymbol{sol}_k$, their corresponding hierarchical plan decomposition $\boldsymbol{\pi}_k$, and their corresponding evaluation $\boldsymbol{eval}_k$.

Next, the generation loop of the algorithm is performed, between Lines 11 and 28, until reaching the stop condition, consistent on testing if a predefined number of iterations is met. In each generation (algorithm iteration), the new set of solutions, named *Children* in Algorithm 2, are created by selecting from *Population* pairs of solution plans expressed at the MA-level (named $\boldsymbol{par}_k$ and $\boldsymbol{par}_{k+1}$), which will undergo crossover, mutation, decomposition, and evaluation (see Lines 18 to 21). Afterwards, the old and the new population are combined in Line 26 to determine the new population of the following generation.

In particular, the pairs of parents selection is performed with *TournamentSelection* ($Population, k_{tournament}$) that implements the k-tournament operator proposed in [19] for constrained multiobjective problems. That is, for each parent, it selects randomly, according to the uniform distribution, $k_{tournament}$ solutions of *Population*, and among them it selects the best one, preferring infeasible solutions with smaller $\varphi(\boldsymbol{\pi})$ to infeasible solutions with a bigger $\varphi(\boldsymbol{\pi})$, feasible solutions (i.e., those with $\varphi(\boldsymbol{\pi}) = 0$) to infeasible ones (i.e., those with $\varphi(\boldsymbol{\pi}) > 0$) and the Pareto dominating feasible solutions to the dominated ones.

Next, the crossover of the two parents ($\boldsymbol{par}_k$ and $\boldsymbol{par}_{k+1}$) is performed with *Crossover* ($\boldsymbol{par}_k, \boldsymbol{par}_{k+1}, p_{crossover}$), that implements a single-point crossover that takes into consideration the expected ending time of the MA-level tasks of each parent. That is, unlike the typical genetic operators for crossover (which select the gene where the crossover should be performed in both parents), we select randomly, according to the uniform distribution and as shown in Figure 6a, the crossover time $t_{crossover}$. Each parent is then divided into a head and tail component at the start time of a task (at the MA-level) closest to $t_{crossover}$ (as marked in the red ellipses), and afterwards the head of one parent and the tail of the other (and vice versa) are concatenated to build the new list of solutions of each child, as shown in Figure 6b. Besides, the probability of crossover $p_{crossover}$ is used to decide, for each pair of parents, if they should undergo the crossover process or if they should be directly copied as new possible solutions.

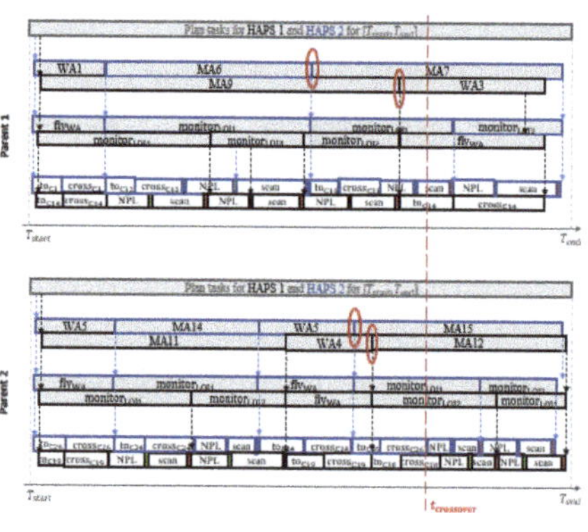

(a) Random selection of a crossover time $t_{crossover}$.

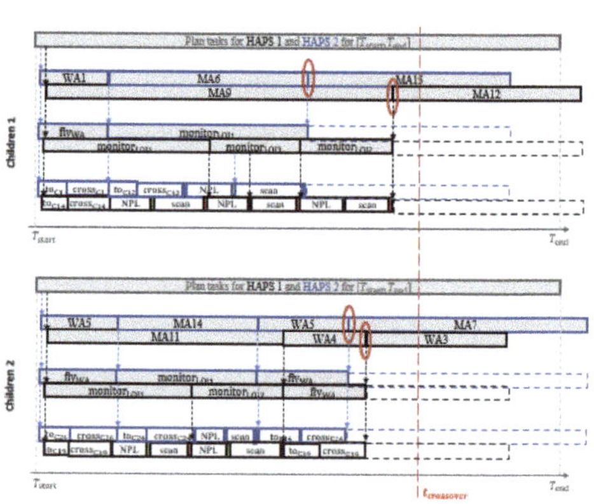

(b) Single-point crossover at the MA-level.

Figure 6. Single-point crossover with a random selection of the temporal crossover point.

After crossover, each child is mutated in Line 20 with $Mutate(\boldsymbol{sol}_l, p_{\mathrm{mut}})$, which uses the probability of mutation p_{mut} to determine, according to the uniform distribution, if each of the MA-tasks in $\boldsymbol{sol}_l$ has to mutate and be changed by any other MA# or WA# task randomly selected at the MA-level.

After mutation, each of the children is decomposed and evaluated with $Decompose\&$ $Evaluate(\boldsymbol{sol}_k)$. Moreover, as crossover preserves the head actions of already decomposed plans (as Figure 6b shows) and mutation does not influence in the timing (duration) of the parts of the plan that are previous to the mutation point, we can use the corresponding invariant decomposed plans of the parents to perform more efficiently the decomposition of the new children. Besides, it is worth noting that the connection constraint $\varphi_{\mathrm{con}}(\pi^{MA})$ can be violated after a crossover or a mutation. Hence, in order to make the decomposition quicker, the sequence of high level tasks ($< o^{\mathrm{MA}}_{h,i+1}, \ldots, o^{\mathrm{MA}}_{h,end} >$) after the last connected one ($o^{\mathrm{MA}}_{h,i}$) are not decomposed into tasks of lower levels, neither will the density distribution for the ending time of their tasks be determined. Finally, we prefer to use crossover and mutation operators that allow to create unconnected high-level (MA) plans to allow them to be reconnected afterwards, eventually, after other crossovers and mutations. By doing so, the planner can sometimes create invalid solutions that are used by the search process to transverse infeasible regions of the search space while moving from one side of the feasible search space to the other. The planner configurations under analysis in the following section will show the importance of this fact.

Once the children population has been completely created, the *Population* and their *Children* are first compared to discard the duplicate solution. Afterwards they are sorted together into nondominated fronts by using the same criteria as in the tournament selection (i.e., their objective functions and constraint criteria are taken into account to prefer feasible to infeasible, solutions that are closer to be feasible to those that are farther to be feasible). Finally, the sorted population is truncated to contain only the best N_P solutions, using the crowding distance, as described in [19], to pick the surviving solutions that belong to the last front that can be admitted into the new population.

At the end of the algorithm, once the generation loop has finished, the planner returns the set of solutions that belong to the first front of the last *Population*. In this front, it is expected to find solutions that fulfill constraints and that are equally good, from the Pareto comparison perspective, regarding the objective functions.

Finally, it is worth noting that although it is not stated in Algorithm 2 for simplicity, all the *Population* of all iterations obtained by the planner (in the initialization and during the generation loop) are also stored to be able to analyze the performance of the planner, over different scenarios, in the following section.

5. Results and Analysis

This section analyzes the performance of the GA-based planner described in this paper for determining the hierarchical task decomposition of a set of HAPSs that carry out realistic monitoring missions in complex time-varying environments with a highly-organized airspace structure. This planner combines the algorithms described in Section 4 as well as the evaluation functions and constraint criteria formally elaborated in Section 3 in order to take into account the realistic mission requirements and constraints described in Section 2.

To highlight the benefits of the planner, different scenarios are used during the performance tests. The subsequent subsections will first introduce the chosen scenarios, followed by a description of the different variants of the planner that are tested (to determine which configurations are better for each scenario), by an interpretation of the graphical representation of the results, and finally, by their in-depth analysis.

The algorithms are implemented in Matlab and tested on a 4-core i7 processor at 1.80 GHz. On average, an iteration takes 15 s and can go up to 30 s under challenging weather conditions or when more HAPSs are involved, due to the constraint evaluations. The computation time is acceptable for the mission at hand, as the planning is meant to be performed prior to the execution (as opposed to real-time planning), and therefore more generous planning time is allowed. Besides, the planner can also be implemented as an "anytime" planner, as the algorithm provides a Pareto front with feasible solutions at each iteration. To accelerate the code in a future release, the evaluation functions could be implemented in C.

5.1. Scenarios

The three scenarios considered in the performance tests of this paper share the mission map depicted in Figure 1 and the HAPSs and mission parameters presented in Appendix A. The scenarios differ in the weather data and/or the number of HAPSs involved. The following paragraphs briefly introduce the settings of each scenario, while Table 2 provides an overview of all of them.

Nominal scenario. In the first scenario, historical weather data of COSMO-DE (predecessor of COSMO-D2) taken from a relatively calm day in April 2018 is used. The weather conditions are considered moderate, with some strong wind before noon time and some cloudy hours. Besides, the mission is performed by two HAPSs, placed initially at WA2 and WA4, that have to monitor the LOIs depicted in Figure 1 taking into account the rewards and coverage information provided in Table A3. This scenario is useful to see how the planner works under good (nominal) weather conditions.

Challenging weather scenario. In this scenario, we use a weather data of the same format as the real weather data considered in the first scenario but synthetically increase the wind to make it stronger. In particular, in some mission areas or corridors, strong wind can occur during more than half of the time of the day. This synthetic scenario is created in order to demonstrate the performance of the planner under challenging weather conditions. Finally, similar to the first scenario, only two HAPSs, placed again initially at WA2 and WA4, are considered in this scenario.

Three HAPSs scenario. To demonstrate the scalability of the planner regarding the number of HAPSs, in this scenario the monitoring mission is performed by three HAPSs, placed initially at WA2, WA4, and WA5. The weather conditions are identical to those of the first scenario.

Finally, it is worth noting that the first scenario (labelled as SC1 hereafter) will be considered the basis to compare against the other two, as the second scenario (SC2) is similar to the first but with worsened weather conditions, while the third scenario (SC3) is the first with an additional HAPS.

Table 2. Scenarios considered for performance tests.

Scenarios	Weather Data	Number of HAPSs
SC1 (Nominal scenario)	Historical (April 2018)	2
SC2 (Challenging weather scenario)	With synthetically increased wind magnitude	2
SC3 (Three HAPSs scenario)	Historical (April 2018)	3

5.2. Planner Configurations

The general input parameters of Algorithm 2 are presented in Table 3. They have been selected after analyzing the behavior of the planner under different combinations of parameters over the presented scenarios.

Besides, several configurations of the GA are considered to optimize the hierarchical task plan and to analyze the performance and benefits of each one for the different scenarios. The three Planner Configurations (PC1, PC2, and PC3) analyzed in the paper are implemented in general according to Algorithm 2 and two of them (PC2 and PC3) contain some slight variations injected into parts of the code to support the following behaviors:

Planner Configuration 1 (PC1). The constraint-handling technique proposed by [19] is used for select the pair of parents in the k-tournament selection (at Line 17 of Algorithm 2) and for recombining the old and new populations (at Line 26). In other words, solutions that fulfill or are closer to fulfilling the constraints are preferred to solutions that do not fulfill or are further to fulfilling them, and among solutions that are equally good regarding the constraints, solutions that Pareto dominate the others regarding the objective functions are preferred to solutions that are Pareto dominated. As this configuration implements the standard constraint-handling techniques of NSGA-II [19], it is also the one described in Section 4.

Planner Configuration 2 (PC2). The constraint-handling criteria are only applied to select the pair of parents in the k-tournament selection in Line 17 of Algorithm 2 and ignored during the recombination of old and new populations. That is, during the recombination step in Line 26 of Algorithm 2, the solutions are sorted by only taking into account the ordering imposed by the Pareto comparison of the objective functions. The motivation of this variation is to have a planner configuration that is less "stringent" with the hierarchical plans that violate the constraints (i.e., that have $\varphi(\pi^{MA}) > 0$), and to give them more chances to be selected for the next generation (or even be selected as parents for the generation of children solutions of the next iteration).

Planner Configuration 3 (PC3). The diversity objective function (OF_{div}) is ignored both during the parents selection and recombination steps. This configuration has been set up to put forth the benefit of considering the diversity (and not only the expected reward or the effort) for planning.

Table 3. Planner configuration parameters.

Planner Parameters	Parameter Values
Crossover probability, $p_{\text{crossover}}$	0.9
Mutation probability, p_{mut}	0.1
Population size, N_{P}	50
Tournament size, $k_{\text{tournament}}$	3
Number of generations, I_{max}	100 (in SC1 and SC3), 60 (in SC2)
Constraint Thresholds	**Threshold Values**
p_{saf}	0.1
p_{coex}	0.3

For readers familiar with the stochastic ranking mechanism for constrained evolutionary optimization presented in [20], it is interesting to highlight that PC1 and PC2 represent the two extreme cases that are obtained when the probability of ignoring the constraints is respectively set to 0 (for PC1) or to 1 (for PC2). That is, during the recombination step, in PC1 the constraints are never ignored while in PC2 the constraints are always ignored. Comparing the behavior of the extreme cases will facilitate the understanding of the effects of taking into account (or ignoring) the constraints in the recombination step.

Finally, for the computation of the expected reward using Equation (7), we assume $L(\mu = \text{success}|s_{h,i}, t_i, o_{h,i}^{MA}, w_{t_i}) = 0.8$, if the cloud coverage of w_{t_i} is smaller than the image coverage required by the mission, as shown in Table A3. Otherwise, $L(\mu = \text{success}|s_{h,i}, t_i, o_{h,i}^{MA}, w_{t_i}) = 0.2$.

5.3. Results Representation

In order to provide an overview on the the weather (wind and cloud coverage) conditions of each scenario, on the time windows where each mission area can be visited, as well as on a representative solution obtained by the planner, we use the graphics displayed in Figures 10–14, whose vertical axes represent, from the bottom to the top, the mission areas (MA#) and waiting areas (WA#), while the horizontal axes represent the hour of the day. Further, the graphics also contain the above-mentioned information on the weather, mission, and plan, which is represented by the following items:

- The light-grey bars represent the time windows with clear sky, while the dark-grey bars signify high cloud coverage above the corresponding MA.
- The light-green bars represent the time windows for the absence of critical weather conditions at the MA or WA, while the dark-green bars signify critical weather conditions for the corresponding MA or WA, e.g., strong wind.
- The light-blue bars represent the time windows where monitoring missions at the corresponding MA are requested (and therefore rewarded), while the dark-blue bars signify the absence of mission request for the corresponding MA.
- Red lines represent the monitoring/fly-by tasks of HAPS1 to be performed on the corresponding MA/WA, according to the representative plan π_1^{MA}. Moreover, the thicker line in the middle marks the median start and end time, while the thinner lines mark the time range from the minimum starting time to the maximum end time of each task.
- Similarly, blue lines represent the tasks at MA-level to be performed by HAPS2 and magenta lines represent tasks at MA-level to be performed by HAPS3.

Taking into account the previous information, the sequence of mission elements (i.e., MA# and WA#) traversed by each HAPS in a representative solution can be observed, along with the weather conditions and the mission time window of each scenario. For example, in Figure 10a we can observe, following the red line, that HAPS1 moves from WA2 (the starting location of HAPS1, which is not represented in the graphic) to MA1, WA1, WA2, MA2, WA2, WA1, MA3, WA1, and MA3. Besides, MA1 is visited when not requested, while MA2 and the two visits to MA3 are within the correct mission time windows. Besides, MA2 is partially visited under cloudy conditions, which can reduce the expected reward obtained by HAPS1, while MA3 are visited under good weather conditions, which provides HAPS1 two times the total reward of MA3.

Besides, in order to analyze the performance of the different configurations of the planner in different scenarios, we store for each scenario-planner configuration pair and for each iteration of the GA, the values of the objective functions (OF_{rew}, OF_{eff} and OF_{div}) and constraint criteria (φ_{saf}, φ_{coex} and φ_{con}) of all the feasible solutions (i.e., $\varphi(\pi) = 0$) of the best Pareto front obtained during the execution of the algorithm. With that information, we represent the following graphs:

- The evolution over iterations of the Mean and Standard Deviation (M&SD) of the values of each objective function of *the feasible solutions that belong to the best front*. Considering *only feasible solutions of the best front* is initially necessary for the three planner configuration, since it is possible that the first Pareto fronts are initially infeasible. Besides, it is always necessary in PC2, since the fronts are obtained by ignoring the constraints, and therefore, the best front obtained using PC2 can contain infeasible solutions. Moreover, this is meaningful since only the final feasible solution plans of the best Pareto front will be presented to the HAPS operator. The M&SD evolution graphs for each objective function are presented in the first row of Figures 8–13 for Scenario 1, 2, and 3, respectively. The mean and standard deviation values of each objective function are represented in different columns of the figures (left column OF_{rew}, middle column OF_{eff}, and right column OF_{div}). Besides, while the mean is depicted over iterations with a bold line, the shadowed area around it represents the standard deviation, using a different color for each planner configuration (blue for PC1, green for PC2, and red for PC3).
- The evolution of the Maximum (Max) value of each objective function obtained among the solutions of the first Pareto front that also fulfill the constraints are plotted in the lower row of graphs of Figures 8–13. These graphs, organized as the previous and using only a line for the Max value, complement the M&SD evolution graphs as they show the objective values of the best solutions with respect to each objective in the Pareto front.

Finally, we also use two additional types of graphs in order to analyze further certain scenarios:

- A 3D representation of the values of the three objective functions of all the solutions of the population versus the values of the objective functions of the solutions of the best Pareto front, at selected iterations of the planner (and a 2D representation of OF_{eff} versus OF_{rew}). This information, represented in Figure 7, marking in red the points associated to the solutions of the best Pareto front and in black the remaining solutions of the population, is used to graphically demonstrate the effectiveness of the planner in evolving and finding solutions of the Pareto front.
- The number of infeasible solutions within the population at each iteration. This information, represented in Figure 9, is used to put forth the advantage of enabling/disabling the constraint handling during the recombination step of PC1 and PC2.

5.4. Comparative Analysis

In the following sections, the results obtained from each scenario and planner configurations, characterized using the types of graphical representation explained in Section 5.3, are analyzed.

5.4.1. Analysis for Scenario 1 (SC1)

For SC1, all the configurations of the planner (PC1-PC3) are tested and their GA are set to run for 100 iterations to illustrate better the convergence behavior of the planner.

Figure 7 summarizes, at three selected iteration counts (in particular at the 1st, 40th, 100th iteration), the evolution of the OF of the population and of the best Pareto front, obtained using PC1. The graphics show: (1) how the number of solutions belonging to the best Pareto front in the population increases as the iteration number grows and (2) how all the solutions move towards the Pareto optimal front, along the axis in the direction of increasing values of the three objective functions. This is the expected behavior of NSGA-II, which is the optimizer that supports the search of the sequence of high-level tasks in our planner. To avoid increasing unnecessarily the length of the paper, it does not include more graphics of this type for the other configurations (PC2 and PC3) or for the remaining scenarios (SC2 and SC3), as they present similar behaviors. Besides, the evolution graphs, used in the rest of the paper are more suitable to provide further insights on the behavior of the planner configurations.

Figure 8a–c show the evolution of the mean and standard deviation of the values of the objective functions (OF_{rew}, OF_{eff} and OF_{div}) of the feasible solutions that belong to the best Pareto front and that are found using the three planner configurations, while Figure 8d–f show the maximum value of each OF. According to Figure 8a–c, the standard deviations on the values of OF_{rew}, OF_{eff} and OF_{div} obtained using PC2 are substantially wider than those obtained with PC1 and PC3. This is due to the fact that the constraint criteria are used only for the parents selection, allowing the MOEA to have a bigger "exploring" capability. Besides, we can also observe that the GA search converges earlier, around iteration 60. Additionally, to show the importance of considering OF_{div}, in SC1 we also test PC3, where OF_{div} is neglected deliberately during the parents selection and recombination steps. Figure 8c,f show how the diversity criterion evolution is worst for the configuration where OF_{div} is neglected (that is, for PC3), while PC1 and PC2 reach similar values (in particular the maximum value of OF_{div} for PC1 is not observed as it is equal to the maximum value of OF_{div} for PC2). This implies that the mission plans obtained with PC3 suffer from having a low diversity, resulting in a more challenging selection process to be performed by the human operator who is responsible of choosing a "well-balanced" plan among the feasible plans of the best Pareto front returned as plan suggestions by the planner. Besides, if we compare PC1 and PC2, we can conclude that for SC1, PC2 produces overall better solutions regarding OF_{rew}, while PC1 produces overall better solutions with respect to OF_{eff}. However, the values of OF_{eff} of the solutions found using PC1 is only marginally better than the values of OF_{eff} of the solutions obtained by PC2, while the values of OF_{rew} of the solutions obtained with PC2 is significantly bigger than the values of of

OF_{rew} of the solutions obtained with PC1. Therefore, we conclude that PC2, with its better exploring capability, is more suitable for the first scenario.

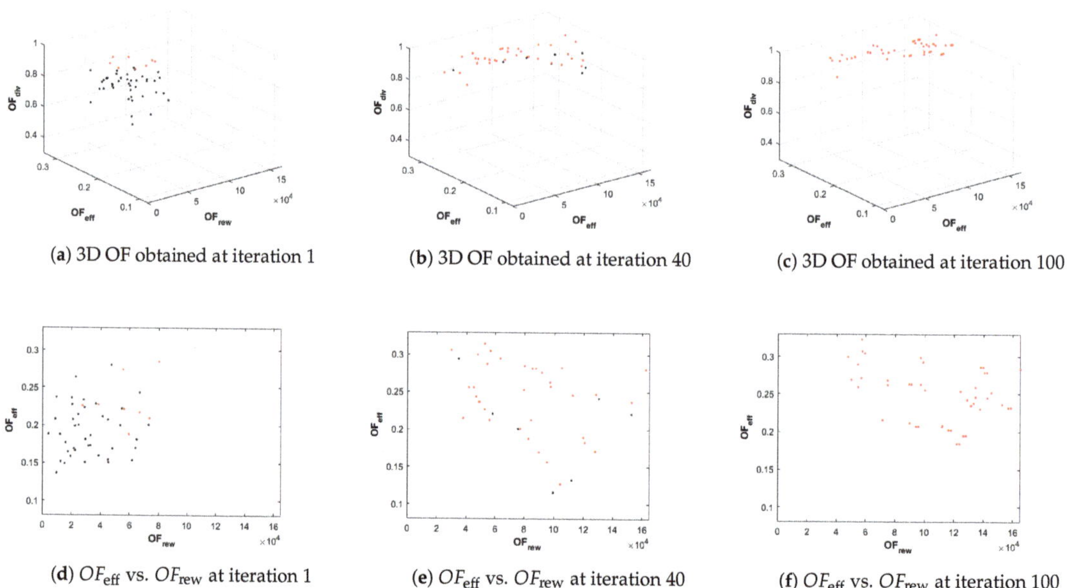

(**a**) 3D OF obtained at iteration 1

(**b**) 3D OF obtained at iteration 40

(**c**) 3D OF obtained at iteration 100

(**d**) OF_{eff} vs. OF_{rew} at iteration 1

(**e**) OF_{eff} vs. OF_{rew} at iteration 40

(**f**) OF_{eff} vs. OF_{rew} at iteration 100

Figure 7. Values of the objective functions of the best Pareto front (marked in red) vs. values of the OFs of the remaining population (marked in black) for SC1 and PC1. The top row of graphics represent in 3D the values of the three OFs, while the second row only shows the values of two of them.

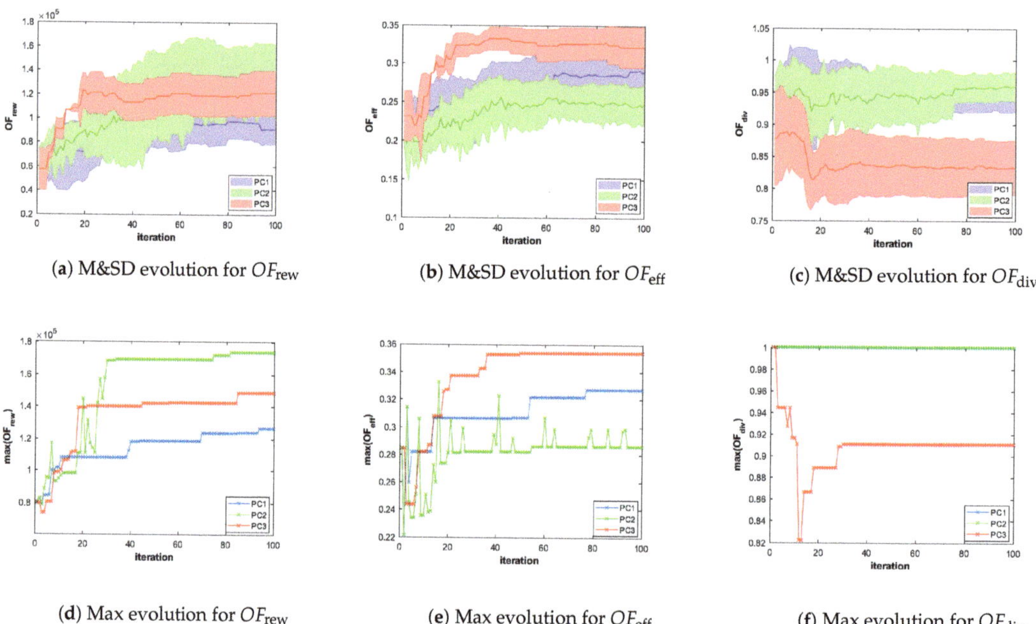

(**a**) M&SD evolution for OF_{rew}

(**b**) M&SD evolution for OF_{eff}

(**c**) M&SD evolution for OF_{div}

(**d**) Max evolution for OF_{rew}

(**e**) Max evolution for OF_{eff}

(**f**) Max evolution for OF_{div}

Figure 8. Evolution graphics of the OFs of the feasible solutions of the best Pareto front for SC1. The top row of graphics shows the evolution of the mean and standard deviation (M&SD) of each OF, while the bottom row shows the evolution of the best (Max) value of each OF.

Figure 9a shows that the number of infeasible individuals in the population is also higher in PC2 than in PC1, which practically excludes all infeasible individuals after eight iterations. This demonstrates the difficulty of PC1 to explore new regions of the space that can be reached with the help of some infeasible solutions obtained after random crossover or mutation operations.

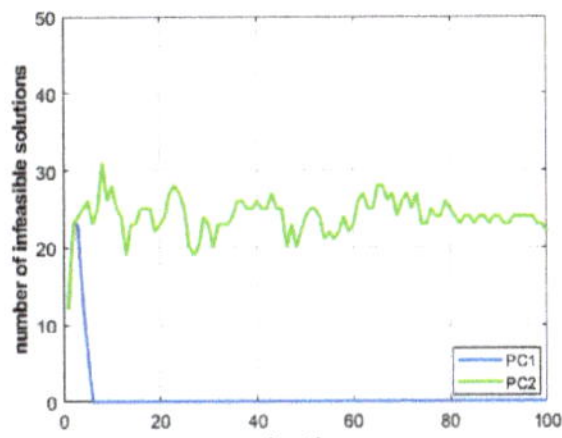
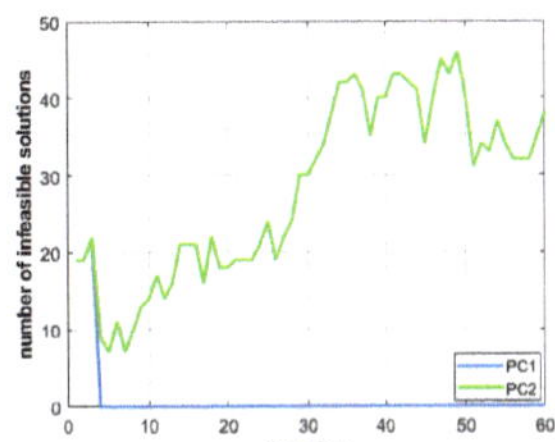

(**a**) Number of infeasible solutions in SC1   (**b**) Number of infeasible solutions in SC2

Figure 9. Infeasible plans in the population for PC1 (blue) and PC2 (green) for SC1 (**a**) and SC2 (**b**).

Figure 10 shows two representative plans of SC1, which have been selected among the feasible plans of the best Pareto front found using PC2 (in the upper figure) and PC3 (in the lower figure). In particular, we have decided to display the plans that have the maximum OF_{rew}. Comparing the plans of both figures, we can observe that the plan found with PC2 (displayed in Figure 10a) has fewer repetitions of the visited MA than the plan found using PC3 (represented in Figure 10b), where HAPS2 stays monitoring only MA10. This happens because OF_{div} is neglected in PC3 during the search of the solutions. Additionally, both graphics show how both HAPSs try to accommodate their visit to the MAs to the requested time windows and clear sky weather conditions in order to increment the overall obtained reward. Lastly, by analyzing Figure 8d we can observe that the plan returned by the planner with maximum OF_{rew} in PC2 has a higher value of OF_{rew} than the one obtained with PC3, because by including the diversity objective function and by ignoring the constraints in the recombination step, the GA configuration used in PC2 is able to explore the search space more efficiently, jumping to search regions that contain solutions of higher rewards.

Finally, it is worth noting that for the remaining scenarios we do not test against PC3, in order to focus the analysis on the comparison of PC1 and PC2, i.e., the variants of planner configurations that use and ignore the constraint criteria during the recombination of the old and new populations.

5.4.2. Analysis for Scenario 2 (SC2)

In the second scenario, weather data of the same format as the real weather data in SC1 are used but with synthetically increased strong wind. We also set the maximum iterations to 60, which was the iteration number in which the GA converges for SC1.

The M&S and Max evolution graphs of the values of the OFs of the feasible solutions belonging to the first Pareto front are shown in Figure 11. The graphics show that the results obtained with respect to the evolution of the objective functions over iterations are comparable in terms of order of magnitude using PC1 and PC2. However, the results obtained with PC2 fluctuate much more than PC1. In fact, the behavior of PC2 is predictable since in this configuration the best Pareto front can obtain both feasible and infeasible solutions, and the feasible solutions, which are the only ones considered for plotting the M&S and Max evolution graphs, can be overtaken by infeasible solutions whose objective function values dominate the objective function values of the feasible solutions.

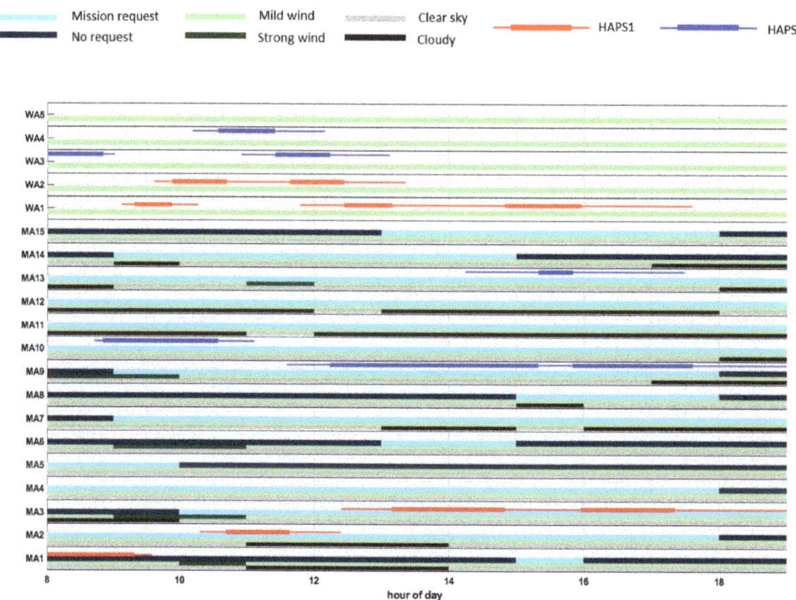

(**a**). MA-level task sequencing of the feasible plan with the largest OF_{rew} obtained using PC2.

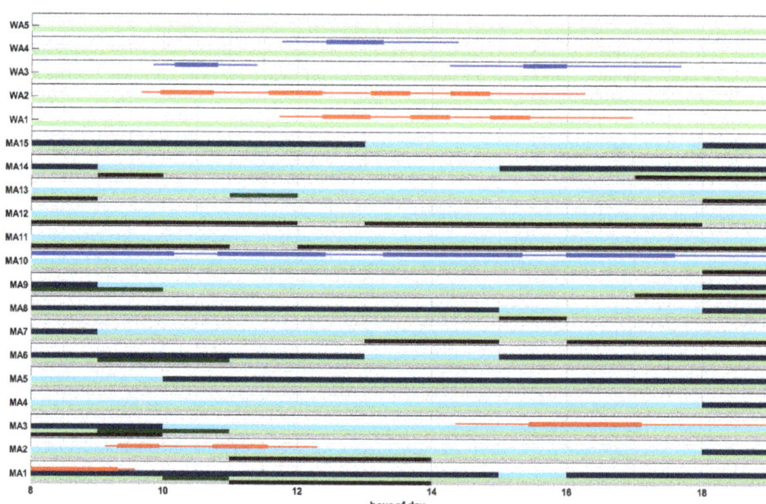

(**b**). MA-level task sequencing of the feasible plan with the largest OF_{rew} obtained using PC3.

Figure 10. Illustrative examples of feasible plans obtained by PC2 and PC3 for the first scenario (SC1).

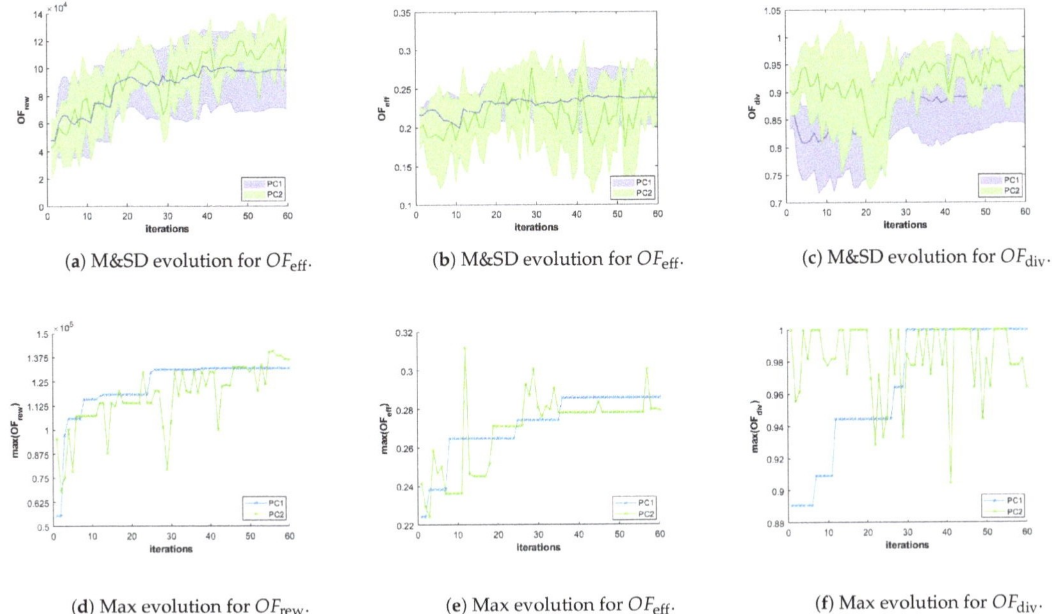

(**a**) M&SD evolution for OF_{eff}.

(**b**) M&SD evolution for OF_{eff}.

(**c**) M&SD evolution for OF_{div}.

(**d**) Max evolution for OF_{rew}.

(**e**) Max evolution for OF_{eff}.

(**f**) Max evolution for OF_{div}.

Figure 11. Evolution graphics of the OFs of the feasible solutions of the best Pareto front for SC2. The top row of graphics shows the evolution of the mean and standard deviation (M&SD) of each OF, while the bottom row shows the evolution of the best (Max) value of each OF.

Although PC2 works well under nominal weather conditions, this planner configuration can be "unstable" under challenging weather conditions, which can facilitate a more frequent violation of the constraints criteria. This behavior can be better explained using Figure 9, where the number of infeasible solutions of the population obtained with PC1 and PC2 for SC1 and SC2 are displayed side-by-side. As already described in the previous subsection, the graphics show how PC1 is much "stricter" against infeasible solutions, as the constraint criteria are used in the recombination step, resulting in a reduction of the "survivability" of infeasible individuals and of the exploring capability of the planner. However, PC2's higher exploring capability appears to be too "lenient" with the infeasible solutions for SC2, allowing an excessive number of them predominate the population in the final iterations. This behavior, which appears in the more constrained scenario imposed by the stronger winds of SC2, is prone to end up having too few feasible individuals remaining in the best Pareto front of the last iteration, thereby losing the best feasible ones identified along the iterations of the algorithm. Therefore, under more challenging weather conditions, PC1 should be the preferred configuration.

Figure 12 depicts the plan that has the highest expected reward among the feasible plans of the best Pareto front found using PC1 over SC2. The figure shows that the mission elements are affected by strong wind (which occupies more than 20% of the time) more often than in SC1. Besides, HAPS1 monitors MA1 and HAPS2 monitors MA6 at time windows that are not requested by the clients, in order to be able to reach other more promising MAs (and due to the fact that a MA cannot be transversed without monitoring its LOIs).

5.4.3. Analysis for Scenario 3 (SC3)

Three HAPSs are used in the third scenario to analyze the scalability of the planner. However, since the search space of the possible solutions has grown (due to the additional HAPS), more iterations of the GA are necessarily. For this reason, we set the maximum iterations of the stop condition to 100, which is also observed to be necessary, as the search takes more iterations to converge according to the evolution graphics of OF_{rew} presented in Figure 13. Besides, since the nominal weather setting for the environment is used in this scenario (as in SC1), Figure 13 shows how PC2 again exhibits higher variability and better performance than PC1, thanks to its higher "exploring" capability.

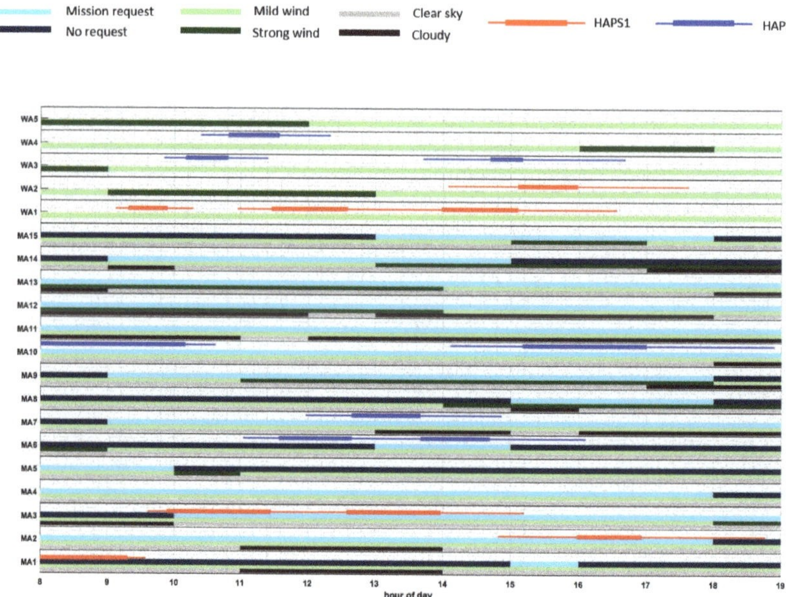

Figure 12. Illustrative example of the feasible MA-level plan with the largest OF_{rew} obtained by PC1 for the second scenario (SC2).

The mission plan with the largest OF_{rew} among the feasible plans of the first Pareto front found using PC2 is illustrated in Figure 14, along with the operation environment and requirements. With the additional HAPS, more MAs can be monitored, compared to missions where only two HAPS operate (whose illustrative plans are presented in Figures 10 and 12). This fact is also observable comparing the evolution of the Max graphs of OF_{rew} of SC1 and SC3, because the expected reward obtained by the plans for SC3 (Figure 13d) is higher than the one obtained for SC1 (Figure 8d). Figure 14 also shows how the coexistence in the same MA of multiple HAPS is tolerated (e.g., the presence of HAPS2 and HAPS3 in MA10), since the constraint criterion φ_{coex} is probabilistically evaluated and violated when the probability of coexistence exceeds a given p_{coex} (which is set to 0.3 in this paper). Changing the value of this parameter, the constraint violation can be "tightened" or "relaxed" as much as desired. This is a novelty of the planner presented in this paper, since the original version presented in [10] implemented a deterministic evaluation of the coexistence criterion where no overlapping of the start and end time range of MA# tasks was allowed.

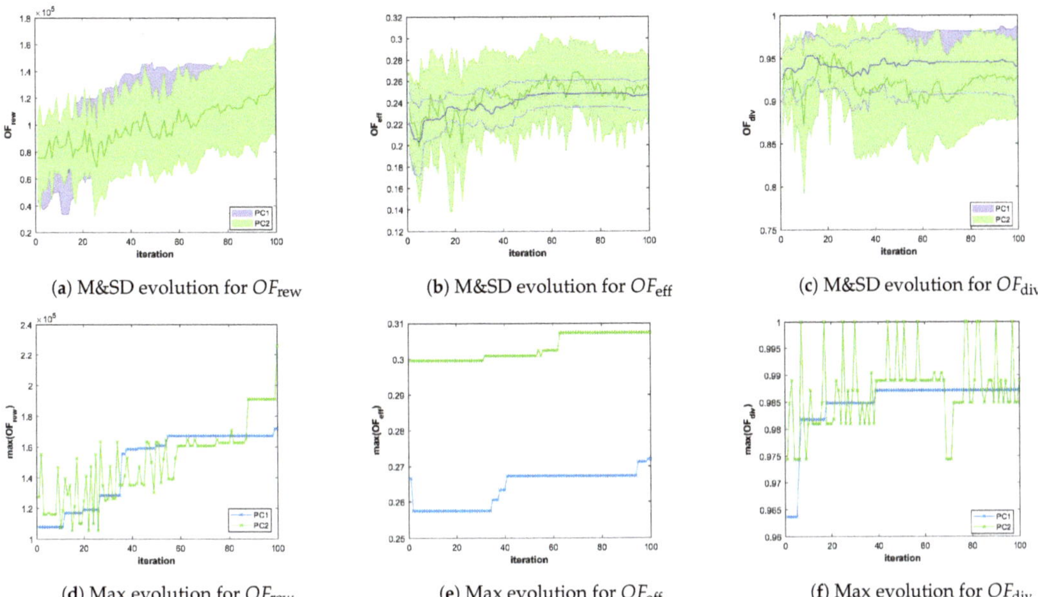

(**a**) M&SD evolution for OF_{rew}

(**b**) M&SD evolution for OF_{eff}

(**c**) M&SD evolution for OF_{div}

(**d**) Max evolution for OF_{rew}

(**e**) Max evolution for OF_{eff}

(**f**) Max evolution for OF_{div}

Figure 13. Evolution graphics of the OFs of the feasible solutions of the best Pareto front for SC3. The top row of graphics shows the evolution of the mean and standard deviation (M&SD) of each OF, while the bottom row shows the evolution of the best (Max) value of each OF.

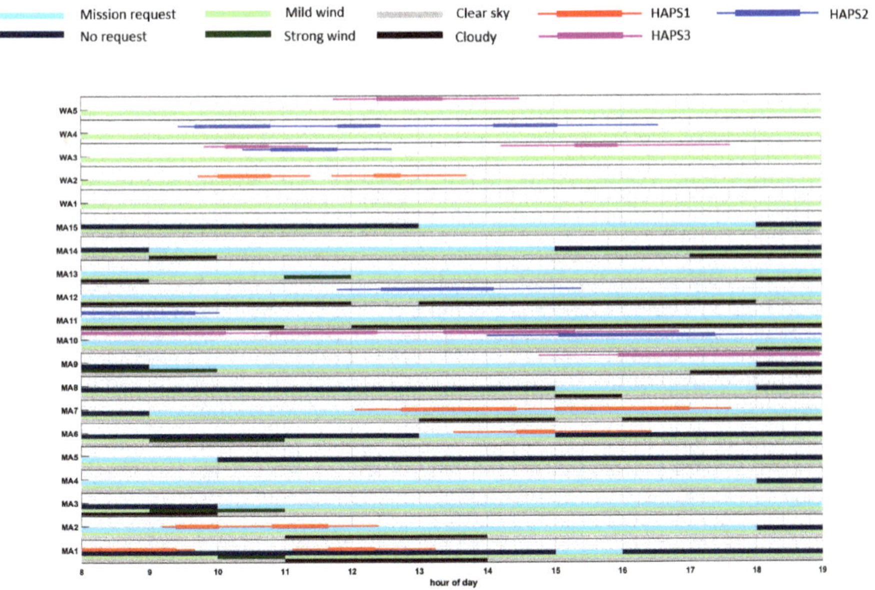

Figure 14. Illustrative example of the feasible MA-level plan with largest OF_{rew} obtained by PC1 for the third scenario (SC3).

6. Related Works

UAVs have recently become a popular alternative for monitoring ground activities [21], mapping [22] or search and rescue missions [23], since the operation of these platforms is more cost-efficient than using manned aerial vehicles while achieving the same purpose [5]. Furthermore, the deployment of these platforms is also more flexible, because numerous UAV platforms are capable of vertical take-off and landing, enabling the deployment in many missions where vast areas for takeoff and landing are scarce. Lastly, the use of unmanned platforms also allows the immediate deployment in risk zones, without compromising the safety of human pilots nor delaying the operations.

Following the development of battery technologies, light-weight but robust material as well as technologies for optimal harvesting of solar energy, the development of unmanned High Altitude Long Endurance (HALE) aerial vehicles has became the focus of giant aeronautics industries [2,24]. Moreover, new path and mission planning strategies for solar powered UAVs are being continuously developed, intended to (1) improve their trajectory by deriving the most energy-efficient flight patterns [25–29], to (2) determine the optimal path for improving operational efficiency in missions meant for communications [30–32], or (3) to track different types of targets [33]. The planning strategies exploited in these works are based on different types of optimization approaches, ranging from nonlinear optimization strategies [26,28–31] to rapidly-exploring random trees [32], the grasshopper optimization algorithm [33] and particle swarm optimization [27]. Nevertheless, the planners presented in these works either (1) optimize the trajectories without considering any aspect that is relevant to the mission or (2) tackle missions which are significantly different from the one proposed in this paper (and hence they consider a different set of requirements and constraints). Furthermore, the planning methods proposed are different than the GA-based one presented in this paper, although the last two (i.e., [27,33]) are also variants of evolutionary algorithms.

Moreover, the HAPSs considered in this work are special types of HALE platforms aimed to be an alternative to satellites for long-term remote sensing while offering more flexibility in its deployment. With the success stories around Kelleher [2], HAPSs are deemed fit for deployment in the near future at larger scale. However, although HAPSs operations can be beneficial, they can be extremely challenging, given the fragility of the platform under critical weather conditions, their lack of maneuverability, and the requirement for plans for long operations, which oblige the consideration of weather parameters that vary over time within the plan horizon [34]. Specific studies on automated planning for HAPS include [10,35], both aiming to reduce the operators' workload. In particular, for a complex mission scenario as the one depicted in Figure 1, Ref. [35] proposes a sequential task and motion planning framework for a collective operation area, simplifying the constraints of the planning problem, while [10] uses a GA to extend the temporal hierarchical task planner for multiple HAPSs. Moreover, the current work extends the planner presented in [10] by (1) including the evaluation of the safety and coexistence constraints with the new probability based functions presented in Sections 3.3.1 and 3.3.2, by (2) substituting the weighted evaluation function used in [10] for the constrained multiobjective Pareto-front evaluation mechanisms of NSGA-II, and by (3) returning the set of the hierarchical plans that form part of the final best Pareto front. Besides, this paper analyzes the behavior of the new planner with new scenario and the influence of the diversity objective function and of different constraint handling techniques within our planner.

Evolutionary algorithms, including variants for solving multiple-objective problems with powerful constraint-handling techniques (such as [19,36]), have often been used for the mission planning of Satellite and UAV operations. For instance, Refs. [37–39] present different GA-based planner for scheduling the observation tasks of different satellites, while [23,40–43] use multiple-objective evolutionary algorithms to solve task planning problems for multiple UAVs engaged in performing monitoring tasks in dissected areas of interest. Although our planner also uses a GA algorithm to determine the best solution plans for a given scenario, it solves a different type of monitoring task mission problem, involving exogenous time-varying events (i.e., weather) and time-dependent mission requirements. Therefore, its evolutionary encoding has a different interpretation and is customized for HAPSs instead of satellites or other types of UAVs. Besides, similar to other works that take into account the uncertainty associated to weather conditions [44–46] or to other elements of the mission (e.g., the target location and movement in search and rescue missions [47,48] or the probability of target detection and destruction in hostile environments [49,50]), in this work the uncertainties are incorporated into the models used to evaluate how probable is that each HAPS is at a mission area at a given time, which affects the outcome of the objective and constraint values.

Finally, it is worth highlighting that although this work uses NSGA-II for a constrained multiobjective optimization, it is only a part of the temporal hierarchical task planner, in which the search for optimal decomposition into an ordered list of nonprimitive tasks poses a combinatorial search problem. With an appropriate encoding of the problem at the task level at which the combinatorial problem prevails, NSGA-II is used for guiding the decomposition into executable tasks within a temporal hierarchical task network with a nested Time-Dependent Multi-Vehicle Routing Problem (TDMVRP). Note also that Hierarchical Task Planning often refers to an Artificial Intelligence (AI) planning paradigm and that although there are some domain-independent frameworks meant for it [51,52]; they do not yet support a nested TDMVRP.

7. Conclusions and Future Work

This paper presents a new approach for planning the tasks that a group of HAPSs must perform to carry out ground monitoring mission in a structured airspace. The new approach returns a Pareto front of feasible hierarchical plans, whose sequence of higher level tasks is determined using a MOEA that optimizes the expected reward to be received by the HAPSs team for monitoring the different LOIs, the diversity of the LOIs visited by the HAPSs and the time that the HAPSs are actually monitoring (and not traversing the airspace). Besides, it also considers multiple constraints, some encoded in the decomposition method of the hierarchical planner, while others validated by measuring the constraint criteria related to the mission safety, the coexistence of HAPSs in the same MA and the connectivity of the plan. The planner also considers, through the evaluation functions and constraint criteria, the uncertainty that the weather conditions impose on the duration of each task (due to the wind vectors) and on the visibility for the mission camera (due to cloud coverage).

The performance of the different configurations of the planner, carefully set up for increasing/decreasing the "survivability" of the infeasible solutions or to disable the diversity requirement, is tested against several scenarios, with varying number of HAPSs and different weather conditions. The quality of the results is scenario-dependent, although it seems advisable to use the second configuration (PC2) for the planner when the HAPSs operate under mild weather conditions and the first configuration (PC1) for challenging weather conditions, as suggested by the results of the performance tests presented in Section 5. Besides, when planning for two HAPSs the number of iterations required by the planner to converge is smaller than when planning for three HAPSs.

In order to further improve the planner, we will consider several possibilities. Firstly, a "softer" constraint-handling method can be used to improve PC1. For this purpose, we are planning to adopt the approach proposed by [20], in which, with a low probability, some infeasible solutions can be ranked better than feasible solutions. This new planner configuration could avoid, for example, the early convergence of PC1 in SC2, while still managing to maintain the right balance between feasible and infeasible solution plans in the population.

Secondly, while the planner is typically customized for solving HAPS mission planning problems, it can be extended for more generic uses. As a matter of fact, temporal hierarchical task planners (without a nested TDMVRP) are gearing toward general implementation [52]. Hence, with careful considerations of the encoding of the chromosomes for generic planning problems and more generic approaches for tuning the planner parameters, the approach presented in this work could be implemented in a "domain-independent" fashion.

Finally, although the underlying mission planner reduces the operator's workload, when the planner suggest many solutions, selecting the one to execute can still be challenging for the operator. Therefore, in order to increase usability of the planner, it can be convenient to provide operators with a set of tools that help them to analyze the solutions of the best Pareto front more easily, by (1) taking into account the explicability of the plans with a visualization interface that can highlight the probable constraint violations, the rewarding mission tasks, the diversity of the clientele pool and the effort; or by (2) designing filter mechanisms that accelerate the selection of the plan of the best Pareto front that better fits the operator preference (e.g., the one with the maximum value of an objective function or the one with the best preference weighting [53]). In a similar line, more interactive functions can be integrated to enable "mixed-initiative planning", which can favor quick local replanning performed by the operator whenever necessary, due to unexpected weather change or to take into account the operator's preferences.

Author Contributions: Conceptualization, J.J.K., E.B.-P., and A.S.; methodology, J.J.K., E.B.-P.; software, J.J.K.; validation, J.J.K.; formal analysis, J.J.K.; writing—original draft preparation, J.J.K.; writing—review and editing, E.B.-P.; visualization, J.J.K.; supervision, E.B.-P., A.S.; project administration, J.J.K., A.S.; funding acquisition, A.S. All authors have read and agreed to the published version of the manuscript.

Funding: This research was funded by Project StraVARIA, a Ludwig Bolköw Campus project. Eva Besada Portas' contributions are funded by the the Spanish National Challenge Grant RTI2018-098962-B-C21.

Institutional Review Board Statement: Not applicable.

Informed Consent Statement: Not applicable.

Data Availability Statement: Not applicable.

Acknowledgments: The authors thank partners from the German Aerospace Center and from ADS GmbH for their support in providing information on HAPS, as well as on methodology to be adopted for validation purposes.

Conflicts of Interest: The authors declare no conflict of interest.

Abbreviations

The following abbreviations are used in this manuscript:

AI	Artificial Intelligence
C	Corridor
EA	Evolutionary Algorithm
EO	Electro-Optical
FL	Flight Level
GA	Genetic Algorithm
GCS	Ground Control Station
HALE	High Altitude Long Endurance
HAPS	High Altitude Pseudo-Satellite
HFR	High-level Flight Rules
LOI	Location Of Interest
MA	Mission Area
MC	Mission Constraint
MOEA	Multi-Objective Evolutionary Algorithm
MR	Mission Requirement
NSGA	Nondominated Sorting Genetic Algorithm

PC	Planner Configuration
TDMVRP	Time Dependent Multi-Vehicle Routing Problem
UAV	Unmanned Aerial Vehicle
WA	Waiting Area
WP	Waypoint

Appendix A. Numerical Details on the HAPS Model and on the Mission Parameters

Table A1 details numerical information on the HAPS that are assumed in this work.

Table A1. Model of the HAPS considered in this work: build, performance, and mission payload.

Build/Fight Performance/Payload	Parameter Values
Weight	100 kg
Wingspan	30 m
Payload	5–10 kg
Battery capacity	15 kWh
Electro-motor maximum propulsive power	1700 W
Operating altitude	18 km
Cruise airspeed at the operating altitude	30 m/s
Endurance	3 months
Ground sampling distance at 18 km	30 cm
hxw of an image	360×3000 m

The dimensions of the mission elements depicted in Figure 1 in form of their longest diagonals in kilometers are given in Table A2.

Table A2. Dimensions of the mission elements depicted in Figure 1.

C	Longest Diagonal [km]	MA/WA	Longest Diagonal [km]	LoI 1	LoI 2	LoI 3
C1	60.78	MA1	27.58	17.10		
C2	37.83	MA2	69.49	19.83		
C3	33.67	MA3	103.11	20.07	44.27	
C4	58.88	MA4	25.32	15.56		
C5	34.22	MA5	52.90	13.25		
C6	17.64	MA6	46.51	21.61		
C7	70.41	MA7	84.03	21.42	25.88	
C8	40.27	MA8	36.62	10.84		
C9	66.19	MA9	133.30	34.36	21.24	38.13
C10	72.10	MA10	123.35	39.96	44.74	
C11	20.52	MA11	74.98	31.62	24.53	
C12	73.35	MA12	73.49	18.28	26.38	
C13	39.34	MA13	56.87	18.62		
C14	88.46	MA14	36.28	14.31		
C15	63.04	MA15	104.92	23.66	36.46	
C16	70.70	WA1	47.90			
C17	71.98	WA2	34.44			
C18	71.38	WA3	46.41			
C19	61.77	WA4	47.96			
C20	42.75	WA5	26.44			
C21	46.67					
C22	39.80					
C23	74.26					
C24	38.15					
C25	106.86					
C26	63.86					

The rewards obtained for the successful monitoring of all LOI(s) of a MA are listed in Table A3, along with the image coverage of the ground required for the monitoring mission to be considered successful.

Table A3. Rewards to be given for each MA ($\times\ 10^3$).

Mission Area	Coverage (%)	Reward (€)	Mission Area	Coverage (%)	Reward (€)
MA1	80	4	MA9	60	13
MA2	80	50	MA10	60	18
MA3	60	100	MA11	70	20
MA4	80	20	MA12	70	10
MA5	80	3	MA13	60	8
MA6	70	5	MA14	90	18
MA7	70	15	MA15	50	9
MA8	80	3			

References

1. Klöckner, A. *A Behavior Trees for Mission Management of High-Altitude Pseudo-Satellites*; Verlag Dr. Hut: Munich, Germany, 2016.
2. Airbus. Zephyr Press Release. Available online: https://www.airbus.com/newsroom/press-releases/en/2018/08/Airbus-Zephyr-Solar-High-Altitude-Pseudo-Satellite-flies-for-longer-than-any-other-aircraft.html (accessed on 19 February 2021).
3. Hunter, S. Safe Operations above FL600. In Proceedings of the 2015 Space Traffic Management Conference, Daytona Beach, FL, USA, 12–13 November 2015.
4. Kiam, J.J. AI-Based Mission Planning for High-Altitude Pseudo-Satellites in Time-Varying Environments. Ph.D. Thesis, University der Bundeswehr Munich, Munich, Germany, 2019.
5. Finnegan, P. *World Civil Unmanned Aerial Systems: Teal Market Profile and Forecast*; TEAL CROUP: Fairfax, VA, USA, 2017.
6. Airbus. *Adverse Weather Operations: Optimum Use of the Weather Radar*; Technical Report; Flight Operations Briefing Notes; Airbus: Blagnac, France, 2007.
7. Leena, P.; Ratnam, M.V.; Murthy, B.K.; Rao, S. Detection of High Frequency Gravity Waves Using High Resolution Radiosonde Observations. *J. Atmos. Sol. Terr. Phys.* **2012**, *77*, 254–259. [CrossRef]
8. EUROCONTROL; EASA. UAS ATM Integration: Integration Operational Concept. EUROCONTROL, November 2018. Available online: https://www.eurocontrol.int/publication/unmanned-aircraft-systems-uas-atm-integration (accessed on 19 February 2021).
9. EUROCONTROL. Advance Flexible Use of Airspace (FUA) Concept. Eurocontrol; 24 July 2015. Available online: https://www.eurocontrol.int/publication/advanced-flexible-use-airspace-afua-concept (accessed on 19 February 2021).
10. Kiam, J.; Besada-Portas, E.; Hehtke, V.; Schulte, A. GA-guided task planning for multiple-HAPS in realistic time-varying operation environments. In Proceedings of the 2019 Genetic and Evolutionary Computation Conference, Prague, Czech Republic, 13–17 July 2019.
11. Müller, R.; Kiam, J.; Mothes, F. Multiphysical simulation of a semi-autonomous solar powered high altitude pseudo-satellite. In Proceedings of the 2018 IEEE Aerospace Conference, Big Sky, MT, USA, 4–11 March 2018.
12. Delauré, B.; Michiels, D.; Lewyckyj, N.; van Achteren, T. The development of a family of lightweight and wide swath UAV camera systems around an innovative dual-sensor on-single-chip detector. *Int. Arch. Photogramm. Remote. Sens. Spat. Inf. Sci.* **2013**, *XL-1/W2*, 101–106.
13. Baldauf, M.; Gebhardt, C.; Theis, S.; Ritter, B.; Schraff, C. *Beschreibung des Operationellen Kürzestfristvorhersagemodells COSMO-D2 und COSMO-D2-EPS und Seiner Ausgabe in die Datenbanken des DWD*; Technical Report; Deutscher Wetterdienst, DWD: Offenbach, Germany, 2018.
14. National Centres for Environmental Information (NOAA). Global Forecast System Weather Download. Available online: https://www.ncdc.noaa.gov/data-access/model-data/model-datasets/global-forcast-system-gfs (accessed on 19 February 2021).
15. European Centre for Medium-Range Weather Forecasts. *IFS Documentation—Cy45r1: Operational Implementation 5 June 2018*; ECMWF: Reading, UK, 2018.
16. Bradley, D.; Gupta, R. On the Distribution of the Sum of n Non-Identically Distributed Uniform Random Variables. *Ann. Inst. Stat. Math.* **2002**, *54*, 689–700. [CrossRef]
17. Boyan, J.; Littman, M. Exact solutions to time-dependent MDPs. In Proceedings of the 13th International Conference on Neural Information Processing Systems (NIPS), Vancouver, BC, Canada, 3–8 December 2000.
18. Simpsons, E. Measurement of Diversity. *Nature* **1949**, *163*, 688. [CrossRef]
19. Deb, K.; Pratap, A.; Agarwal, S.; Meyarivan, T. A Fast and Elitist Multiobjective Genetic Algorithm: NSGA-II. *IEEE Trans. Evol. Comput.* **2002**, *6*, 182–197. [CrossRef]

20. Runarsson, T.P.; Yao, X. Stochastic Ranking for Constrained Evolutionary Optimization. *IEEE Trans. Evol. Comput.* **2000**, *4*, 284–294. [CrossRef]

21. Linchant, J.; Lisein, J.; Semeki, J.; Lejeune, P.; Vermeulen, C. Are Unmanned Aircraft Systems (UASs) the Future of Wildlife Monitoring? A Review of Accomplishments and Challenges. *Mammal Rev.* **2015**, *45*, 239–252. [CrossRef]

22. Remondino, F.; Barazzetti, L.; Nex, F.; Scaioni, M.; Sarazzi, D. UAV Photogrammetry for Mapping and 3D Modeling—Current Status and Future Perspectives. *Int. Arch. Photogramm. Remote Sens. Spat. Inf. Sci.* **2011**, *38*, C22. [CrossRef]

23. Wang, Y.; Kirubarajan, T.; Tharmarasa, R.; Jassemi-Zargani, R.; Kashyap, N. Multiperiod Coverage Path Planning and Scheduling for Airborne Surveillance. *IEEE Trans. Aerosp. Electron. Syst.* **2018**, *54*, 2257–2273. [CrossRef]

24. Aurora. Odysseus Press Realase. Available online: https://www.aurora.aero/odysseus-high-altitude-pseudo-satellite-haps/ (accessed on 19 February 2021).

25. Spangelo, S.; Gilbert, E. Power Optimization of Solar-powered Aircraft with Specified Closed Ground Tracks. *J. Aircr.* **2012**, *50*, 232–238. [CrossRef]

26. Hosseini, S.; Dai, R.; Mesbahi, M. Optimal path planning and power allocation for a long endurance solar-powered UAV. In Proceedings of the 2013 American Control Conference, Washington, DC, USA, 17–19 June 2013.

27. Gao, X.; Hou, Z.; Guo, Z.; Chen, X.; Chen, X. Joint Optimization of Battery Mass and Flight Trajectory for High-Altitude Solar-powered Aircraft. *Proc. Inst. Mech. Eng. Part G J. Aerosp. Eng.* **2014**, *228*, 2439–2451. [CrossRef]

28. Lee, J.; Yu, K. Optimal Path Planning of Solar-powered UAV Using Gravitational Potential Energy. *IEEE Trans. Aerosp. Electron. Syst.* **2017**, *53*, 1442–1451. [CrossRef]

29. Bolandhemmat, H.; Thomsen, B.; Marriott, J. Energy-optimized trajectory planning for High Altitude Long Endurance (HALE) aircraft. In Proceedings of the 18th European Control Conference, Naples, Italy, 25–28 June 2019.

30. Guerra-Padilla, G.; Kim, K.; Yu, K. *Flight Path Planning and Signal Behavior Analysis of a LALE Solar-Powered UAV for Communication Relay*; Asia Pacific International Symposium on Aerospace Technology: Gold Coast, Australia, 2019.

31. Guerra-Padilla, G.; Kim, K.; Park, S.; Yu, K. Flight Path Planning of Solar-powered UAV for Sustainable Communication Relay. *IEEE Robot. Autom. Lett.* **2020**, *5*, 6772–6779. [CrossRef]

32. Huang, H.; Savkin, A.V.; Ni, W. Energy-efficient 3D Navigation of a Solar-powered UAV for Secure Communication in the Presence of Eavesdroppers and No-fly Zones. *Energies* **2020**, *13*, 1445. [CrossRef]

33. Wu, J.; Wang, H.; Li, N.; Yao, P.; Huang, Y.; Su, Z.; Yu, Y. Distributed Trajectory Optimization for Multiple Solar-powered UAVs Target Tracking in Urban Environment by Adaptive Grasshopper Optimization Algorithm. *Aerosp. Sci. Technol.* **2017**, *70*, 497–510. [CrossRef]

34. Kiam, J.; Schulte, A. Multilateral quality mission planning for solar-powered long-endurance UAV. In Proceedings of the 2017 IEEE Aerospace Conference, Big Sky, MT, USA, 4–11 March 2017.

35. Kiam, J.J.; Schulte, A. Multilateral mission planning in a time-varying vector field with dynamic constraints. In Proceedings of the 2018 IEEE International Conference on Systems, Man, and Cybernetics (SMC), Miyazaki, Japan, 7–10 October 2018.

36. Fonseca, C.; Fleming, P. Genetic Algorithms for multiobjective optimization: Formulation, discussion and generalization. In Proceedings of the Fifth International Conference on Genetic Algorithms, Urbana-Champaign, IL, USA, 17–21 June 1993.

37. Liu, W.; Li-Gang, L. Mission Planning of Space Astronomical Satellite Based on Improved Genetic Algorithm. *Comput. Simul.* **2014**, *31*, 54–58.

38. Chen, H.; Du, C.; Li, J.; Jing, N.; Wang, L. An approach of satellite periodic continuous observation task scheduling based on evolutionary computation. In Proceedings of the Genetic and Evolutionary Computation Conference Companion (GECCO), Berlin, Germany, 15–19 July 2017.

39. Long, J.; Li, C.; Zhu, L.; Chen, S.; Liu, J. An Efficient Task Autonomous Planning Method for Small Satellites. *Information* **2018**, *9*, 181. [CrossRef]

40. Ramirez-Atienza, C.; Ser, J.; Camacho, D. Weighted Strategies to Guide a Multi-objective Evolutionary Algorithm for Multi-UAV Mission Planning. *Swarm Evol. Comput.* **2019**, *44*, 480–495. [CrossRef]

41. Wang, Z.; Liu, L.; Long, T.; Wen, Y. Multi-UAV Reconnaissance Task Allocation for Heterogeneous Targets Using an Opposition-based Genetic Algorithm with Double-chromosome Encoding. *Chin. J. Aeronaut.* **2018**, *31*, 339–350. [CrossRef]

42. Liu, J.; Wang, W.; Li, X.; Wang, T.; Bai1, S.; Wang, Y. Solving a Multi-objective Mission Planning Problem for UAV Swarms with an Improved NSGA-III Algorithm. *Int. J. Comput. Intell. Syst.* **2018**, *11*, 1067–1081. [CrossRef]

43. Wilhelm, J.; Rojas, J.; Eberhart, G.; Perhinschi, M. Heterogeneous Aerial Platform Adaptive Mission Planning Using Genetic Algorithms. *Unmanned Syst.* **2017**, *5*, 19–30. [CrossRef]

44. Zhang, B.; Tang, L.; Roemer, M. Probabilistic Weather Forecasting Analysis for Unmanned Aerial Vehicle Path Planning. *J. Guid. Control Dyn.* **2014**, *37*, 309–312. [CrossRef]

45. Thibbotuwawa, A.; Bocewicz, G.; Radzki, G.; Nielsen, P.; Banaszak, Z. UAV Mission Planning Resistant to Weather Uncertainty. *Sensors* **2020**, *20*, 515. [CrossRef]

46. Luo, H.; Liang, Z.; Zhu, M.; Hu, X.; Wang, G. Integrated Optimization of Unmanned Aerial Vehicle Task Allocation and Path Planning under Steady Wind. *PLoS ONE* **2018**, *13*, e0194690. [CrossRef]

47. Perez-Carabaza, S.; Besada-Portas, E.; Lopez-Orozco, J.; de la Cruz, J. A real world multi-UAV evolutionary planner for minimum time target detection. In Proceedings of the 18th Conference on Genetic and Evolutionary Computation, Denver, CO, USA, 20–24 July 2016.

48. Perez-Carabaza, S.; Besada-Portas, E.; Lopez-Orozco, J.; Pajares, G. Minimum Time Search in Real-World Scenarios Using Multiple UAVs with Onboard Orientable Cameras. *J. Sens.* **2019**, *2019*, 7673859. [CrossRef]
49. Besada-Portas, E.; de la Torre, L.; de la Cruz, J.; Andres-Toro, B. Evolutionary Trajectory Planner for Multiple UAVs in Realistic Scenarios. *IEEE Trans. Robot.* **2010**, *26*, 619–634. [CrossRef]
50. Yang, P.; Tang, K.; Lozano, J.A.; Cao, X. Path Planning for Single Unmanned Aerial Vehicle by Separately Evolving Waypoints. *IEEE Trans. Robot.* **2015**, *31*, 1130–1146. [CrossRef]
51. Fdez-Olivares, J.; Castillo, L.; Garcıa-Pérez, O.; Palao, F. Bringing users and planning technology together. Experiences in SIADEX. In Proceedings of the 2006 International Conference on Autonomous Planning and Scheduling (ICAPS), Cumbria, UK, 6–10 June 2006.
52. Dvorak, F.; Bit-Monnot, A.; Ingrand, F.; Ghallab, M. A flexible ANML actor and planner in robotics. In Proceedings of the 2nd ICAPS Workshop on Planning and Robotics (ICAPS-PlanRob), Portsmouth, NH, USA, 22 June 2014.
53. Hehtke, V.; Kiam, J.; Schulte, A. *An Autonomous Mission Management System to Assist Decision Making of a HALE Operator*; Deutscher Luft- und Raumfahrtkongress: Munich, Germany, 2017.

Article

Monitoring Winter Stress Vulnerability of High-Latitude Understory Vegetation Using Intraspecific Trait Variability and Remote Sensing Approaches

Elmar Ritz [1], Jarle W. Bjerke [2] and Hans Tømmervik [2,*]

[1]   Institute for Water and River Basin Management, Department of Aquatic Environmental Engineering, Karlsruhe Institute of Technology, Gotthard-Franz-Str. 3, 76131 Karlsruhe, Germany; elmar.ritz@kit.edu

[2]   Norwegian Institute for Nature Research, FRAM–High North Research Centre for Climate and the Environment, P.O. Box 6606 Langnes, NO-9296 Tromsø, Norway; jarle.bjerke@nina.no

*   Correspondence: hans.tommervik@nina.no

Received: 4 March 2020; Accepted: 6 April 2020; Published: 8 April 2020

Abstract: In this study, we focused on three species that have proven to be vulnerable to winter stress: *Empetrum nigrum*, *Vaccinium vitis-idaea* and *Hylocomium splendens*. Our objective was to determine plant traits suitable for monitoring plant stress as well as trait shifts during spring. To this end, we used a combination of active and passive handheld normalized difference vegetation index (NDVI) sensors, RGB indices derived from ordinary cameras, an optical chlorophyll and flavonol sensor (Dualex), and common plant traits that are sensitive to winter stress, i.e. height, specific leaf area (SLA). Our results indicate that NDVI is a good predictor for plant stress, as it correlates well with height ($r = 0.70$, $p < 0.001$) and chlorophyll content ($r = 0.63$, $p < 0.001$). NDVI is also related to soil depth ($r = 0.45$, $p < 0.001$) as well as to plant stress levels based on observations in the field ($r = -0.60$, $p < 0.001$). Flavonol content and SLA remained relatively stable during spring. Our results confirm a multi-method approach using NDVI data from the Sentinel-2 satellite and active near-remote sensing devices to determine the contribution of understory vegetation to the total ecosystem greenness. We identified low soil depth to be the major stressor for understory vegetation in the studied plots. The RGB indices were good proxies to detect plant stress (e.g. Channel G%: $r = -0.77$, $p < 0.001$) and showed high correlation with NDVI ($r = 0.75$, $p < 0.001$). Ordinary cameras and modified cameras with the infrared filter removed were found to perform equally well.

Keywords: climate change; evergreen plants; extreme events; flavonol and chlorophyll sensor (Dualex); greenness indices; mosses; near-remote sensing active and passive NDVI sensors; Sentinel-2; subarctic vegetation damage

---

1. Introduction

Global warming will affect arctic and subarctic regions more than any other area in the world [1]. It is expected to increase the productivity of subarctic and arctic ecosystems [2–4]. Increasing productivity and biomass is generally known as 'greening' [5]. Major drivers are a longer growing season and increasing summer warming [6]. However, negative trends in productivity and biomass, known as 'browning', have also been reported [6]. For the Arctic as a whole, trends are complex, as Myers-Smith et al. state: "Figures vary from 42% greening and 2.5% browning from 1982 to 2014 in the GIMMS3g AVHRR dataset to 20% greening and 4% browning from 2000 to 2016 in Landsat data, and to estimates of 13% greening and 1% browning for the MODIS trends calculated for 1,000 random points in the tundra polygon from 2000 to 2018." ([7], p. 107).

In the subarctic region of Scandinavia, i.e. Norway, Finland, and Sweden north of the Arctic Circle, the main drivers of browning are winter warming events and pest outbreaks [8]. Winter warming events can melt the insulating snow cover that normally protects photosynthetic short-statured organisms overwintering with aboveground tissue (e.g. prostrate shrubs, cushion plants, bryophytes, and lichens) from the harsh ambient winter weather conditions. After a few thaw days, ground vegetation becomes exposed to ambient air and hibernation is interrupted, thus reducing the protection of photosynthetic organisms against frost, which may easily lead to freezing damage upon return of normal winter weather [9]. Soil communities, including both micro-arthropods and bacteria, can also be severely affected [10]. Overall, a warmer winter climate changes species compositions and reduces carbon cycling [11]. In subarctic and arctic regions evergreen plants in particular are sensitive to changing winter climate and reduced snow cover [12]. This includes the widespread dwarf shrubs *Empetrum nigrum* L., *Vaccinium vitis-idaea* L., *Cassiope tetragona* (L.) D.Don, and *Calluna vulgaris* (L.) Hull, as well as the tall coniferous shrub *Juniperus communis* L. [6,11,13]. Bryophytes, such as the widespread feathermoss *Hylocomium splendens* (Hedw.) Schimp., deciduous shrubs, such as *Vaccinium myrtillus* L., evergreen horsetails (*Equisetum* spp.), as well as small cushion plants show reduced growth following exposure to winter warming [9,11,13,14]. The other major factor causing browning are pest outbreaks. Recently, increasing frequency and intensity of outbreaks of leaf-defoliating geometrid moths led to massive canopy defoliation of their preferred host tree *Betula pubescens* Ehrh. and understory plants [13,15,16]. Overall, multiple stress events are main drivers of browning. Given the high focus on climate change-induced changes in northern primary productivity, it is important to develop easy and reliable methods for assessment of plant vitality.

For a long time, satellites have monitored the global vegetation status [2,3,17]. Spectral sensors operated near the target vegetation are increasingly applied for assessing the plant status [18,19]. However, near-remote time series of the plant status are still uncommon, which is partly due to the need for expensive equipment, for example spectroradiometers [20]. In recent years, several new and low-cost active and passive proximal sensors were developed. This includes sensors measuring the normalized difference vegetation index (NDVI). NDVI is a radiometric measure of the amount of radiation ($\approx$~400–700 nm) absorbed by vegetation during photosynthesis. It is calculated from contrasting reflectance at near-infrared (NIR) and red bands [21,22].

NDVI has been widely used in studies of phenology, productivity, biomass, and disturbance monitoring, as it has proven to be a good proxy of the vegetation's photosynthetic activity [19,23]. NDVI works well for subarctic ecosystem monitoring and is widely used on different scales and as a vegetation marker [24–26].

Previously, modified cameras—with the infrared filter removed—were found to be good NDVI surrogates. In such cameras, the NDVI proxy is commonly calculated by using the enhanced red channel and the blue channel (BNDVI) [27]. However, a combination of the enhanced red channel and the green channel might also be of interest due to a strong linear correlation with the chlorophyll content (GNDVI) [28]. Additionally, ordinary cameras were increasingly applied for vegetation analysis and phenology studies in recent years. Greenness indices based on ordinary RGB images from such cameras are promising NDVI substitutes [28,29], even for high-arctic vegetation [30].

In subarctic forests, the contribution of understory vegetation (i.e. dwarf shrubs, herbs, graminoids, bryophytes and lichens) to the total ecosystem productivity is similar to that of trees [31]. Moreover, biodiversity of vascular plants at high latitudes is relatively low, which makes research into dominant species and their vulnerability to environmental change even more important [32]. We hypothesized that in situ estimates of plant damage would be correlated to optical measurements of plant greenness, but that greenness indices would vary in their explanatory power. Our second hypothesis was that plant stress would vary over short distances in a rolling subarctic landscape and that this would be detectable both by near-remote sensing measurements and by Sentinel. To this end, we combined near-remote sensing approaches with classical determination of plant traits of understory vegetation to address the following research questions:

(1)   What is the range of intraspecific variability of common traits of dwarf shrubs and mosses in subarctic spring?

(2)   Which traits are reliable indicators of plant stress?

(3)   How do the indices derived from ordinary and modified RGB cameras correlate with common plant traits?

To answer these questions, we made analyses in a widespread subarctic heath ecosystem, focusing on vegetation plots dominated by two evergreen dwarf shrubs and a mat-forming moss.

2. Materials and Methods

2.1. Site Description

We selected plots in wind-exposed areas where snow cover generally is shallow and plants are more susceptible to winter frost-thaw stress. Eighteen plots (1 m × 1 m) were assessed within a total area of approx. 1 km^2 in Tromsø, Troms County, northern Norway (Figure S1, Table S1, Figure S2b,c). Satellite upscaling was conducted in homogeneous plots within a wider area in Tromsø (Table S2), from 8–12 June 2017, corresponding to days of year (DOY) 159–163. The three areas (a total of ca. 0.5 ha) for the upscaling approach were not congruent to the eighteen field plots (a total of 18 m^2). We chose three separate areas in order to avoid trails, snow patches, unvegetated ground and rocky steep slopes (as exemplified in Figure S2a). Unvegetated ground was estimated to be around 10% in the area shown in Figure S2a. The other two areas had around 5% of unvegetated ground. Satellite upscaling was performed when snow patches became sparse, but before budburst of the deciduous trees in the heath. The study focuses on the evergreen dwarf shrub species *Empetrum nigrum* and *Vaccinium vitis-idaea* and the mat-forming moss *Hylocomium splendens*. These species are abundant, co-occur in boreal ecosystems, and are linked to browning [9,14,33]. Our continuous monitoring of plant vitality in the study area shows that these species have not been exposed to severely stressful events since 2012, as reported in Bjerke et al. [15]. Minor damage rates were recorded in more recent years, then mostly restricted to wind-exposed sites with little snow accumulation (unpublished observations). Thus, plant traits in the study area were expected to vary naturally along microclimatic gradients. Plots with different stress levels (or health states) were established for each species to monitor natural intraspecific trait variabilities. The studied plots were dominated by (with number of plots in parentheses): *V. vitis-idaea* (two), *H. splendens* (three), *E. nigrum* (nine), and mixed plots of *E. nigrum* with a lower layer of *H. splendens* (four).

2.2. Data Collection

Five greenness measurements were made in each plot (*n* = 87) from DOY 130 to DOY 180. Weather conditions varied between days of measurements. For greenness measurements we used four handheld spectral devices (Table 1). The Mapir NDVI camera was only accessible in the last two sampling cycles (*n* = 35). All passive spectral devices were applied 1.5 to 2.0 m above the plot for photographing. We avoided photographing direct light reflectance in the calibration target, and also avoided overexposure. The active Greenseeker sensor measurements were acquired 60 cm above ground.

Measurements of epidermal chlorophyll and flavonol content were conducted with the optical Dualex 4 scientific instrument (Force-A, Orsay, France). The readings of this instrument show a linear relationship to chlorophyll concentrations calculated from extractions. Readings are given in µg cm^{-2}, and the measuring wavelength for chlorophyll is a ratio of transmittance at 710 and 850 nm [34]. Within each plot, we sampled at five different spots. The sampling dates were the same as for the greenness measurements. Measurements were made on the newest, fully developed segment of *H. splendens* and on shoot tips of *E. nigrum*. Eight *V. vitis-idaea* leaves of one plant were measured per sample, starting with the upper (= newer) leaves. Hence, we measured physiological traits in different health states on three plots per species (nine in total). All measurements were performed with the adaxial setting of the device. High correlations with the abaxial side are found [35]. Measurements of *V. vitis-idaea* leaves showed reliable results. However, for the shoots of *H. splendens*, stable and reproducible results could only be achieved when three shoots were stacked and fixed with a transparent tape (Figure S3b). The same process was used for the shoots of *E. nigrum* (Figure S3a). At least eight chlorophyll measurements were performed on each sample and readings were then averaged and divided by the number of stacks in the tape. This results in 40 measurements per sampling date and plot. Palta [36] identified leaf anatomy, leaf veins, the presence of other pigments, and leaf thickness as main causes of large variations in chlorophyll meter readings. Hence, we decided to take the specific leaf area (SLA) into account. SLA is also of additional value to determine growth and plant stress [37,38]. After the chlorophyll measurements had been performed, a 6 mm circle was punched out of the prepared samples (Figure S3) [39]. First, fresh weight of the samples was measured. Then, the samples were dried for 24 h at 70 °C before dry weight was measured. This provided information on moisture content and SLA. Plant height was measured as median height above ground. Plant height of *H. splendens* refers to the thickness of the moss layer. We assessed stress levels for each plot in the first and last sampling periods. The stress estimate is a bare-eye classification of visibly dead or dying leaves versus healthy leaves of evergreens within small plots. Leaves that are dead or dying are brown, while healthy leaves are green. The stress estimate thus ranges from 0 to 100%, and it has turned out to be closely correlated to NDVI [13] and CO_2 fluxes [40].

2.3. Data Processing

2.3.1. Greenness Indices

We applied four different devices and extracted six different greenness indices from these (Table 1). The calibration methods applied are also listed. The Greenseeker did not need a calibration; internal tests suggest that measurements are not dependent on environmental changes [41].

The spectral properties of the devices we used for NDVI calculation are listed in Table 2. The active Greenseeker device has very similar bandwidths and -peaks to the Sentinel-2 bands 4 and 7. The spectral properties of the passive Mapir camera are closer to the "normal" NDVI calculated with Sentinel-2 bands 4 and 8.

Table 1. Background information on the greenness indices used in this study. Market prices refer to price levels in 2019.

Name	Equation	Device	Market Price	Calibration	Comments	Source
Greenseeker NDVI	Automatically calculated NDVI output; range 0–1 [1]	GreenSeeker handheld crop sensor (Trimble Inc., Sunnyvale, CA, USA)	ca. 700 $	Not needed	Active sensor	[41,42]
Mapir NDVI	$\frac{NIR-Red}{NIR+Red}$	Survey2 Camera – NDVI (Mapir Inc., San Diego, CA, USA)	ca. 400 $	Mapir target	Passive sensor	[43]
BNDVI	$\frac{Red-Blue}{Red+Blue}$	Modified Canon Eos 450D (no IR filter) (MaxMax, LDP LLC, Carlstadt, CA, USA)	Body + conversion = 250 $	White balance (WB) on gray area [2]	Widely used NDVI surrogate	[44]
GNDVI	$\frac{Red-Green}{Red+Green}$	Same as for BNDVI	Same as for BNDVI	White balance (WB) on gray area [2]	More linear correlation with chlorophyll than NDVI	[28,29]
GRVI	$\frac{Green-Red}{Green+Red}$	α7 (ILCE-7) (Sony Corp., Tokyo, Japan)	ca. 700 $, but could be any RGB camera	3-step gray card [2], when EV = −0.7	RGB index	[45]
Channel G%	$\frac{Green}{Red+Green+Blue}$	Same as for GRVI	Same as for GRVI	3-step gray card and white balance on white area [2] when EV = −0.7	RGB index	[46,47]

[1] According to the manufacturer's documents, even non-chlorophyll-containing surfaces, such as soil, have small NDVI values. Therefore, values below 0.15 are rarely measured. Likewise, normal NDVI, with a range from −1 to 1, shows NDVI values between 0.00 and 0.12 for scarce vegetation or bare soil. [2] Ordinary gray card with (N2, N5, N9.5) with an accuracy of about 5% on the Macbeth color space.

Table 2. Comparison of spectral properties of the various NDVI devices listed in Table 1. Numbers in parentheses show the bandwidth in nanometers.

	Mapir	Greenseeker	Sentinel-2	Sentinel-2	Sentinel-2
	Survey 2 NDVI	Handheld Crop Sensor	Band 4	Band 7	Band 8
Red band	660 nm (50)	660 nm (25)	665 nm (30)		
NIR band	850 nm (70)	780 nm (25)		783 nm (20)	833 nm (106)
Spatial resolution	16 MP camera	an oval depending on the height of the sensor: at 60 cm, length is 25 cm	10 m	20 m	10 m

2.3.2. Analysis of the Data

For the calibration of the RGB indices as well as of BNDVI and GNDVI, an ordinary gray card was applied. The card is printed on Teslin Synthetic (greywhitebalancecolorcard, Northfleet, UK]. According to the manufacturer, it has an accuracy of 5% on the Macbeth color space. To decide whether a three-step reflectance calibration on black, gray, and white is superior to a normal white balance, the Channel G% index was calculated for both calibrations. For the Mapir camera (Mapir Inc., San Diego, CA, USA) conversion and calibration were performed with the Mapir calibration target V1 and the QGIS software plug-in version 1.1.2. (Mapir Inc.). All other calibrations and NDVI calculations were performed in WINCAM pro 2013a (Regent Instruments Inc., Quebec City, QC, Canada). Further processing was done in EXCEL (Microsoft Corp., Redmond, WA, USA), while statistical analyses were performed with SPSS 25.0. (IBM Corp., Armonk, NY, USA). Additional, logistic curve fitting was analyzed using the Excel add-on Xlfit version 5.3.1.3 (ID Business Solutions Ltd., Guildford, UK). We calculated all correlations with a two-tailed Pearson´s testimony. Percentiles are weighted averages. Satellite NDVI data were retrieved from ESA's Sentinel-2 Open Access Hub [https://scihub.copernicus.eu]. Sentinel-2 satellite images were analyzed using ESA's SNAP software version 5.0.8 with the integrated Sentinel-2 Toolbox (ESA, Common Service Section, Rome, Italy). Atmospherically corrected 2A products from DOY 159 and DOY 163 in 2017 were used. Cloud cover was below 1% and products were analyzed with a resampled spatial resolution of 20 m. Downscaling was done with the "mean method," which calculates the output as mean of every source pixel value. The downscaling was performed to compare the different bands of Sentinel-2 for a pixelwise NDVI comparison, and also to reduce small-scale effects, like imprecise GPS coordinates (up to 3 m). To compare spaceborne and handheld NDVI data, at least four GPS waypoints were taken per area and NDVI values between the waypoints were measured with the Greenseeker handheld sensor (Trimble Inc., Sunnyvale, CA, USA, see Table S2). The sensor can also be used for measurement over a larger area. Then, it calculates an average of the scanned area. The Sentinel-2 image pixels corresponding to the GPS waypoints were identified and values were compared with the Greenseeker data.

3. Results

3.1. Descriptive Statistics

Descriptive statistics of the vegetation indices and plant traits are listed in Table 3. Due to relatively large sampling sizes ($n > 36$) for all traits on plot level we can assume that data is normally distributed. The combination of three different species in one dataset (Table 3), might lead to less normal distributed data. According a Kolmogorov-Smirnov test, all datasets of Table 1 are normally distributed ($p > 0.05$), except for Channel G%, Plant height, Stress level, Soil depth, SLA and Flav. However, a species-specific normal distribution is achieved for Plant height, NBI, SLA and Flav. For Channel G% the species-specific normal distribution does not hold for *V. vitis-idaea*, which was monitored on only two plots with highly contrasting stress levels. Stress level in total is not expected to be normally distributed due to plot selection by contrasting health states. Moreover, Soil depth is also not expected to be normally distributed.

Table 3. Descriptive statistics of the monitored plant traits on plot level.

Name	Mean	Std. Error	Std. Deviation	N	5% PCTL	Median	95% PCTL	Skewness	Kurtosis
Greenseeker NDVI	0.64	0.0099	0.092	88	0.48	0.63	0.80	0.018	−0.701
Mapir NDVI	0.69	0.0104	0.062	35	0.56	0.69	0.78	−0.611	0.246
BNDVI	0.86	0.0096	0.090	88	0.68	0.87	0.99	−0.569	−0.233
Channel G%	0.40	0.0050	0.047	87	0.48	0.39	0.80	0.848	0.283
GRVI	−0.098	0.0087	0.081	88	−0.212	−0.108	0.072	0.659	0.349
GNDVI	0.47	0.0085	0.079	88	0.34	0.47	0.60	−0.395	0.393
Plant height	9.46	0.0638	4.508	50	4.00	8.25	19.00	0.748	−0.430
Stress level	32.07	4.6593	27.956	36	0.85	30.00	93.63	1.009	0.274
Soil depth	13.73	0.7238	6.790	88	4.00	12.00	25.00	0.267	−1.301
SLA	0.149	0.0151	0.093	38	0.059	0.103	0.343	0.993	−0.238
Chl	23.46	0.7151	4.401	38	15.28	22.92	32.78	−0.014	0.617
Flav	1.01	0.0672	0.414	38	0.61	0.81	1.86	1.065	−0.420
NBI	25.31	0.0151	7.241	38	14.34	25.93	36.04	−0.018	−1.166

3.2. Comparison of Different Vegetation Indices and Plant Traits

We assessed the use of the six greenness indices and the different calibration methods. Accuracy was not found to be improved by using a relatively cheap 3-step reflectance target instead of an ordinary gray card for white balance (Table S3). However, it was important to avoid overexposure of the calibration target. Comparing the greenness indices (Table 4): Mapir NDVI and Greenseeker NDVI were significantly correlated ($r = 0.951$, $p < 0.001$; Figure 1a), while ordinary RGB indices showed a much lower correlation, albeit still significant. NDVI to BNDVI ($r = 0.779$, $p < 0.001$) correlated slightly better than Channel G% to NDVI ($r = 0.749$, $p < 0.001$) and NDVI to GRVI ($r = 0.689$, $p < 0.001$). Greenseeker NDVI showed the best correlation with the chlorophyll content ($r = 0.634$; $p < 0.001$), while other indices, such as BNDVI, showed rather low correlations with the chlorophyll content ($r = 0.433$, $p < 0.01$; Table 4). Figure 1 illustrates the results for different species and how they correlate with the other indices. Comparison of BNDVI and Greenseeker NDVI (Figure 1b) reveals several high BNDVI values around 1 and increased deviation of lower values. Moreover, spaceborne NDVI data from the Sentinel-2 satellites based on band 4/7/8 (Figure 1e,f) are highly correlated with ground-sampled NDVI values ($r = 0.956$ and $r = 0.968$, $p < 0.001$, $n = 14$) obtained using the Greenseeker device. In principle, this allows for a significant upscaling from ground to space.

Joint analyses of all plant species resulted in variable correlations between the greenness indices and other plant traits, i.e. SLA, chlorophyll content, and plant height (Table 4). The nitrogen index (NBI) shows a good correlation with soil depth ($r = 0.685$, $p < 0.001$), indicating that the nutrients may be limited by shallow soil depths. The NDVI and the Channel G% indices allow for an assumption of plant height, as correlations are good (NDVI: $r = 0.703$, $p < 0.001$; Channel G%: $r = 0.515$, $p < 0.001$), even when comparing across functional groups, i.e. by considering mosses and dwarf shrubs together. Species-specific correlations are listed in Tables S5–S8. Specifically, the correlation (Figure 2b) between SLA and chlorophyll ($r = -0.718$, $p < 0.001$) is almost solely driven by the moss *H. splendens*, which shows a strong correlation when analyzed separately ($r = -0.745$, $p < 0.01$), whereas *V. vitis-idaea* and *E. nigrum* showed no significant correlation. A similar case is the SLA to flavonol correlation ($r = -0.512$, $p = 0.001$; Table 4). In this case, *E. nigrum* is the only species showing a significant correlation when analyzed species-wise ($r = -0.589$, $p < 0.05$), while the two other species showed no significant correlation.

Table 4. Correlations between greenness indices and plant traits at plot level. Root Mean Squared Error (RMSE) of the linear regressions (slope, intercept) is computed for significant correlations with *r* > 0.4. Chl = chlorophyll content, Flav = flavonol absorbance, and NBI = nitrogen balance index, i.e. the ratio between Chl and Flav.

		BNDVI	Channel G%	GRVI	GNDVI	Plant Height	Stress Level	Soil Depth	SLA	Chl	Flav	NBI
NDVI	Correlation	0.779	0.749	0.689	0.440	0.703	−0.600	0.454	−0.451	0.634	0.260	0.190
	Sig.	0.000	0.000	0.000	0.000	0.000	0.000	0.000	0.004	0.000	0.116	0.254
	RMSE	0.0571	0.0314	0.0592	0.0716	3.2404	22.682	6.086	0.0841	3.4572		
	Slope/	0.762/	0.379/	0.606/	0.377/	36.892/	−188.3/	33.324/	−0.446/	29.706/		
	Intercept	0.376	0.159	−0.485	0.232	−14.313	152.9	−7.562	0.436	4.320		
	N	88	87	88	88	50	36	88	38	38	38	38
BNDVI	Correlation		0.730	0.641	0.547	0.468	−0.654	0.246	−0.232	0.433	−0.029	0.327
	Sig.		0.000	0.000	0.000	0.001	0.000	0.021	0.161	0.007	0.863	0.045
	RMSE		0.0324	0.0627	0.0668	4.026	21.469			4.027		
	slope/		0.378/	0.576/	0.480/	26.478/	−233.09/			19.167/		
	intercept		0.076	−0.595	0.059	−13.720	228.97			7.033		
	N		87	88	88	50	36	88	38	38	38	38
Channel G%	Correlation			0.909	0.454	0.515	−0.768	0.395	−0.141	0.376	−0.051	0.354
	Sig.			0.000	0.000	0.000	0.000	0.000	0.397	0.020	0.759	0.029
	RMSE			0.0340	0.0714	3.946	18.176					
	Slope/			1.565/	0.164/	49.185/	−419.85/					
	Intercept			−0.728	0.768	−10.268	203.61					
	N			87	87	49	36	87	38	38	38	38
GRVI	Correlation				0.338	0.554	−0.745	0.393	−0.124	0.387	0.019	0.301
	Sig.				0.001	0.000	0.000	0.000	0.459	0.016	0.908	0.066
	RMSE					3.7909	18.909					
	slope/					30.426/	−246.65/					
	intercept					12.371	11.770					
	N				88	50	36	88	38	38	38	38
GNDVI	Correlation					0.045	−0.220	0.070	−0.324	0.226	−0.120	0.294
	Sig.					0.758	0.196	0.515	0.047	0.172	0.472	0.074
	N					50	36	88	38	38	38	38
Plant height	Correlation						−0.553	0.425	−0.357	0.365	0.383	−0.160
	Sig.						0.001	0.002	0.103	0.095	0.079	0.477
	RMSE						22.129	6.3474				
	slope/						−3.085/	0.654/				
	intercept						58.926	7.765				
	N						34	50	22	22	22	22
Stress level	Correlation							−0.336	−0.110	−0.232	−0.090	−0.021
	Sig.							0.045	0.707	0.425	0.760	0.944
	N							36	14	14	14	14
Soil depth	Correlation								−0.025	0.103	−0.441	0.685
	Sig.								0.883	0.539	0.006	0.000
	RMSE										0.3768	5.3469
	slope/										−0.027/	0.721/
	intercept										1.385	15.182
	N								38	38	38	38
SLA	Correlation									−0.718	−0.512	0.086
	Sig.									0.000	0.001	0.607
	RMSE									3.1102	0.3605	
	slope/									−34.048/	−2.282/	
	intercept									28.541	1.353	
	N									38	38	38
Chl	Correlation										0.541	0.045
	Sig.										0.000	0.790
	RMSE										0.3530	
	slope/										0.051/	
	intercept										−0.180	
	N										38	38
Flav	Correlation											−0.787
	Sig.											0.000
	RMSE											4.5306
	slope/											−13.759/
	intercept											39.239
	N											38

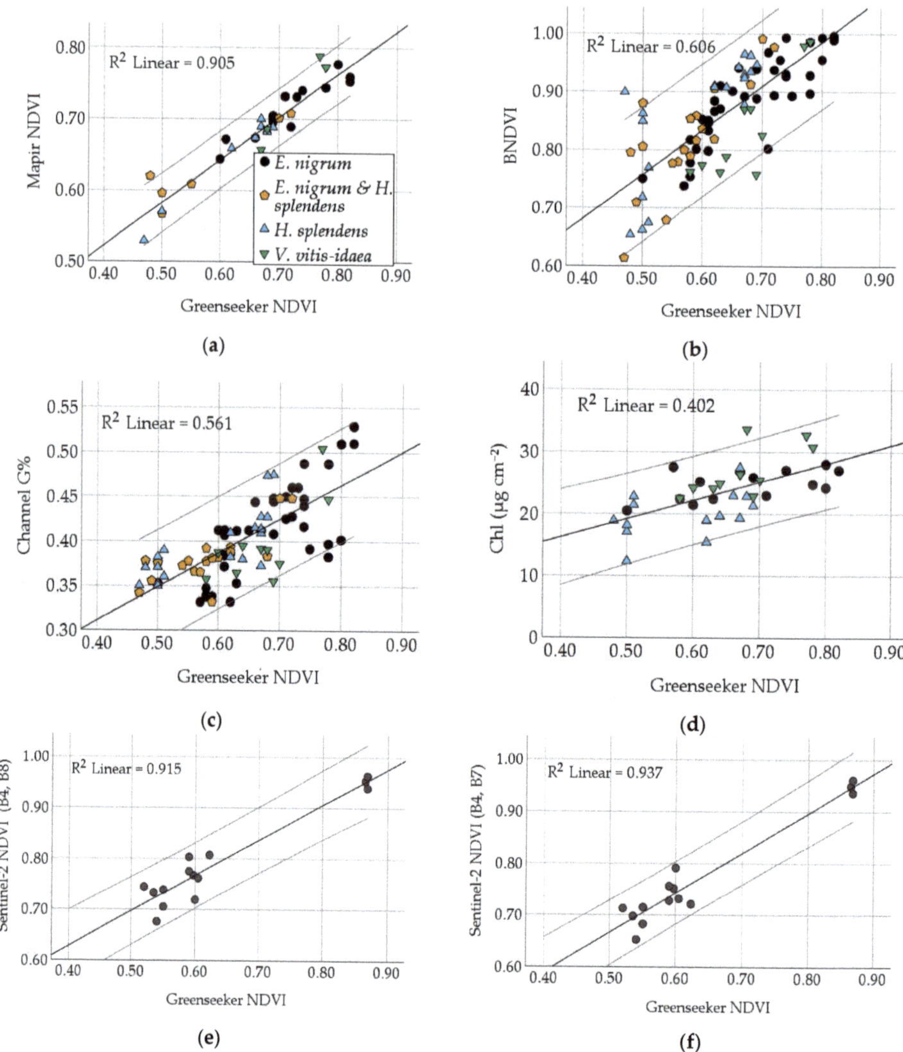

Figure 1. Species are marked with different symbols (see legend). 95% confidence intervals are grey. All correlations are significant ($p < 0.001$). The panels show the relationships between: (**a**) The active Greenseeker and the passive Mapir NDVI; (**b**) the BNDVI from the modified camera and NDVI; (**c**) Channel G% from the ordinary camera and NDVI; (**d**) chlorophyll content (Chl) and NDVI; (**e**) and (**f**) represent our satellite upscaling; comparing ground-based Greenseeker NDVI and Sentinel-2 NDVI, (note $n = 14$). NDVI is calculated with the named bands (= B). Equations and RMSE are shown in Table S4.

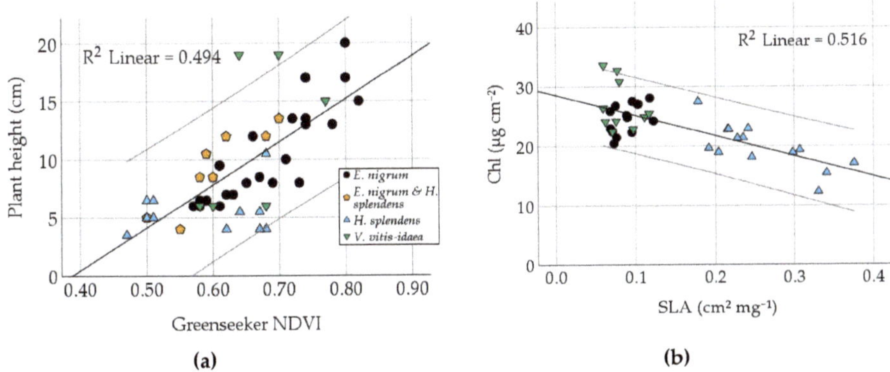

Figure 2. Relationship of two trait correlations from Table 4. Species are marked. 95% confidence intervals are grey. Relation between (**a**) median plant height and NDVI, (**b**) chlorophyll content and SLA. Equations and RMSE are shown in Table S4.

3.3. Intraspecific Trait Variablility

Table 5 illustrates the overall trait variability during the study. Plant traits varied in the course of the study, with species showing contrasting responses (Figure 3). For example, Chl increased in *V. vitis-idaea,* but was stable in *E. nigrum* and *H. splendens* (Figure 3a), while the flavonol content was rather constant over time, but significantly higher in *V. vitis-idaea* than in the other two species (Figure 3b). In early June, however, all studied species showed a sudden decline in chlorophyll content (Figure 3a; DOY 152), ($F_{4,14} = 7.945$, $p < 0.001$), which coincided with the temperature dropping almost to freezing point and light snowfall during DOYs 150–151 (Figure S4). The post-hoc Bonferroni test confirms significant differences in chlorophyll content on the sampling dates before (DOY 138) and after (DOY 172) the temperature drop ($p = 0.020$ and $p = 0.015$). NDVI varied considerably within species (long boxes in Figure 3c) and only *V. vitis-idaea* showed a temporal trend in NDVI, which coincided with an increase in chlorophyll content (compare Figure 3a,c). SLA was constant over time, but significantly higher in the moss than in the two dwarf shrubs (Figure 3d).

Table 5. Descriptive statistics of the monitored plant traits on species level.

Name	Specie	Mean	N	Std. Deviation	Std. Error	Median	5% PCTL	25% PCTL	75% PCTL	95% PCTL
Greenseeker NDVI	*E. nigrum*	0.65	59	0.094	0.012	0.62	0.49	0.58	0.72	0.82
	H. splendens	0.59	19	0.086	0.020	0.62	0.47	0.50	0.67	0.68 (90%)
	V. vitis-idaea	0.67	10	0.066	0.020	0.67	0.58	0.62	0.72	0.78 (90%)
Channel G%	*E. nigrum*	0.405	58	0.050	0.006	0.393	0.333	0.372	0.444	0.511
	H. splendens	0.396	19	0.037	0.008	0.383	0.350	0.383	0.415	0.474 (90%)
	V. vitis-idaea	0.397	10	0.046	0.015	0.389	0.355	0.363	0.408	0.497 (90%)
Plant height	*E. nigrum*	10.4	33	3.975	0.692	9.5	4,7	7.0	13.3	17.9
	H. splendens	5.5	11	1.955	0.589	5.0	3.5	4.0	6.5	9.7 (90%)
	V. vitis-idaea	11.8	6	6.55	2.676	10.5	6.0	6.0	19.0	-
SLA	*E. nigrum*	0.089	69	0.022	0.0027	0.085	0.057	0.072	0.105	0.130
	H. splendens	0.257	69	0.086	0.0104	0.246	0.164	0.189	0.291	0.411
	V. vitis-idaea	0.080	61	0.019	0.0024	0.079	0.052	0.066	0.092	0.117
Chl	*E. nigrum*	24.72	70	4.267	0.510	25.26	16.25	21.44	27.57	30.67
	H. splendens	19.88	69	5.122	0.617	20.01	11.90	15.59	23.88	28.54
	V. vitis-idaea	28.25	61	6.491	0.831	26.55	20.21	23.63	31.39	40.96
Flav	*E. nigrum*	0.833	69	0.164	0.020	0.818	0.542	0.727	0.943	1.145
	H. splendens	0.736	67	0.126	0.015	0.728	0.519	0.642	0.842	0.942
	V. vitis-idaea	1.671	61	0.232	0.030	1.706	1.297	1.558	1.843	1.944

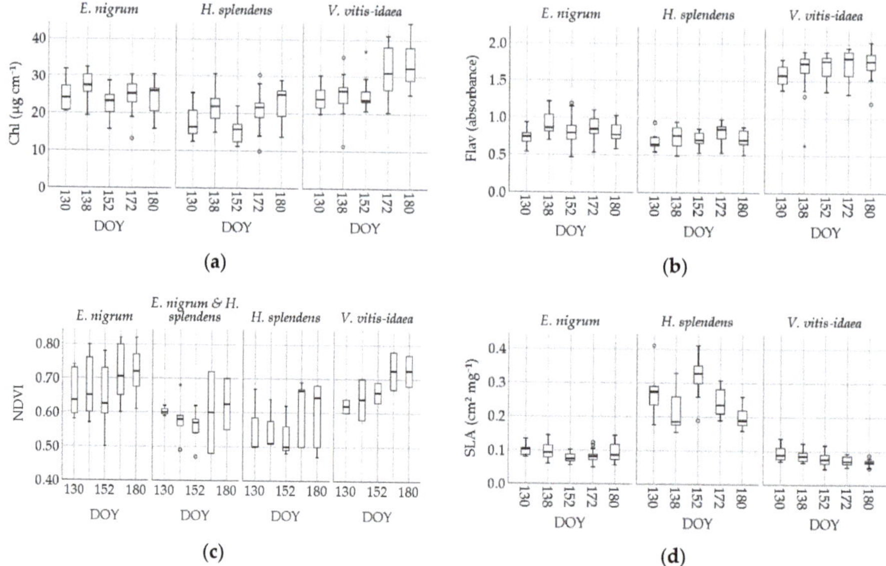

Figure 3. Intraspecific trait variability and its changes during the spring season for the species *Empterum nigrum*, *Vaccinium vitis-idaea*, and *Hylocomium splendens*. All species were monitored in varying health states, while each value represents one plot dominated by the named species. (**a**) Chlorophyll content (Chl); (**b**) flavonols (Flav); relative absorbance values; (**c**) Greenseeker NDVI; (**d**) SLA. These results were obtained at a moisture content (percent of wet weight) ranging from 55% to 75% for *H. splendens*, 63 to 68% for *V. vitis-idaea*, and 63 to 71% for *E. nigrum*.

3.4. Suitable Traits for Stress Monitoring

Our stress level estimate (Table 4) correlated with NDVI ($r = -0.600$, $p < 0.001$) and BNDVI ($r = -0.654$, $p < 0.001$), while the RGB indices performed best; $r = 0.768$, $p < 0.001$ for the Channel G% and $r = -0.745$, $p < 0.001$ for the GRVI, both as linear functions. Allowing a logistic relationship, a higher correlation level is obtained $r = -0.833$, $p < 0.001$ (Figure 4a) and $r = -0.650$, $p < 0.001$ (Figure 4b). Stress level estimates were also significantly correlated with plant height ($r = -0.553$, $p = 0.001$), as well as with soil depth ($r = -0.336$, $p < 0.05$). For litter, the relation with stress is reasonable, but not significant ($r = -0.419$, $p = 0.084$, $n = 18$). Neither flavonol absorbance nor chlorophyll content or the NBI readings could be related to the stress estimate (Table 4). No correlation was found for SLA and stress level ($n = 14$).

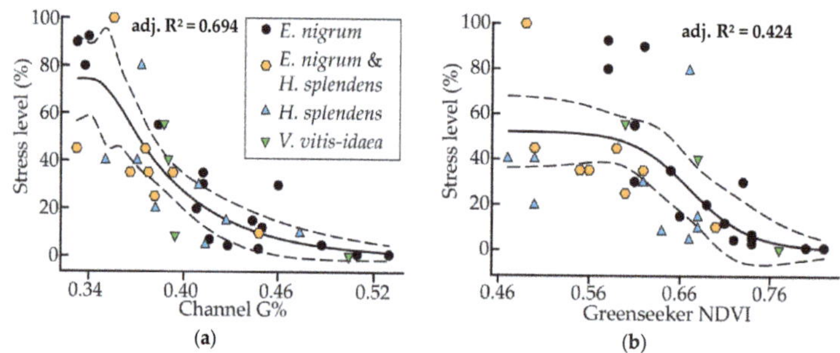

Figure 4. Comparison of stress level to greenness indices. 95% confidence intervals are dashed. Relation between (**a**) stress level and the RGB index Channel G%; (**b**) stress level and Greenseeker NDVI. Equations are shown in Table S4.

4. Discussion

The range of intraspecific trait variability (1st research question), is attributed well during the study. Some plant traits remained relatively stable during spring (SLA, Plant height, Flav), while others showed more variations during the season and to environmental circumstances (NDVI, ChannelG% and Chl). The small leaves of *E. nigrum* and shoots of *H. splendens* made the SLA measurements challenging. However, the infrequently used method that we decided to apply seemed to work well, as our results are comparable to SLA values retrieved in previous studies [48–50]. To our knowledge, our plots did not suffer from any major stress (browning) events during the last 3 years prior to our measurements, except that *Vaccinium myrtillus* in the area had been partly defoliated by larvae of geometrid moths [51], but this species was rare or absent in our plots. In the early growing season, plants are especially vulnerable to winter-related stress and are showing accumulated stress responses from the previous years [13,14]. Hence, we monitored the natural range of trait variabilities from start of the growing season (DOY 130) onwards. Chlorophyll content was dropping significantly when temperature fell to almost freezing point. However, more research is needed to validate this result. It might be that the slight snowfall, or both parameters jointly, instigated the decline in chlorophyll concentrations.

The second research question was to assess whether any of the studied plant traits are suitable for stress monitoring. Our data show that the stress level differed between plots; we found that plant height was related to soil depth and that soil depth was also related to NBI. Although we did not find any significant correlation between plant height and NBI, we assume that soil depth is a limiting factor for this ecosystem. Lower soil depth affects water and nutrient availability as well as soil temperature [52] and is also associated with areas of low snow accumulation during winter [13]. This is supported by the fact that the stress level decreased with increasing soil depth and that NDVI increased with increasing soil depth.

In general, the flavonol content is associated with plant stress reactions [53]. However, we could not relate the flavonol absorbance to our stress level estimates. As the Dualex device estimates the flavonol content from spectral properties, it might not be able to measure the relevant flavonols in relation to the types of stress occurring in these subarctic plants. Dualex flavonol measurements are performed at the wavelengths 375 nm (UV-A) and 650 nm (red) [34]. This results in screening of mainly kaempferol, quercetin, and myricetin [53]. Our results are in agreement with Lefebvre et al. [54], who concluded that the Dualex device could not accurately predict the flavonol content in the three alpine plants they studied.

Concerning the third research question, our results show that ordinary RGB cameras may be used as NDVI surrogates and that they reflect various plant traits well. They performed equally well as modified cameras (with the infrared filter removed) for near-remote sensing approaches in the subarctic ecosystem. We found that a normal gray card, as used by professional photographers, was sufficient for the calibration process. Based on our findings, we recommend a simple white balance. Even if correlations to NDVI were slightly higher for BNDVI ($r = 0.779$) than for RGB greenness indices (0.689–0.749), one of the main strengths of the RGB cameras is that they are easier to operate than the modified devices. Sonnentag et al. [46] showed that different RGB cameras produce comparable results and that the choice of file format is not that important. Also, Nijland et al. [55] identified band separation and dynamic range as main problems when using converted cameras and therefore recommended the use of true color imaging. Another aspect is that the distribution of RGB cameras via smartphones is enormous and might be valuable for citizen science projects or app development [56]. In general, our greenness measurements are in agreement with existing reports on phenology at higher latitudes [30,57].

Moreover, the Channel G% index performed better than NDVI in characterizing some plant traits. This includes the stress level which showed a stronger correlation to Channel G% (linear: $r = -0.768$ vs. $r = -0.600$; logistic: $r = -0.833$ vs. $r = -0.651$) and NBI which showed a significant correlation ($r = -0.354$) to Channel G%, but not to NDVI. Consequently, the Channel G% index is of additional value for screening plant stress (2nd research question). The significant correlation between RGB indices and chlorophyll meter readings ($r = 0.38$, $p < 0.05$) also implies that the RGB-based indices could be potential NDVI surrogates (see Table 4). In contrast to previous studies [28,29,58], our GNDVI data did not show any significant correlation with chlorophyll content or other plant traits. Correlations between chlorophyll and NDVI showed reasonable results [29], indicating that chlorophyll measurements are valid in spite of the untypical leaf structures of *H. splendens* and *E. nigrum*.

We found a high correlation between spaceborne NDVI and ground-sampled NDVI measured by the active Greenseeker device (maximum $r = 0.968$ for the Sentinel-2 NDVI calculated with bands 4 and 7). Nevertheless, despite of the strongly significant correlation, it is based only on 14 data points, implying that relationships have to be handled with care. It is a higher correlation than retrieved in previous studies, where near-remotely sensed NDVI data were compared to NDVI from Sentinel-2 and Landsat 8 [59,60]. A likely reason for the very strong correlation is that this study was carried out in a very open subarctic woodland (in parts nearly treeless and then considered as heath) where understory vegetation contributes very much to the NDVI detected by the satellites. We did not find major differences in the correlations, even when spectral properties (bandwidth and wavelength peaks) were not similar. This strongly suggests that active sensors can be used for validation of spaceborne data, for example, from Sentinel-2.

5. Conclusions

The objective of this study was to assess the applicability of common plant traits and near-remote sensing approaches as tools to monitor the health state of dominant understory subarctic vegetation types that previously were shown to be vulnerable to winter climate change and other types of stress. In order to determine intraspecific trait variability, species were monitored in different health states. Due to this screening we are able to better validate the effect size of a browning event on the studied species. As the study was set in an area not recently damaged by stressful events, the different stress levels could be explained by differences in soil depth, which again act as a surrogate for several potential stressful elements, including moisture and nutrient deficits during the growing season and little snow protection during winter. Channel G% was the best RGB-based index in our study, and we recommend the use of this index. Finally, we found promising results by combining spaceborne Sentinel-2 data with the active near-remote sensor for measurements of NDVI. This could be a useful tool for upscaling the role of understory vegetation to the total NDVI measured by satellites in regions where browning occurs. Further research is recommended on the satellite upscaling, but also on the measured chlorophyll drop following a rapid midsummer temperature decrease to freezing point. Finally, we recommend following the same plots after a stressful weather event, to report direct as well as long-term changes in situ.

Supplementary Materials: The following are available online at http://www.mdpi.com/1424-8220/20/7/2102/s1, **Figure S1:** Study site of the 18 plots at the northern parts of Tromsøya (Norway). Exact coordinates are given in Table S1. **Figure S2:** Typical vegetation plots within the study area. (**a**) Detail view of one of the satellite upscaling areas. The photo was taken on the day of the upscaling experiment. Note: the snow patch on the left, which made plot selection difficult. The leafless trees show that this is from early spring prior to budbreak. (**b**) RGB image of a healthy plot (1 m^2) of *E. nigrum*. (**c**) RGB image of a stressed plot of *E. nigrum*. **Figure S3:** Illustration of the working process for the different species. After measuring at least eight times with the chlorophyll meter (Chl, Flav, NBI), middle parts are hole-punched and cut and stored in sealed, numbered glasses to determine moisture content and SLA. **Figure S4:** Weather statistics for Tromsø from November 2016 to July 2017. The red line shows the mean value of daily temperature. The black line shows the average temperature from 1989-2018. Blue bars indicate the snow depth. Snow depth and long-time temperature data are from the Tromsø weather station (SN90450) located about 5.6 km south of the study area, while daily mean temperature is from the Stakkevollan weather station (SN90495) located about 900 m south of the main cluster of field plots. **Table S1:** Plot descriptions, including coordinates, stress level estimates, plant height, soil depth, litter, slope, and vegetation assessment. **Table S2:** Coordinates of the waypoints for satellite referencing. Waypoint coordinates are different from plot coordinates. **Table S3:** Plot-level relationships between different greenness indices and chlorophyll content for different calibration methods. Wb = white balance, 3-step = 3-step reflectance calibration. **Table S4:** Equations corresponding to the linear regressions in Figures 1 and 2. **Table S5:** Correlation table for plots dominated by *E. nigrum*. Correlations are computed by Pearson's correlation, while significance is two-tailed. **Table S6:** Correlation table for plots dominated by *Vaccinium vitis-idaea*. Correlations are computed by Pearson's correlation, while significance is two-tailed., **Table S7:** Correlation table for plots dominated by *Hylocomium splendens*. Correlations are computed by Pearson's correlation, while significance is two-tailed. **Table S8:** Correlation table for mixed plots of *Empetrum nigrum* and *Hylocomium splendens*. Correlations are computed by Pearson's correlation, while significance is two-tailed.

Author Contributions: Conceptualization, E.R., J.W.B. and HT.; methodology, E.R., H.T. and J.W.B.; formal analysis, E.R., J.W.B. and H.T.; writing—original draft preparation, E.R, J.W.B. and H.T.; writing—review and editing, E.R, J.W.B. and H.T.; project administration, J.W.B.; funding acquisition, J.W.B. All authors have read and agree to the published version of the manuscript.

Funding: Funding was provided by the Research Council of Norway grant 287402.

Conflicts of Interest: The authors declare no conflict of interest.

References

1. Pachauri, R.K.; Allen, M.R.; Barros, V.R.; Broome, J.; Cramer, W.; Christ, R.; Church, J.A.; Clarke, L.; Dahe, Q.; Dasgupta, P.; et al. *Climate Change 2014: Synthesis Report*; Pachauri, R.K., Mayer, L., Eds.; Intergovernmental Panel on Climate Change: Geneva, Switzerland, 2015.

2. Ju, J.; Masek, J.G. The vegetation greenness trend in Canada and US Alaska from 1984–2012 Landsat data. *Remote Sens. Environ.* **2016**, *176*, 1–16. [CrossRef]

3. Myneni, R.B.; Keeling, C.D.; Tucker, C.J.; Asrar, G.; Nemani, R.R. Increased plant growth in the northern high latitudes from 1981 to 1991. *Nature* **1997**, *386*, 698–702. [CrossRef]

4. Verbyla, D. The greening and browning of Alaska based on 1982–2003 satellite data. *Glob. Ecol. Biogeogr.* **2008**, *17*, 547–555. [CrossRef]

5. Jia, G.J.; Epstein, H.E.; Walker, D.A. Greening of arctic Alaska, 1981–2001. *Geophys. Res. Lett.* **2003**, *30*. [CrossRef]

6. Epstein, H.E.; Bhatt, U.S.; Raynolds, M.K.; Walker, D.A.; Forbes, B.C.; Phoenix, G.K.; Bjerke, J.W.; Tømmervik, H.; Karlsen, S.-R.; Myneni, R.; et al. Tundra greenness. *Bull. Am. Meteorol. Soc.* **2019**, *100*, S163–S168.

7. Myers-Smith, I.H.; Kerby, J.T.; Phoenix, G.K.; Bjerke, J.W.; Epstein, H.E.; Assmann, J.J.; John, C.; Andreu-Hayles, L.; Angers-Blondin, S.; Beck, P.S.A.; et al. Complexity revealed in the greening of the Arctic. *Nat. Clim. Chang.* **2020**, *10*, 106–117. [CrossRef]

8. Phoenix, G.K.; Bjerke, J.W. Arctic browning: Extreme events and trends reversing arctic greening. *Glob. Chang. Biol.* **2016**, *22*, 2960–2962. [CrossRef]

9. Bokhorst, S.F.; Bjerke, J.W.; Bowles, F.W.; MELILLO, J.; Callaghan, T.V.; Phoenix, G.K. Impacts of extreme winter warming in the sub-Arctic: Growing season responses of dwarf shrub heathland. *Glob. Chang. Biol.* **2008**, *14*, 2603–2612. [CrossRef]

10. Sorensen, P.O.; Finzi, A.C.; Giasson, M.A.; Reinmann, A.B.; Sanders-DeMott, R.; Templer, P.H. Winter soil freeze-thaw cycles lead to reductions in soil microbial biomass and activity not compensated for by soil warming. *Soil. Biol. Biochem.* **2018**, *116*, 39–47. [CrossRef]

11. Bokhorst, S.F.; Bjerke, J.W.; Street, L.E.; Callaghan, T.V.; Phoenix, G.K. Impacts of multiple extreme winter warming events on sub-Arctic heathland: Phenology, reproduction, growth, and CO2 flux responses. *Glob. Chang. Biol.* **2011**, *17*, 2817–2830. [CrossRef]

12. Bokhorst, S.F.; Phoenix, G.K.; Berg, M.P.; Callaghan, T.V.; Kirby-Lambert, C.; Bjerke, J.W. Climatic and biotic extreme events moderate long-term responses of above- and belowground sub-Arctic heathland communities to climate change. *Glob. Chang. Biol.* **2015**, *21*, 4063–4075. [CrossRef] [PubMed]

13. Bjerke, J.W.; Treharne, R.; Vikhamar-Schuler, D.; Karlsen, S.R.; Ravolainen, V.; Bokhorst, S.F.; Phoenix, G.K.; Bochenek, Z.; Tømmervik, H. Understanding the drivers of extensive plant damage in boreal and Arctic ecosystems: Insights from field surveys in the aftermath of damage. *Sci. Total Environ.* **2017**, *599*, 1965–1976. [CrossRef] [PubMed]

14. Bjerke, J.W.; Bokhorst, S.F.; Callaghan, T.V.; Phoenix, G.K.; Dorrepaal, E. Persistent reduction of segment growth and photosynthesis in a widespread and important sub-Arctic moss species after cessation of three years of experimental winter warming. *Funct. Ecol.* **2017**, *31*, 127–134. [CrossRef]

15. Bjerke, J.W.; Rune Karlsen, S.; Arild Høgda, K.; Malnes, E.; Jepsen, J.U.; Lovibond, S.; Vikhamar-Schuler, D.; Tømmervik, H. Record-low primary productivity and high plant damage in the Nordic Arctic Region in 2012 caused by multiple weather events and pest outbreaks. *Environ. Res. Lett.* **2014**, *9*, 84006. [CrossRef]

16. Tømmervik, H.; Bjerke, J.W.; Park, T.; Hanssen, F.; Myneni, R.B. Legacies of historical exploitation of natural resources are more important than summer warming for recent biomass increases in a boreal–Arctic transition region. *Ecosystems* **2019**, *113*, 11770. [CrossRef]

17. Xu, L.; Myneni, R.B.; Chapin III, F.S.; Callaghan, T.V.; Pinzon, J.E.; Tucker, C.J.; Zhu, Z.; Bi, J.; Ciais, P.; Tømmervik, H.; et al. Temperature and vegetation seasonality diminishment over northern lands. *Nat. Clim. Chang.* **2013**. [CrossRef]

18. Kitić, G.; Tagarakis, A.; Cselyuszka, N.; Panić, M.; Birgermajer, S.; Sakulski, D.; Matović, J. A new low-cost portable multispectral optical device for precise plant status assessment. *Comput. Electron. Agric.* **2019**, *162*, 300–308. [CrossRef]

19. Bokhorst, S.F.; Tømmervik, H.; Callaghan, T.V.; Phoenix, G.K.; Bjerke, J.W. Vegetation recovery following extreme winter warming events in the sub-Arctic estimated using NDVI from remote sensing and handheld passive proximal sensors. *Environ. Exp. Bot.* **2012**, *81*, 18–25. [CrossRef]

20. Street, L.E.; Shaver G., R.; Williams, M.; Van Wijk, M.T. What is the relationship between changes in canopy leaf area and changes in photosynthetic CO 2 flux in arctic ecosystems? *J. Ecol.* **2007**, *95*, 139–150. [CrossRef]

21. Rouse, J.; Haas, H.R.; Schell, A.J.; Deering, W.D. Monitoring vegetation systems in the Great Plains with ERTS. In Proceedings of the 3rd Earth Resource Technology Satellite-1 Symposium, Washington, DC, USA, 10–14 December 1973.

22. Tucker, C.J. Red and photographic infrared linear combinations for monitoring vegetation. *Remote Sens. Environ.* **1979**, *8*, 127–150. [CrossRef]

23. Pettorelli, N.; Vik, J.O.; Mysterud, A.; Gaillard, J.-M.; Tucker, C.J.; Stenseth, N.C. Using the satellite-derived NDVI to assess ecological responses to environmental change. *Trends Ecol. Evol.* **2005**, *20*, 503–510. [CrossRef] [PubMed]

24. Boelman, N.T.; Stieglitz, M.; Rueth, H.M.; Sommerkorn, M.; Griffin, K.L.; Shaver, G.R.; Gamon, J.A. Response of NDVI, biomass, and ecosystem gas exchange to long-term warming and fertilization in wet sedge tundra. *Oecologia* **2003**, *135*, 414–421. [CrossRef] [PubMed]

25. Jepsen, J.U.; Hagen, S.B.; Høgda, K.A.; Ims, R.A.; Karlsen, S.R.; Tømmervik, H.; Yoccoz, N.G. Monitoring the spatio-temporal dynamics of geometrid moth outbreaks in birch forest using MODIS-NDVI data. *Remote Sens. Environ.* **2009**, *113*, 1939–1947. [CrossRef]

26. Tømmervik, H.; Karlsen, S.-R.; Nilsen, L.; Johansen, B.; Storvold, R.; Zmarz, A.; Becker, P.S.; Johansen, K.-S.; Hogda, K.-A.; Goetz, S.; et al. Use of Unmanned Aircraft Systems (UAS) in a Multi-Scale Vegetation Index Study of Arctic Plant Communities in Adventdalen on Svalbard. Available online: https://www.nina.no/Portals/NINA/Bilder%20og%20dokumenter/Prosjekter/ArcticBiomass/Tommervik_et_2014_UAS_RPAS.pdf (accessed on 26 January 2020).

27. Leblanc, S.G.; Chen, W.; Maloley, M.; Humphreys, E.; Elliott, C. NDVI Digital Camera for Monitoring Arctic Vegetation. In Proceedings of the 2014 IEEE International Geoscience & Remote Sensing Symposium and 35th Canadian Symposium on Remote Sensing (IGARSS 2014/35th CSRS), Quebec, Canada, 13–18 July 2014.

28. Gitelson, A.A.; Merzlyak, M.N. Remote estimation of chlorophyll content in higher plant leaves. *Int. J. Remote Sens.* **1997**, *18*, 2691–2697. [CrossRef]

29. Hunt, E.R.; Doraiswamy, P.C.; McMurtrey, J.E.; Daughtry, C.S.T.; Perry, E.M.; Akhmedov, B. A visible band index for remote sensing leaf chlorophyll content at the canopy scale. *Int. J. Appl. Earth Obs. Geoinf.* **2013**, *21*, 103–112. [CrossRef]

30. Anderson, H.; Nilsen, L.; Tømmervik, H.; Karlsen, S.; Nagai, S.; Cooper, E.J. Using Ordinary Digital Cameras in Place of Near-Infrared Sensors to Derive Vegetation Indices for Phenology Studies of High Arctic Vegetation. *Remote Sens.* **2016**, *8*, 847. [CrossRef]

31. Nilsson, M.-C.; Wardle, D.A. Understory vegetation as a forest ecosystem driver: Evidence from the northern Swedish boreal forest. *Front Ecol. Environ.* **2005**, *3*. [CrossRef]

32. Lomolino, M.V.; Riddle, B.R.; Whittaker, R.J.; Brown, J.H. *Biogeography*, 4th ed.; Sinauer: Sunderland, MA, USA, 2010; p. 662.

33. Bjerke, J.W.; Bokhorst, S.F.; Zielke, M.; Callaghan, T.V.; Bowles, F.W.; Phoenix, G.K. Contrasting sensitivity to extreme winter warming events of dominant sub-Arctic heathland bryophyte and lichen species. *J. Ecol.* **2011**, *99*, 1481–1488. [CrossRef]

34. Cerovic, Z.G.; Masdoumier, G.; Ghozlen, N.B.; Latouche, G. A new optical leaf-clip meter for simultaneous non-destructive assessment of leaf chlorophyll and epidermal flavonoids. *Physiol. Plant* **2012**, *146*, 251–260. [CrossRef]

35. Cartelat, A.; Cerovic, Z.G.; Goulas, Y.; Meyer, S.; Lelarge, C.; Prioul, J.-L.; Barbottin, A.; Jeuffroy, M.-H.; Gate, P.; Agati, G.; et al. Optically assessed contents of leaf polyphenolics and chlorophyll as indicators of nitrogen deficiency in wheat (*Triticum aestivum* L.). *Field Crops Res.* **2005**, *91*, 35–49. [CrossRef]

36. Palta, J.P. Leaf chlorophyll content. *Remote Sensing Reviews* **1990**, *5*, 207–213. [CrossRef]

37. Pierce, L.L.; Running, S.W.; Walker, J. Regional-Scale Relationships of Leaf Area Index to Specific Leaf Area and Leaf Nitrogen Content. *Ecol. Appl.* **1994**, *4*, 313–321. [CrossRef]

38. Liu, M.; Wang, Z.; Li, S.; Lü, X.; Wang, X.; Han, X. Changes in specific leaf area of dominant plants in temperate grasslands along a 2500-km transect in northern China. *Sci. Rep.* **2017**, *7*, 10780. [CrossRef] [PubMed]

39. Cornelissen, J.H.C.; Lavorel, S.; Garnier, E.; Díaz, S.; Buchmann, N.; Gurvich, D.E.; Reich, P.B.; Steege, H.t.; Morgan, H.D.; Van der Heijden, M.G.A.; et al. A handbook of protocols for standardised and easy measurement of plant functional traits worldwide. *Aust. J. Bot.* **2003**, *51*, 335. [CrossRef]

40. Treharne, R.; Bjerke, J.W.; Tømmervik, H.; Stendardi, L.; Phoenix, G.K. Arctic browning: Impacts of extreme climatic events on heathland ecosystem CO2 fluxes. *Glob. Chang. Biol.* **2019**, *25*, 489–503. [CrossRef] [PubMed]

41. Martin, D.; López, J.; Lan, Y. Laboratory evaluation of the GreenSeeker™ hand-held optical sensor to variations in orientation and height above canopy. *Int. J. Agric. Biol. Eng.* **2012**, 43–47. [CrossRef]

42. Verhulst, N.; Govaerts, B.; Sayre, K.D.; Deckers, J.; François, I.M.; Dendooven, L. Using NDVI and soil quality analysis to assess influence of agronomic management on within-plot spatial variability and factors limiting production. *Plant Soil* **2009**, *317*, 41–59. [CrossRef]

43. Koucká, L.; Kopačková, V.; Fárová, K.; Gojda, M. UAV Mapping of an Archaeological Site Using RGB and NIR High-Resolution Data. *Proceedings* **2018**, *2*, 351. [CrossRef]

44. MaxMax. Manufacturer Homepage; Product Specifications. Available online: https://maxmax.com/ndvi_camera_technical.htm (accessed on 6 September 2017).

45. Bannari, A.; Morin, D.; Bonn, F.; Huete, A.R. A review of vegetation indices. *Remote Sensing Reviews* **1995**, *13*, 95–120. [CrossRef]

46. Sonnentag, O.; Hufkens, K.; Teshera-Sterne, C.; Young, A.M.; Friedl, M.; Braswell, B.H.; Milliman, T.; O'Keefe, J.; Richardson, A.D. Digital repeat photography for phenological research in forest ecosystems. *Agric. For. Meteorol.* **2012**, *152*, 159–177. [CrossRef]

47. Richardson, A.D.; Jenkins, J.P.; Braswell, B.H.; Hollinger, D.Y.; Ollinger, S.V.; Smith, M.-L. Use of digital webcam images to track spring green-up in a deciduous broadleaf forest. *Oecologia* **2007**, *152*, 323–334. [CrossRef] [PubMed]

48. Suzuki, S.; Kudo, G. Short-term effects of simulated environmental change on phenology, leaf traits, and shoot growth of alpine plants on a temperate mountain, northern Japan. *Glob. Chang. Biol.* **1997**, *3*, 108–115. [CrossRef]

49. Pensa, M.; Karu, H.; Luud, A.; Kund, K. Within-species correlations in leaf traits of three boreal plant species along a latitudinal gradient. *Plant Ecol.* **2010**, *208*, 155–166. [CrossRef]

50. Van Wijk, M.T.; Williams, M.; Shaver, G.R. Tight coupling between leaf area index and foliage N content in arctic plant communities. *Oecologia* **2005**, *142*, 421–427. [CrossRef] [PubMed]

51. Bjerke, J.W.; Wierzbinski, G.; Tømmervik, H.; Phoenix, G.K.; Bokhorst, S. Stress-induced secondary leaves of a boreal deciduous shrub (*Vaccinium myrtillus*) overwinter then regain activity the following growing season. *Nord. J. Bot.* **2018**, *36*, e01894. [CrossRef]

52. Tybirk, K.; Nilsson, M.-C.; Michelsen, A.; Kristensen, H.L.; Shevtsova, A.; Strandberg, M.T.; Johansson, M.; Nielsen, K.E.; Riis-Nielsen, T.; Strandberg, B.; et al. Nordic Empetrum dominated ecosystems: Function and susceptibility to environmental changes. *Ambio* **2000**, *29*, 90–97. [CrossRef]

53. Cerovic, Z.G.; Ounis, A.; Cartelat, A.; Latouche, G.; Goulas, Y.; Meyer, S.; Moya, I. The use of chlorophyll fluorescence excitation spectra for the non-destructive in situ assessment of UV-absorbing compounds in leaves. *Plant Cell Environ.* **2002**, *25*, 1663–1676. [CrossRef]

54. Lefebvre, T.; Millery-Vigues, A.; Gallet, C. Does leaf optical absorbance reflect the polyphenol content of alpine plants along an elevational gradient? *Alp. Bot.* **2016**, *126*, 177–185. [CrossRef]

55. Nijland, W.; De Jong, R.; De Jong, S.M.; Wulder, M.A.; Bater, C.W.; Coops, N.C. Monitoring plant condition and phenology using infrared sensitive consumer grade digital cameras. *Agric. For. Meteorol.* **2014**, *184*, 98–106. [CrossRef]

56. Leeuw, T.; Boss, E. The HydroColor app: Above water measurements of remote sensing reflectance and turbidity using a smartphone camera. *Sensors* **2018**, *18*, 256. [CrossRef]

57. Westergaard-Nielsen, A.; Lund, M.; Hansen, B.U.; Tamstorf, M.P. Camera derived vegetation greenness index as proxy for gross primary production in a low Arctic wetland area. *ISPRS Int. J. Geoinf.* **2013**, *86*, 89–99. [CrossRef]

58. Hunt, E.R.; Hively, W.D.; McCarty, G.W.; Daughtry, C.S.T.; Forrestal, P.J.; Kratochvil, R.J.; Carr, J.L.; Allen, N.F.; Fox-Rabinovitz, J.R.; Miller, C.D. NIR-Green-Blue High-Resolution Digital Images for Assessment of Winter Cover Crop Biomass. *GIScience Remote Sens.* **2011**, *48*, 86–98.

59. Manna, S.; Raychaudhuri, B. Retrieval of leaf area index and stress conditions for Sundarban mangroves using Sentinel-2 data. *Int. J. Remote Sens.* **2019**, 1–21. [CrossRef]

60. Barbanti, L.; Adroher, J.; Damian, J.M.; Di Virgilio, N.; Falsone, G.; Zucchelli, M.; Martelli, R. Assessing wheat spatial variation based on proximal and remote spectral vegetation indices and soil properties. *Ital. J. Agronomy* **2018**, *13*, 21. [CrossRef]

 *sensors*

Article

Assessing the Trend of the Trophic State of Lake Ladoga Based on Multi-Year (1997–2019) CMEMS GlobColour-Merged CHL-OC5 Satellite Observations

Augustine-Moses Gaavwase Gbagir [1],* and Alfred Colpaert [2]

[1]   School of Forest Sciences, University of Eastern Finland, Yliopistokatu 7, Borealis Building A 3rd Floor, 80101 Joensuu, Finland

[2]   Department of Geographical and Historical Studies, University of Eastern Finland, Yliopistokatu 7, Metria-Building, P.O. Box 111, FI-80101 Joensuu, Finland; alfred.colpaert@uef.fi

*   Correspondence: augustine.gbagir@uef.fi

Received: 25 September 2020; Accepted: 27 November 2020; Published: 1 December 2020

Abstract: The trophic state of Lake Ladoga was studied during the period 1997–2019, using the Copernicus Marine Environmental Monitoring Service (CMEMS) GlobColour-merged chlorophyll-a OC5 algorithm (GlobColour CHL-OC5) satellite observations. Lake Ladoga, in general, is mesotrophic but certain parts of the lake have been eutrophic since the 1960s due to the discharge of wastewater from industrial, urban, and agricultural sources. Since then, many ecological assessments of the Lake's state have been made. These studies have indicated that various changes are taking place in the lake and continuous monitoring of the lake is essential to update the current knowledge of its state. The aim of this study was to assess the long-term trend in chl-a in Lake Ladoga. The results showed a gradual reduction in chl-a concentration, indicating a moderate improvement. Chl-a concentrations (minimum-maximum values) varied spatially. The shallow southern shores did not show any improvement while the situation in the north is much better. The shore areas around the functioning paper mill at Pitkäranta and city of Sortavala still show high chl-a values. These findings provide a general reference on the current trophic state of Lake Ladoga that could contribute to improve policy and management strategies. It is assumed that the present warming trend of surface water may result in phytoplankton growth increase, thus partly offsetting a decrease in nutrient load. Precipitation is thought to be increasing, but the influence on water quality is less clear. Future studies could assess the current chemical composition to determine the state of water quality of Lake Ladoga.

Keywords: Lake Ladoga; CMEMS GlobColour CHL-OC5; eutrophication; water quality assessment; pulp and paper mill; climate change; ecological status; remote sensing; phytoplankton and chlorophyll-a; chemical wastewater pollution

---

1. Introduction

Assessing the trophic state of Lake Ladoga is essential because of its past history of severe water pollution [1,2]. The history of the Great Lake Ladoga begins after the melting of the ice of the last glaciation, when Ladoga became separated from the so-called Yoldia Sea about 10,000 years ago [3]. The earliest drainage of the lake was to the Gulf of Finland in the direction of the region now known as the city of Vyborg through the threshold of Vetokallio near Heinjoki [3]. Post-glacial uplift, which was, and still is, faster in the north, tilted Lake Ladoga causing a transgression of its southern shallow shoreline, leading to the creation of the current Neva River connection about 3300 years ago [3]. For the same reason, about 9500 years ago, a connection was established from Lake Onega to Ladoga via the River Svir [3]. The birth of the Neva caused the surface of Ladoga to drop by about 12 m, producing the present shape of the lake [3]. The Lake was oligotrophic (mostly free from nutrients) before the 1950s and mesotrophic (moderate nutrients) in the 1970s but some areas of the lake (mostly coastal locations) have become eutrophic due to anthropogenic activities [1,4]. These anthropogenic activities were mainly industrial enterprises (e.g., pulp and paper mills, aluminum) and agriculture [1,5]. Consequently, the discharge of wastewater (containing high amounts of phosphorus and nitrogen) into the lake from these industries resulted in the degradation of water quality [1,2]. The lake has a large catchment (Figure 1), and thus receives huge amounts of water containing anthropogenic nutrients [2,6]. An important part of these waste waters discharge into the shallow Volkhov Bay (southern coastal area) which receives its waters from the Volkhov sub-catchment (West Ladoga), and Syas River (from the South) [5]. Also, extensive agricultural activities resulted in the washing of phosphorus and nitrogen into the lake [1,6]. It is well documented that the lake contains toxic chemical substances, pesticides, other harmful material as well as large amounts of sediment [5,6]. Eventually, Lake Ladoga flows through River Neva near St-Petersburg into the Gulf of Finland [7]. The lake is the main source of clean water not only for local communities, but also for the city of St. Petersburg (over five million inhabitants), thus there is much pressure to maintain good water quality. Also, the importance of the lake for recreational use is steadily increasing [6,8].

Thus, continuous monitoring of the ecological status and water quality of this large lake is essential to understand and assess changes in its environment. Monitoring activities have been on-going for several decades [1,3,5,6]. In this paper, ecological status means the assessment of water eutrophication, through changes in phytoplankton levels. The use of phytoplankton biomass and chlorophyll-a (chl-a) concentration is a common method employed to study ecological status in water bodies [9,10]. This method is successful because chl-a correlates well with phytoplankton biomass abundance [9,11] and is used as an indicator for eutrophication [10]. Currently, low-cost instruments exist for measuring in situ chl-a in water bodies [12–14]. On the other hand, the use of these low-cost instruments still rely on the use of expensive research vessels and time consuming field surveys. However, the development of remote sensing and satellite-based imagery serves a wider audience [10,15]. Although estimates of chl-a based on remote sensing are generally less accurate than laboratory measurements, the approach provides a cost-effective and sufficiently reliable regional picture of a phenomenon that can be used for environmental monitoring and management needs. By using remote sensing data, the amount of chl-a can be calculated using empirical formulas based on the correlation between the violet-green light reflected from the water surface and the measured amounts of chl-a [16]. In addition, remote sensing methods provide historical time series of observations, for example, Landsat satellite data have been available since 1972 and MODIS data since 2000. Pozdnyakov et al. [1] have used SeaWiFS satellite images to study the biochemical properties of Ladoga from 1998–2004. Pozdnyakov et al. [1], were also able to validate the consistency of remote sensing derived chl-a with in situ chl-a measurements in Lake Ladoga. Studies using temporal remote sensing data to study the ecological status of Lake Ladoga are quite few and none have been done after 2010 (e.g., Pozdnyakov et al. [1]). The complex nature of coastal and lake water bodies [excessive colored dissolved organic matter (CDOM) and mineral particles] makes it difficult to use remote sensing to assess chl-a concentrations (so called Case II water bodies), thus many empirical, local, algorithms have been developed [1]. Developing different algorithms to suit each local water type around the globe is a dilemma. Thus, the main goal of this study was to use a global remote sensing data product that has considered the complex nature of different water types. In this study, we utilized the merged chlorophyll-a (chl-a) product derived from SeaWiFS, MERIS, MODIS Aqua, VIIR, and OLCI satellite sensors (1997–2019) to characterize the general trend of chl-a in Lake Ladoga. The novelty in our approach is to use geostatistical analytics of remote sensing data to produce a historical and synoptic picture of the state of Lake Ladoga. We anticipate that the results of this long-term assessment will contribute to the current information on the ecological status of Lake Ladoga. This additional information may be beneficial to on-going policy frameworks and management strategies of Lake Ladoga.

2. Materials and Methods

2.1. Study Area

Lake Ladoga is the largest lake in Europe having a surface area of 17,765 km^2 [17], with an average depth of 48.3 m [17] and maximum depth of 230 m (Figure 1) [6,17]. The lake has a volume of 858 km^3 [17], a water retention time of 12 years, and a coastline of 1570 km [17,18]. The lake has a total catchment area of 282,200 km^2, 20% of the area is located in Finland [6] (Figure 1b). The deepest parts of the lake are in the North (60–200 m), and the shallowest parts (10–50 m) in the southern parts of the basin [4]. The northern shores of the lake are rugged and rocky, while the western, southern, and eastern shores are shallow with fine-grained sediments [4]. The East and South contain extensive sandy beaches as well [4]. Water from Finland flows via the River Vuoksi from the west into Lake Ladoga while from the East, the water from Lake Onega drain via the River Svir [4]. From the South, the waters from Lake Ilmen flow into Ladoga via the River Volkhov [4]. Also, the River Syas and dozens of smaller streams and rivers flow into Ladoga as well [5]. Eventually, Lake Ladoga flows through River Neva near St-Petersburg into the Gulf of Finland [7].

Mean air temperature in the area is +3.2 °C [18,19]. The coldest month (February) has a mean temperature of −8.8 °C, while the warmest month (July) has +16.3 °C [19]. The mean annual precipitation in the area is about 475 mm [19].

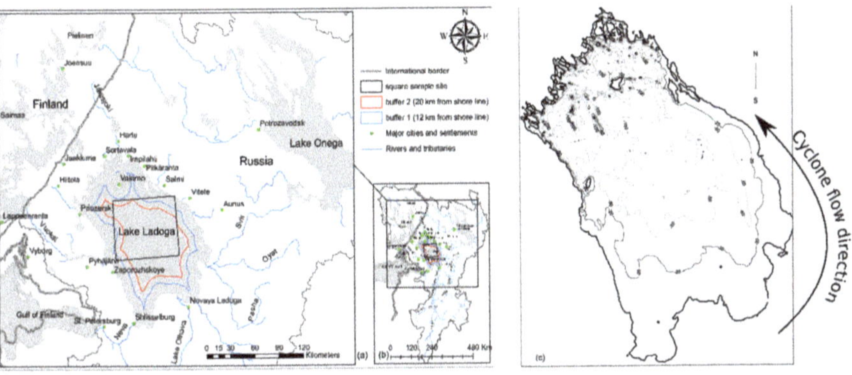

Figure 1. Shows Lake Ladoga, the study area. In (**a**), zoomed in with Ladoga in the center, Lake Onega (upper right), and Gulf of Finland (lower left corner), buffer 1, buffer 2, square are sample sites. In (**b**), Lake Ladoga drainage basin. In (**c**), is the bathymetric map of Lake Ladoga (image adapted and modified from Subetto et al. 1998 [20]).

Along Lake Ladoga's shores there are several cities which include: Sortavala, Priozersk, Shlisselburg, Novaya Ladoga, and Pitkäranta (Figure 1a), with a combined population of about 100,000 people. Also, there is a long history of industrial activity, mainly wood processing industries. Some closed pulp and paper mills were located near Priozersk, and Harlu, Läskelä. Currently, there are still functioning pulp mills in Pitkäranta in the North-East and near the town of Novaya Ladoga in Syasstroy (south). Lake Ladoga is inhabited by several endangered species, as the Ladoga Lake Salmon and the Ladoga Ringed Seal [18].

2.2. Data

We used the merged (4 km × 4 km) global monthly chlorophyll (chl-a) product (1997–2019) from GlobColour [21]. The merged chlorophyll-a OC5 algorithm product (CHL-OC5) is available from September 1997 to present and is updated annually [21]. The global monthly chl-a is a "Cloud Free" level four (L4) product created by merging data from four satellite sensors, the Sea-Viewing

Wide Field-of-View Sensor (SeaWiFS), Medium Resolution Imaging Spectrometer (MERIS), Moderate Resolution Imaging Spectrometer (MODIS), Visible Infrared Imaging Radiometer Suite (VIIRS), and Ocean and Land Color Instrument (OLCI) [21]. The GlobColour products were downloaded from the Copernicus Marine Environmental Monitoring Service (CMEMS) website: http://marine.copernicus.eu/services-portfolio/access-to-products/.

CMEMS CHL-OC5 dataset has been created using the OC5-algorithm which has been shown to work very well for Case II water bodies and has been validated using a large number of global in situ data from sea and lake areas [21,22]. We tested the dataset for lake Ladoga using data from Letanskaya and Protopopova [23], giving a reasonble fit, the CHL-OC5 dataset seemed to overestimate chl-a values in the lower range (below 5 mg m^{-3}), but for the higher ranges (over 15 mg m^{-3}) the data was consistent with the in situ values. As our main interst was in the trend of water chlorophyl-a content and not in absolute values, we used the CMEMS CHL-OC5 dataset in our study, with the notion that the values are not exact values, but relative indicators derived from a global dataset.

The datasets are for the months of June, July, August, September, and October. The downloaded chl-a data was processed further using the SeaWiFS Data Analysis System (SeaDAS) software [24]. SeaDAS is a free and opensource software developed and maintained by Ocean Biology Processing Group (OBPG) [24]. In SeaDAS, the land and water mask were created, and the no data layer added. The appropriate vector data was added, and the final images exported and compiled using Inkscape [25].

Additional processing of the chl-a was done with R software (R Core Team, 2020, [26], with raster package [27], to extract the mean, minimum and maximum values of chl-a. The results were then written to Excel format with the R package writexl [28]. Also, for each year (1997–2019), we calculated the (seasonal) mean of means for the five months (June, July, August, September, and October). Hence, in this study we used season to represent the calculated mean of the five months. We calculated the mean over the whole lake as a unit and within three sub-sample areas (buffer 1, buffer 1, and central part of the lake). Buffers 1 and 2 were twelve and twenty kilometers, respectively, from the shoreline (Figure 1). When calculating the season means, we excluded 1997 (only September) in the trend analysis to prevent bias in the result.

3. Results

3.1. General Trend of Chl-a Concentration in Lake Ladoga: 1997–2019

The general trend of seasonal chl-a concentration in Lake Ladoga is negative (Figure 2, wl). The negative trend is gradual (slope = −0.12416 and R^2 = 0.4069). This indicates a moderate improvement (decline in chl-a). In general, 2009 had the highest chl-a concentrations (>10 mg m^{-3}) while the lowest value was measured in 2017 (<5 mg m^{-3}). Similarly, the trend in the sub-sampled locations (s1, s2 and s3), was negative but non-significant ((buffer1: slope = −0.2962 and R^2 = 0.4212); (buffer2: slope = −0.3039 and R^2 = 0.3991); (center square: slope = −0.2974 and R^2 = 0. 433)). Water surface temperature (SST) had no observable trend in the lake during the study period.

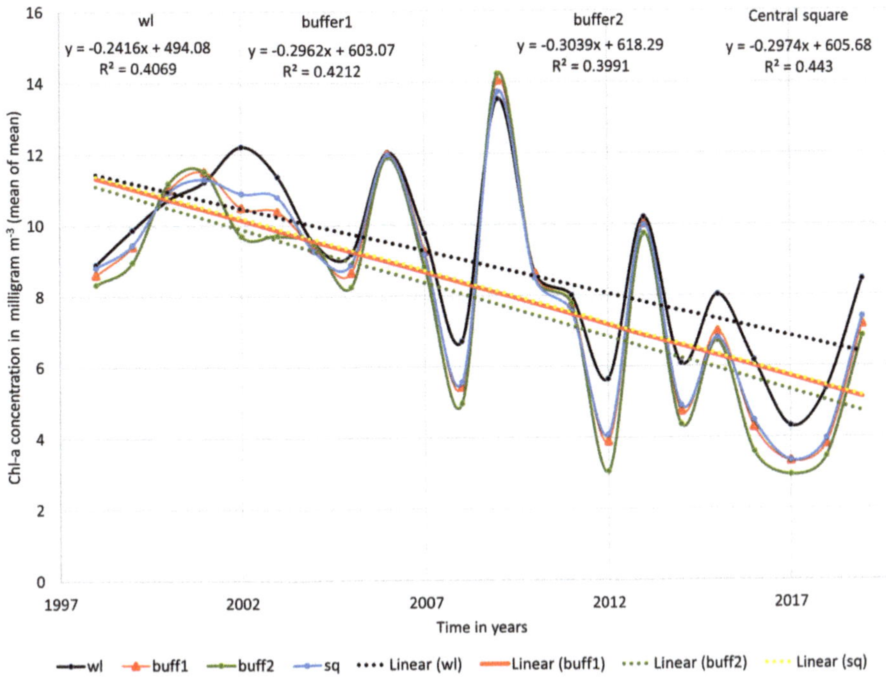

Figure 2. The average seasonal trend of chlorophyll-a concentration (mg m^{-3}) 1998–2019. wl = whole Ladoga lake, buffer1 = 12 km from shore line, buffer2 = 20 km from shore line, and central square = the sample site in the middle of the lake.

3.2. Spatial Distribution of Chl-a Concentration Across Lake Ladoga

Figures 3–7 show the pixel-wise variability of chl-a concentration across the lake. For June, wide distribution of chl-a concentration (moderate to high values) was observed for 1998, 2000, 2002, 2003, 2006 and 2009 (Figure 3). The high values were along the shallow southern shoreline and moderate values were within the central part.

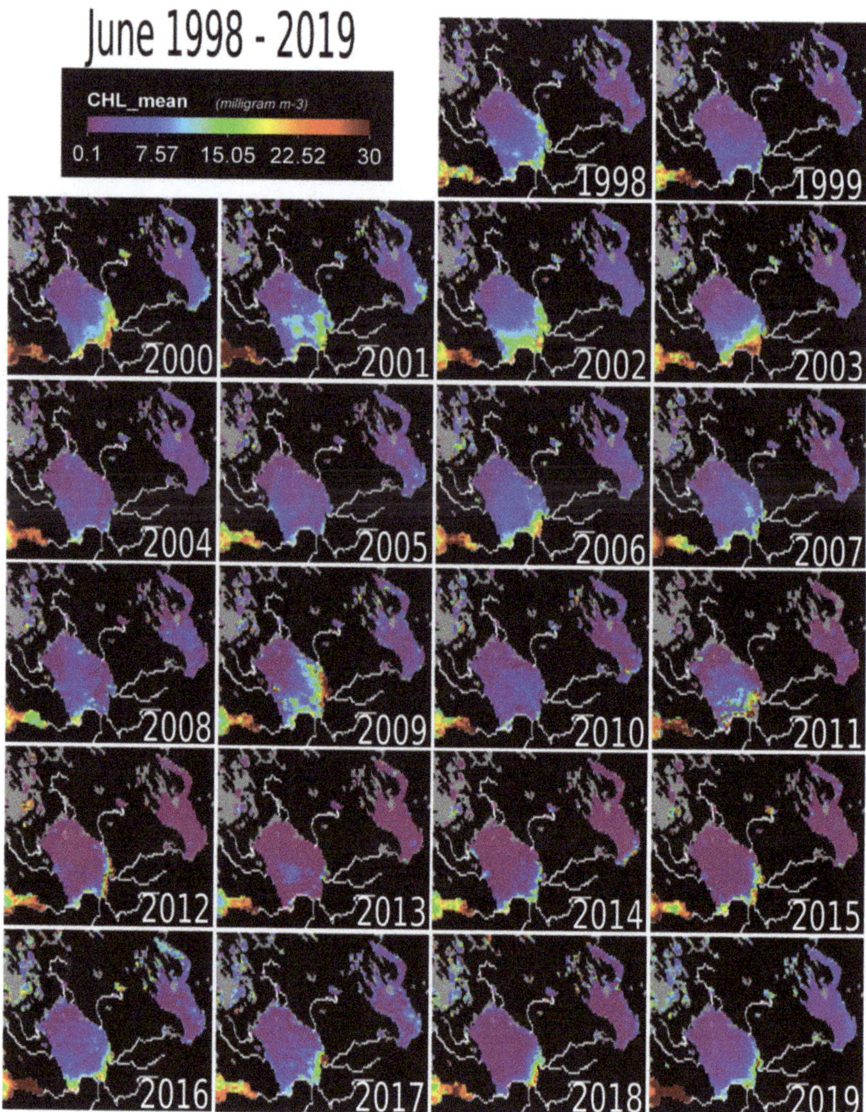

Figure 3. The map of chlorophyll-a concentration at Lake Ladoga for the month of June (1998–2019). Partially displayed as reference points are the Gulf of Finland (Southwest) and Lake Onega (Northeast).

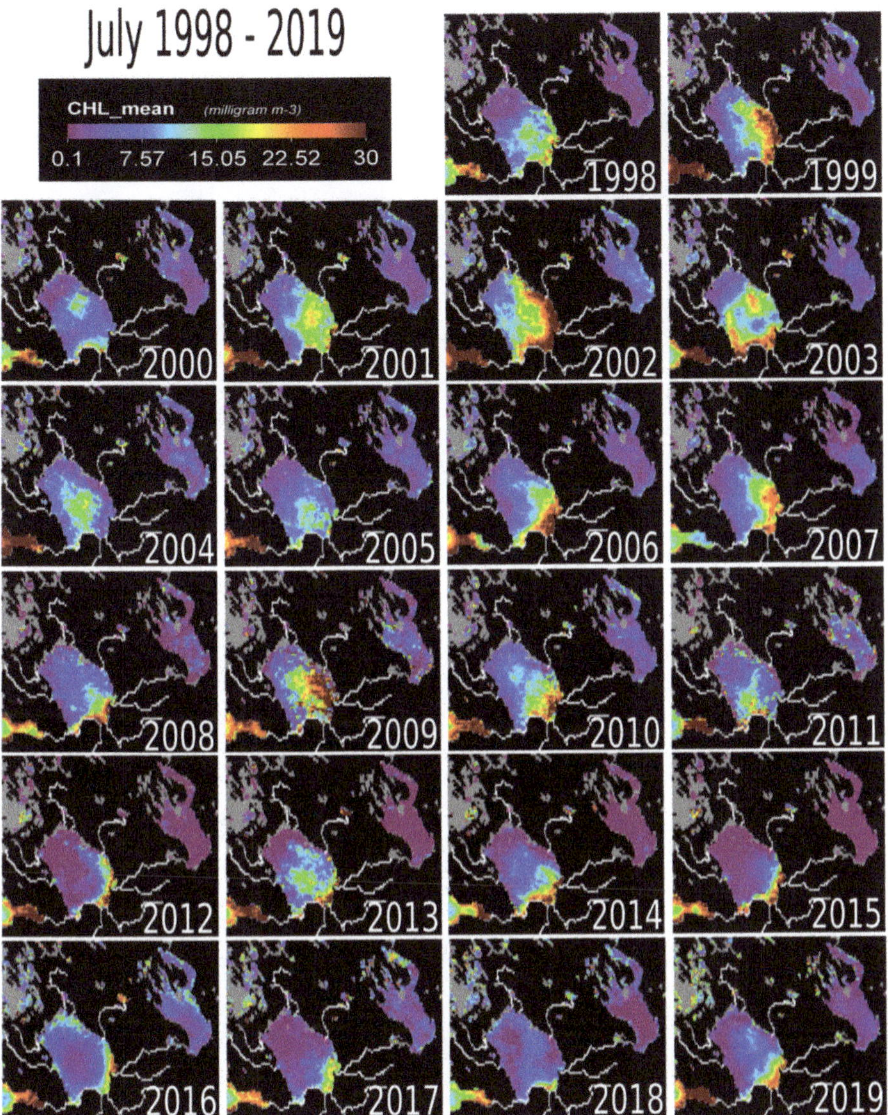

Figure 4. The map of chlorophyll-a concentration at Lake Ladoga for the month of July (1998–2019). Partially displayed as reference points are the Gulf of Finland (Southwest) and Lake Onega (Northeast).

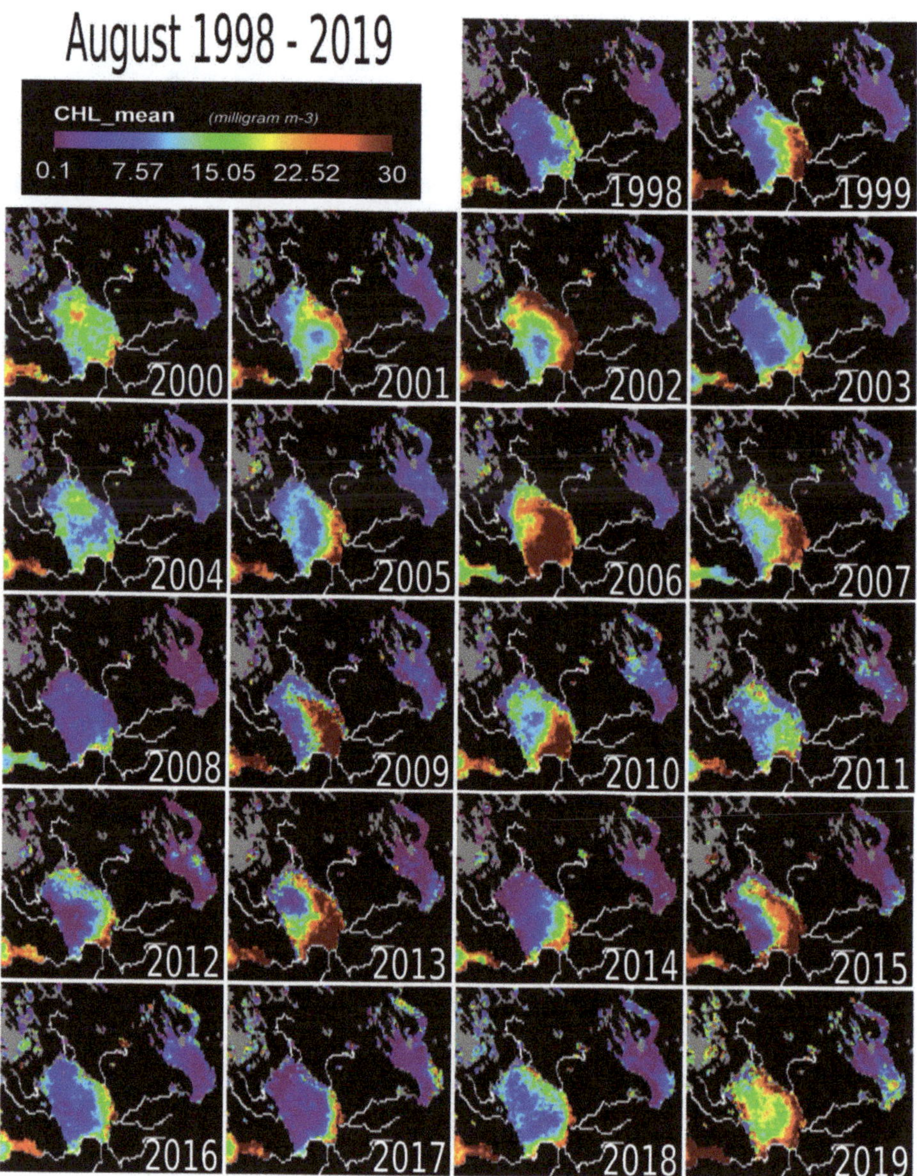

Figure 5. The map of chlorophyll-a concentration at Lake Ladoga for the month of August (1998–2019). Partially displayed as reference points are the Gulf of Finland (Southwest) and Lake Onega (Northeast).

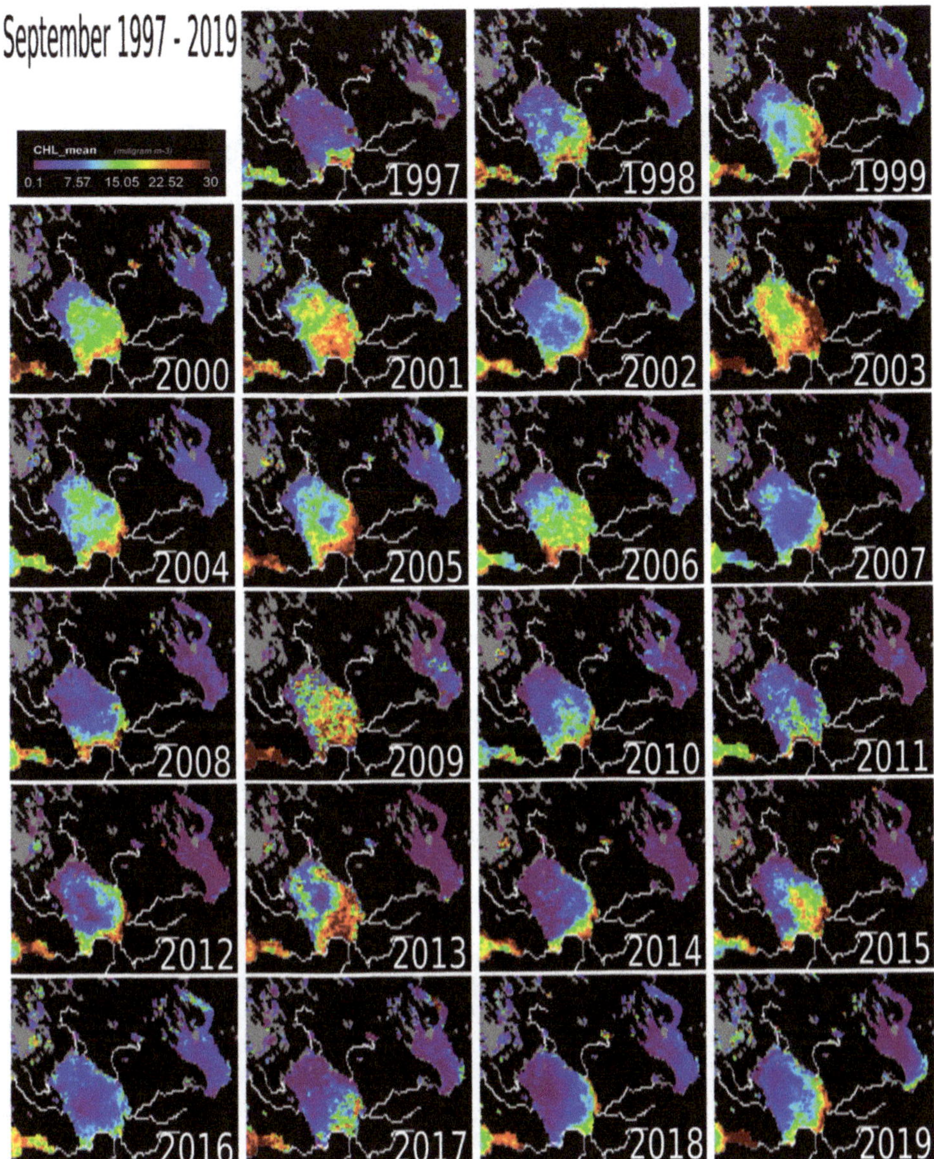

Figure 6. The map of chlorophyll-a concentration at Lake Ladoga for the month of September (1997–2019). Partially displayed as reference points are the Gulf of Finland (Southwest) and Lake Onega (Northeast).

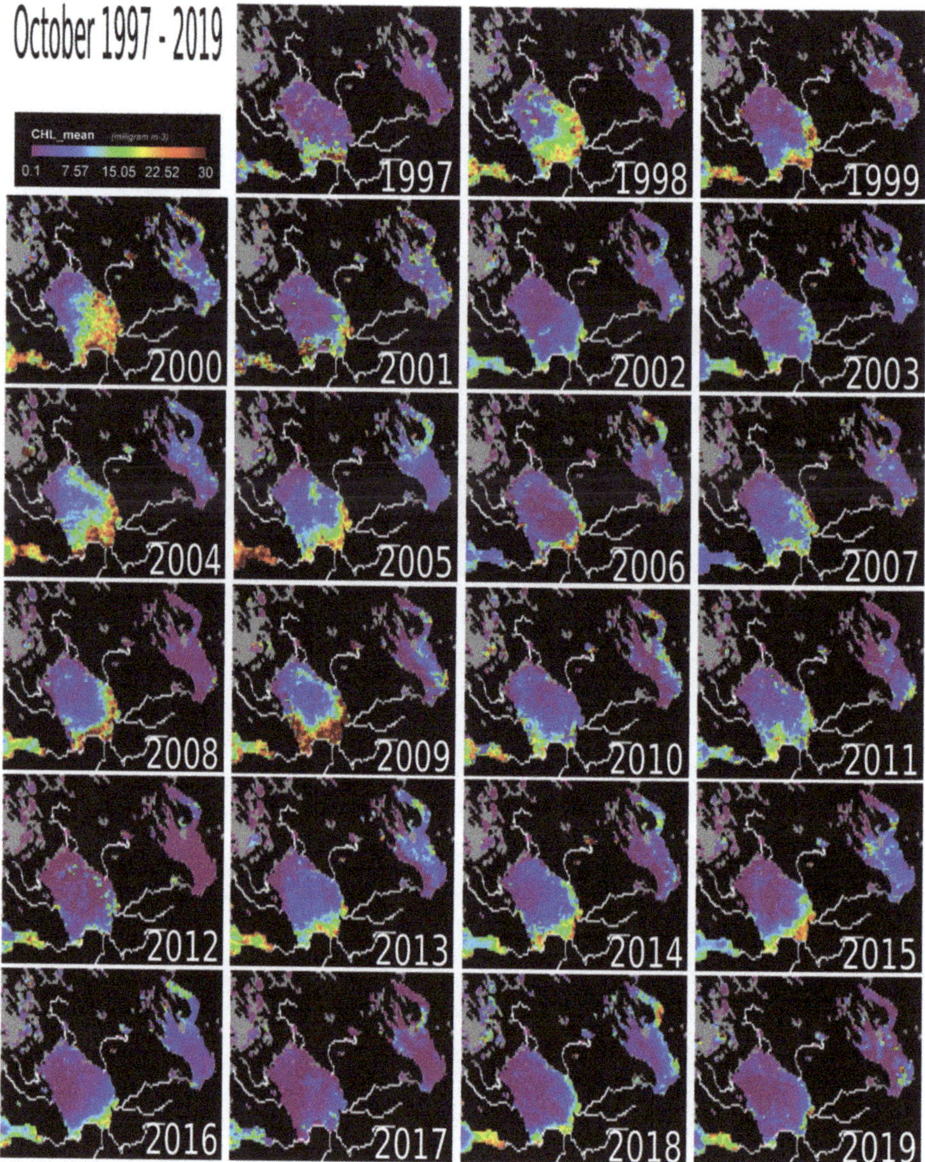

Figure 7. The map of chlorophyll-a concentration at Lake Ladoga for the month of October (1997–2019). Partially displayed as reference points are the Gulf of Finland (Southwest) and Lake Onega (Northeast).

In July, higher chl-a concentrations spread across the lake (Figure 4). During this month, high concentration values were measured in 2002, 2003, 2006, 2009, 2010. On the other hand, 2012 and 2015, appear almost free from high chl-a values except on the southern coast.

The highest chl-a values in August were measured in 2002 (20.70 mg m^{-3}), 2003 (21.61 mg m^{-3}), 2006 (25.34 mg m^{-3}), 2007 (18.82 mg m^{-3}), and 2019 (19.76 mg m^{-3}) (Figure 6). The year 2017 had very low values while 2019 had moderate to high values across the entire lake.

September–October marks the beginning of the winter cooling period causing a drop in chl-a concentrations (Figures 6 and 7). The entire lake still had detectable chl-a values in September, except in 2014, 2017 and 2018. The yearly temporal and spatial distribution of water temperature across Lake Ladoga is driven mainly by the thermal bar phenomenon [1,3], which is even noticeable from satellite images [1]. The thermal bar is a common characteristic of large lakes [1]. In Lake Ladoga, the thermal bar is formed in spring and early summer [1,3]. The thermal bar is the column of water formed between the cold (deep central) and warmer (coastal) waters of a lake [17]. It is the location of vigorous water mixing caused by downwelling of water with a temperature of four degrees Celsius [3]. This is the result of the physical property of water, that has its maximum density at +4 °C, thus heavier rather than colder or warmer water, causes it to sink. The thermal bar effectively acts as a wall between the coastal and inner water masses of a lake, inhibiting mixing of the two water masses. This, in turn, only allows for near shore cyclonic currents caused by the prevalent winds and the Coriolis effect. Consequently, there is a bloom in phytoplankton biomass (chl-a producer) along the coast. The thermal bar disappears around the end of June and beginning of July when the lake's water temperature is homogeneously above +4 °C, usually up to the depth of 50–70 m [1,3]. As the thermal bar disappears, temperature no longer plays a major role in chl-a concentrations in the lake [1,3]. The cyclonic current transports the phytoplankton to the north of Lake Ladoga. It is important to note that during August, the effect of the thermal bar has completely disappeared [1]. Before the disappearance of the thermal bar, the water flow and movement of phytoplankton is cyclonic (counterclockwise) in nature (South-East to North) [3], mainly along the shoreline (Figure 1c). The cyclonic nature of the water movement is driven by the topography, bathymetry, selective westerly winds [1], and the rotation of the earth (i.e., the so-called Coriolis phenomenon). There are also anticyclonic eddy currents in Lake Ladoga [29,30]. In general, anti-cyclonic currents are caused by weak bottom friction and southerly winds [29–31]. Most of the observed low chl-a concentrations are in the north and central parts of the lake while the southern coastal areas remain unchanged. From our results and in-depth literature review, the following factors were identified to be driving the on-going chl-a changes taking place in Lake Ladoga: (i) Temperature, (ii) impact from industry, external load, and littoral settlements. These coastal areas are locations of high industrial activity (e.g., paper mills), settlements and from external sources [1,3].

4. Discussion

As mentioned in Section 3.1 (first paragraph), we observed a gradual and negative trend in chl-a distribution, indicating a moderate improvement of water quality.

4.1. Effect of Temperature and Eutrophication on Chl-a Concentrations

During the month of June, high chl-a concentrations are observed only along the southern–southwestern coastal areas while the deeper parts of the lake remain clean (Figure 3), due to the presence of the thermal bar. These coastal areas have major influxes of nutrients from rivers like the Syas, Volkov and Svir, carrying municipal, industrial and agricultural waste [1,3].

The action of the southern summer winds causing anti-cyclonic eddies moves and distributes the phytoplankton mass towards the central part of the lake [1]. The temperature along the shallow coastal areas of the lake is now homogeneous [1]. Consequently, eutrophication by nutrients originating from bottom sediments could possibly be a factor contributing to chl-a concentrations [1,3]. These eutrophic sediments are thought to originate from high nutrient fall-out between the 1950s and 1990s, the era of intense eutrophication [32]. However, the present role of eutrophication of bottom sediments is arguably not a major factor [1]. Previous studies analyzing water samples from Lake Ladoga revealed that the coastal areas in the North, Southwest, and South were very eutrophic [33]. Also, the southern part of the basin has high concentrations of dissolved organic matter (DOM), primarily from external water load, excretion from green vegetation and decay of macrophytes [1].

The cooling cycle in Lake Ladoga starts in September, causing reduction in phytoplankton biomass and consequently a decline in chl-a values. An important aspect to consider here is the impact of climate change on Lake Ladoga. In general, the last two decades have been exceptionally warm, in fact, the warmest period in the entire history of climate measurement. Sharov et al. [34], reported a temperature increase of +1.5 degrees Celsius between 1950 and 2010 in Lake Ladoga [34]. Likewise, years 2000 and 2015 were exceptionally warm [18]. Consequently, this has increased the duration of the lake's ice-free period from 210 days to 230 days [34–36]. A recent study by Karetnikov et al. [17], also reported increases in temperature giving rise to shorter ice periods. As a result, the increase in available sunshine (less ice) and higher surface water temperatures promote the growth of algae. A similar observation (rising water surface temperature and fewer ice days) have been reported in Lake Peipsi [37]. The impact of climate change on temperature increase promoting algal growth has also been reported in other regions as well [38,39].

4.2. Impact from Industries, Coastal Settlements, and External Load

The decline in chl-a (an indicator of improved water quality) is largely due to the closing of paper mills that led to less wastewater being discharged into Lake Ladoga [40]. High chl-a areas have been observed around functioning paper mills and coastal cities, these sources discharge a considerable amount of nutrient rich wastewater [5]. The closure of several mills led to a significant reduction in primary phytoplankton (chl-a) production and phosphorus concentrations [5]. The anthropogenic activities of the high-density settlements around Lake Ladoga could also be sources for nutrients driving algal blooms [41]. Also, by analyzing comprehensive water samples from Lake Ladoga, Holopainen et. al. [33], found parts of the lake to be eutrophic. They found high chl-a concentrations near Sortavala Bay (26 mg/m^3 in August) and Pitkäranta (functioning paper mill) in the northern shores. High chl-a values were also in the southwest corner of Ladoga, in front of Zaporozhskoye (Metsäpirtti) (11.9 mg/m^3) and in front of the city of Novaya Ladoga (20–25 mg/m^3). The paper and pulp mills in Priozersk (Käkisalmi), Harlu and Läskelä were already established during the Finnish period (before 1945) and continued to operate during the Soviet and Russian era until their closure in the late 1980s. The Pitkäranta pulp mill was also established as early as the 1920s, and its production continues today. An active pulp mill is located near the town of Novaya Ladoga, on the south coast of Ladoga. Here a major accident occurred in 1998 when 700,000 cubic meters of toxic sludge spilled into the Syas River, about a kilometer from the shores of Lake Ladoga [5]. Phosphorus and nitrogen emissions from agriculture have also been a major nutrient source, but agriculture has declined since 1990.

External load from the large catchment area is still a major factor keeping chl-a levels elevated [6] (Figure 1b). The effect of the Volkov Bay (southern Lake Ladoga) is also obvious (Figures 3–7), as the Volkov River is the largest external load on Ladoga. For example, high phosphorus concentrations of 210 µg/L were found in the 1980s as compared to 46 µg/L in the 1950s and 1960s [5]. The sources of these phosphorus values were from a large number of industrial plants (594) and agricultural enterprises (680) in the watershed [42,43]. However, due to the closure of factories in the 1990s [5,40] phosphorus concentrations in the Volkhov Bay were found to have dropped to 120 µg/l from 210 µg/l [5]. Especially the closure of paper mills has been a contributing factor to the improved water quality of Lake Ladoga [1–3,5].

A similar trend of water pollution has been reported at the pulp and paper mills located at the shores of Lakes Onega and Imandra [44–46]. Both of these locations (Kondopoga Bay and Imandra Bolshaya Bay), have seen increased nutrient loads, other toxic chemicals and degraded water quality [45]. However, in comparison to Ladoga, Onega and Imandra, are considerably cleaner (Figures 3–7). To put this into perspective, the chl-a values of Lake Onega (1999–2010) in Lahti, Petrozavodsk varied between 1 and 7 mg/L without a clear trend (with an average of 2–3 mg/L), clearly indicating low pollution levels [34]. Also, Onega's burden is only one urban center, the city of Petrozavodsk. On the other hand, recents studies in Onega have indicated significant water browning especially around Petrozavodsk Bay [47].

The pollution and degradation of water quality from wastewater discharged from pulp and paper mills, urban settlements and agricultural activities have been documented in other countries as well.

For example, in Finland's pulp and papers mills located along Lake Päijänne polluted the lake's water, thus degrading the water quality [48].

On the other hand, it is well documented that, better waste management practices by paper mills, settlements and agricultural activities have improved the water quality of rivers and lakes [49,50]. For example, the Stora Enso Veitsiluoto Mills at Kemi, northern Finland [51,52] and the Kaukas paper mill on southern Lake Saimaa have been able to improve local water quality through better water management processes [53]. It is worthy to note that the closing of paper mills was a major factor to improved water quality in the absence of better waste management practices, e.g., [51,52,54].

5. Conclusions

This study assessed the trophic state of Lake Ladoga during the last 23 years (1997–2019). Geostatistical tools were used to analyze remote sensing data for this purpose. Our analysis reveals a slight decline in chl-a, suggesting there is a moderate improvement in the state of the lake. This study observed that the southernmost and shallowest part of the water body was the most problematic with high observable concentrations of chl-a. This southern area receives nutrients both from the Syas catchment and from bottom sediments causing high phytoplankton growth. Observable differences were seen in the deeper parts of Lake Ladoga in the northern part of the basin. In this northern part, reduced chl-a concentrations were observed, although there were local differences, especially on the north-eastern shore off the Pitkäranta factory site. Also, on the northern shore of Sortavala, there were exceptions as well which could be due to the discharge of municipal sewage.

It is important to note that reducing the nutrient load in Lake Ladoga due to anthropogenic activities is a slow and complicated process. These processes include effects of climate change, mainly warming and ice reduction, both of which contribute to eutrophication of the lake. The impact of climate change on rainfall is not certain, but it is generally assumed that rainfall will increase, thus, increasing the external nutrient load of Lake Ladoga. Furthermore, deforestation in the catchment area also increases nutrient leaching from the soil, and it is therefore uncertain whether the current good trend in the decrease of chl-a will continue. Despite the slight improve of the state of Lake Ladoga reported in this study, it appears that primary production is still relatively high. Our conclusion is in line with previous studies of the lake [6].

In this study, after the visual assessment of satellite imageries and an in-depth literature review, we are of the opinion that the current nutrient load and chemical waste influx in Lake Ladoga is less now, as compared to 23 years ago. We are of the opinion that to further decrease chl-a concentrations in Lake Ladoga, sewage from municipalities should be treated before entering the lake, agricultural practices have to be adjusted to reduce washing of fertilizers, the nutrient load from fish farming should be reduced and traditional fishing should be increased. These measures could potentially improve the water quality of the lake.

Author Contributions: Conceptualization, A.-M.G.G. and A.C.; methodology, A.-M.G.G. and A.C.; software, A.-M.G.G. and A.C.; formal analysis, A.-M.G.G.; investigation, A.-M.G.G. and A.C.; resources, A.C.; data curation, A.-M.G.G. and A.C.; writing—original draft preparation, A.-M.G.G.; writing—review and editing, A.-M.G.G. and A.C.; visualization, A.-M.G.G.; supervision, A.C.; project administration, A.C.; funding acquisition, A.C. All authors have read and agreed to the published version of the manuscript.

Funding: This research was part of the Kone foundation funded Green Belt History project.

Acknowledgments: We thank the University of Eastern Finland for material assistance, and two anonymous reviewers for their valuable comments.

References

1. Pozdnyakov, D.V.; Korosov, A.A.; Petrova, N.A.; Grassl, H. Multi-year satellite observations of Lake Ladoga's biogeochemical dynamics in relation to the lake's trophic status. *J. Great Lakes Res.* **2013**, *39*, 34–45.

2. Moiseenko, T.I.; Sharov, A.N. The retrospective analysis of aquatic ecosystem modification of Russian large lakes under antropogenic impacts. In *Ecotoxicology around the Globe*; Nova Science Publishers, Inc.: Hauppauge, NY, USA, 2011; pp. 309–324.

3. Rukhovets, L.; Filatov, N. *Ladoga and Onego—Great European Lakes*; Springer: Berlin/Heidelberg, Germany, 2010; ISBN 9783540681441.

4. Rumyantsev, V.; Viljanen, M.; Slepukhina, T. The present state of Lake Ladoga, Russia—A review. *Boreal Environ. Res.* **1999**, *4*, 201–214.

5. Naumenko, M.A.; Avinsky, V.A.; Barbashova, M.A.; Guzivaty, V.V.; Karetnikov, S.G.; Kapustina, L.L.; Letanskaya, G.I.; Raspletina, G.F.; Raspopov, I.M.; Rychkova, M.A.; et al. Current ecological state of the Volkhov Bay of the Ladoga Lake. *Ecol. Chem.* **2000**, *9*, 75–87.

6. Kondratyev, S.; Gronskaya, T.; Ignatieva, N.; Blinova, I.; Telesh, I.; Yefremova, L. Assessment of present state of water resources of Lake Ladoga and its drainage basin using Sustainable Development indicators. *Ecol. Indic.* **2002**, *2*, 79–92.

7. Malachovskij, D.B.; Delusin, I.V.; Gej, N.A.; Dginoridzse, R.N. Evidence from the Neva River Valley, Russia, of the Holocene history of Lake Ladoga. *Fennia* **1996**, *174*, 113–123.

8. Sevastiyanov, D.V.; Colpaert, A.; Korostelyov, E.; Mulyava, O.; Shitova, L. Management of tourism and recreation possibilities for the sustainable development of the north-western border region in Russia. *Nord. Geogr. Publ.* **2014**, *43*, 27–38.

9. Harvey, E.T.; Kratzer, S.; Philipson, P. Satellite-based water quality monitoring for improved spatial and temporal retrieval of chlorophyll-a in coastal waters. *Remote Sens. Environ.* **2015**, *158*, 417–430.

10. Pozdnyakov, D.V.; Johannessen, O.M.; Korosov, A.A.; Pettersson, L.H.; Grassl, H.; Miles, M.W. Satellite evidence of ecosystem changes in the White Sea: A semi-enclosed arctic marginal shelf sea. *Geophys. Res. Lett.* **2007**, *34*, 1–4.

11. Marcelli, M.; Piermattei, V.; Madonia, A.; Mainardi, U. Design and application of new low-cost instruments for marine environmental research. *Sensors* **2014**, *14*, 23348–23364.

12. Piermattei, V.; Madonia, A.; Bonamano, S.; Martellucci, R.; Bruzzone, G.; Ferretti, R.; Odetti, A.; Azzaro, M.; Zappalà, G.; Marcelli, M. Cost-effective technologies to study the arctic ocean environment. *Sensors* **2018**, *18*, 2257.

13. Xing, X.; Morel, A.; Claustre, H.; Antoine, D.; D'Ortenzio, F.; Poteau, A.; Mignot, A. Combined processing and mutual interpretation of radiometry and fluorimetry from autonomous profiling Bio-Argo floats: Chlorophyll a retrieval. *J. Geophys. Res. Ocean.* **2011**, *116*, 1–14.

14. Gómez, J.A.D.; Alonso, C.A.; García, A.A. Remote sensing as a tool for monitoring water quality parameters for Mediterranean Lakes of European Union water framework directive (WFD) and as a system of surveillance of cyanobacterial harmful algae blooms (SCyanoHABs). *Environ. Monit. Assess.* **2011**, *181*, 317–334. [PubMed]

15. Hu, C.; Lee, Z.; Franz, B. Chlorophyll a algorithms for oligotrophic oceans: A novel approach based on three-band reflectance difference. *J. Geophys. Res. Ocean.* **2012**, *117*, 1–25.

16. Karetnikov, S.; Leppäranta, M.; Montonen, A. A time series of over 100 years of ice seasons on Lake Ladoga. *J. Great Lakes Res.* **2017**, *43*, 979–988.

17. Karetnikov, S.; Naumenko, M. Lake Ladoga ice phenology: Mean condition and extremes during the last 65 years. *Hydrol. Process.* **2011**, *25*, 2859–2867.

18. Sagitov, R.; Zavarzin, A.; Ieshko, E.; Pogrebov, V.; Baranov, B.; Fokin, Y.; Ussenkov, S.; Kurashov, E.; Kiyko, O.; Vasilevich, V.; et al. Ladoga. Climate. Available online: http://ladoga.krc.karelia.ru/environ/climate/index.shtml (accessed on 23 September 2020).

19. Subetto, D.A.; Davydova, N.N.; Rybalko, A.E. Contribution to the lithostratigraphy and history of Lake Ladoga. *Palaeogeogr. Palaeoclim. Palaeoecol.* **1998**, *140*, 113–119.

20. Garnesson, P.; Mangin, A.; D'Andon, O.F.; Demaria, J.; Bretagnon, M. The CMEMS GlobColour chlorophyll a product based on satellite observation: Multi-sensor merging and flagging strategies. *Ocean Sci.* **2019**, *15*, 819–830.

21. GC-PL-NIVA-FVR-01. *ESA DUE GlobColour Global Ocean Colour for Carbon Cycle Research Full Validation Report*. 14 December 2007. Available online: https://www.yumpu.com/en/document/read/5924619/validation-report-globcolour-project (accessed on 20 November 2020).

22. Letanskaya, G.I.; Protopopova, E.V. The current state of phytoplankton in Lake Ladoga (2005–2009). *Inl. Water Biol.* **2012**, *5*, 310–316.

23. OBPG Ocean Biology Processing Group (OBPG). SeaDAS 7.5.3 Science Software, NASA Goddard Space Flight Center, Ocean Ecology Laboratory. 2019. Available online: https://seadas.gsfc.nasa.gov/about/ (accessed on 26 September 2019).

24. Inkscape: An Open-Source Vector Graphics Editor. Available online: https://inkscape.org/2019 (accessed on 23 March 2019).

25. *R Core Team 2020: A Language and Environment for Statistical Computing*; R Foundation for Statistical Computing: Vienna, Austria, 2020; Available online: http//www.R-project.org/ (accessed on 11 September 2020).

26. Robert, J. Hijmans Raster: Geographic Data Analysis and Modeling. R Package Version 3.0-12. 2020. 2020. Available online: https://rdrr.io/cran/raster/ (accessed on 11 September 2020).

27. Ooms, J. Writexl: Export Data Frames to Excel "xlsx" Format. R Package Version 1.2 2019. 2019. Available online: https://CRAN.R-project.org/package=writexl (accessed on 11 September 2020).

28. Naumenko, M.A.; Karetnikov, S.G.; Tikhomirov, A.I. Main features of the thermal regime of Lake Ladoga during the ice-free period. *Hydrobiologia* **1996**, *322*, 69–73.

29. Naumenko, M.; Karetnikov, S.; Guzivaty, V. Thermal regime of Lake Ladoga as a typical dimictic lake. **2007**, *7*, 63–70.

30. Beletsky, D.; Saylor, J.H.; Schwab, D.J. Mean circulation in the Great Lakes. *J. Great Lakes Res.* **1999**, *25*, 78–93. [CrossRef]

31. Natalia, V. Ignatieva Distribution and release of sedimentary phosphorus in Lake Ladoga. In *A Case Approach to Perioperative Drug-Drug Interactions*; Springer: Berlin/Heidelberg, Germany, 1996; Volume 322, pp. 129–136.

32. Holopainen, A.L.; Huttunen, P.; Letanskaya, G.I.; Protopopova, E.V. The trophic state of Lake Ladoga as indicated by late summer phytoplankton. *Hydrobiologia* **1996**, *322*, 9–16. [CrossRef]

33. Sharov, A.N.; Berezina, N.A.; Nazarova, L.E.; Poliakova, T.N.; Chekryzheva, T.A. Links between biota and climate-related variables in the Baltic region using Lake Onega as an example. *Oceanologia* **2014**, *56*, 291–306. [CrossRef]

34. Filazzola, A.; Blagrave, K.; Imrit, M.A.; Sharma, S. Climate Change Drives Increases in Extreme Events for Lake Ice in the Northern Hemisphere. *Geophys. Res. Lett.* **2020**, *47*, e2020GL089608. [CrossRef]

35. Sharma, S.; Blagrave, K.; Magnuson, J.J.; O'Reilly, C.M.; Oliver, S.; Batt, R.D.; Magee, M.R.; Straile, D.; Weyhenmeyer, G.A.; Winslow, L.; et al. Widespread loss of lake ice around the Northern Hemisphere in a warming world. *Nat. Clim. Chang.* **2019**, *9*, 227–231. [CrossRef]

36. Öğlü, B.; Möls, T.; Kaart, T.; Cremona, F.; Kangur, K. Parameterization of surface water temperature and long-term trends in Europe's fourth largest lake shows recent and rapid warming in winter. *Limnologica* **2020**, *82*. [CrossRef]

37. Šorf, M.; Davidson, T.A.; Brucet, S.; Menezes, R.F.; Søndergaard, M.; Lauridsen, T.L.; Landkildehus, F.; Liboriussen, L.; Jeppesen, E. Zooplankton response to climate warming: A mesocosm experiment at contrasting temperatures and nutrient levels. *Hydrobiologia* **2015**, *742*, 185–203. [CrossRef]

38. Li, J.; Tian, L.; Song, Q.; Sun, Z.; Yu, H.; Xing, Q. Temporal variation of chlorophyll-a concentrations in highly dynamicwaters from unattended sensors and remote sensing observations. *Sensors* **2018**, *18*, 2699. [CrossRef]

39. Feng, J.F.; Zhu, L. Changing trends and relationship between global ocean chlorophyll and sea surface temperature. *Procedia Environ. Sci.* **2012**, *13*, 626–631. [CrossRef]

40. Letanskaya, G.I. Phytoplankton monitoring of Lake Ladoga. In Proceedings of the Third International Lake Ladoga Symposium, Joensuu, Finland, 23–27 August 1999; pp. 114–121.

41. Isachenko, G.A. Lake Ladoga Region: Human impacts and recent environmental changes. *First Int. Lake Ladoga Symp.* **1996**, *322*, 217–221.

42. Rusanov, A.G.; Stanislavskaya, E.V.; Ács, É. Periphytic algal assemblages along environmental gradients in the rivers of the Lake Ladoga basin, Northwestern Russia: Implication for the water quality assessment. *Hydrobiologia* **2012**, *695*, 305–327. [CrossRef]

43. Drabkova, V.G.; Rumyantsev, V.A.; Sergeeva, L.V.; Slepukhina, T.D. Ecological problems of Lake Ladoga: Causes and solutions. *First Int. Lake Ladoga Symp.* **1996**, *322*, 1–7.

44. Moiseenko, T.; Sharov, A.; Voinov, A.; Shalabodov, A. Long-Term Changes in the Large Lake Ecosystems Under Pollution: The Case of the North-East European Lakes. *Geogr. Environ. Sustain.* **2012**, *5*, 67–83. [CrossRef]

45. Moiseenko, T.; Sharov, A. Large Russian lakes ladoga, onega, and imandra under strong pollution and in the period of revitalization: A review. *Geoscience* **2019**, *9*, 492. [CrossRef]

46. Moiseenko, T.I.; Sharov, A.N.; Vandish, O.I.; Kudryavtseva, L.P.; Gashkina, N.A.; Rose, C. Long-term modification of Arctic lake ecosystems: Reference condition, degradation under toxic impacts and recovery (case study Imandra Lakes, Russia). *Limnologica* **2009**, *39*, 1–13. [CrossRef]

47. Kalinkina, N.; Tekanova, E.; Korosov, A.; Zobkov, M.; Ryzhakov, A. What is the extent of water brownification in Lake Onego, Russia? *J. Great Lakes Res.* **2020**. [CrossRef]

48. Meriläinen, J.J.; Hynynen, J.; Palomäki, A.; Veijola, H.; Witick, A.; Mäntykoski, K.; Granberg, K.; Lehtinen, K. Pulp and paper mill pollution and subsequent ecosystem recovery of a large boreal lake in Finland: A palaeolimnological analysis. *J. Paleolimnol.* **2001**, *26*, 11–35. [CrossRef]

49. Soszka, H.; Gołub, M.; Kolada, A.; Cydzik, D. Chlorophyll a based assessment of Polish lakes. *Verh. Internet. Verein. Limonol.* **2008**, *30*, 416–418. [CrossRef]

50. Räike, A.; Pietiläinen, O.P.; Rekolainen, S.; Kauppila, P.; Pitkänen, H.; Niemi, J.; Raateland, A.; Vuorenmaa, J. Trends of phosphorus, nitrogen and chlorophyll a concentrations in Finnish rivers and lakes in 1975–2000. *Sci. Total Environ.* **2003**, *310*, 47–59. [CrossRef]

51. Pöykiö, R.; Nurmesniemi, H.; Kivilinna, V.A. EOX concentrations in sediment in the part of the Bothnian Bay affected by effluents from the pulp and paper mills at Kemi, Northern Finland. *Environ. Monit. Assess.* **2008**, *139*, 183–194. [CrossRef]

52. Nurmesniemi, H.; Pöykiö, R.; Keiski, R.L. A case study of waste management at the Northern Finnish pulp and paper mill complex of Stora Enso Veitsiluoto Mills. *Waste Manag.* **2007**, *27*, 1939–1948. [CrossRef] [PubMed]

53. Kaplin, C.; Hemming, J.; Holmbom, B. Improved water quality by process renewal in a pulp and paper mill. *Boreal Environ. Res.* **1997**, *2*, 239–246.

54. Pöykiö, R.; Taskila, E.; Perämämäki, P.; Nurmesniemi, H.; Kivililinna, V.-A.; Kuokkanen, T.; Virta, P. Sediment, Perch (Perca fluviatilis L.) and bottom fauna as indicators of effluent discharged from the pulp and paper mill complex at Kemi, northern Finland. *Water Air Soil Pollut.* **2004**, *158*, 325–343. [CrossRef]

Publisher's Note: MDPI stays neutral with regard to jurisdictional claims in published maps and institutional affiliations.

Article

Generating Time-Series LAI Estimates of Maize Using Combined Methods Based on Multispectral UAV Observations and WOFOST Model

Zhiqiang Cheng [1,2,3], Jihua Meng [2,*], Jiali Shang [4], Jiangui Liu [4], Jianxi Huang [5], Yanyou Qiao [2], Budong Qian [4], Qi Jing [4], Taifeng Dong [4] and Lihong Yu [6]

[1] Institute of Geography, Fujian Normal University, Fuzhou 350007, China; chengzq@fjnu.edu.cn
[2] Key Laboratory of Digital Earth, Aerospace Information Research Institute, Chinese Academy of Sciences, Beijing 100101, China; qiaoyy@radi.ac.cn
[3] Key Laboratory of Humid Subtropical Eco-Geographical Process (Fujian Normal University), Ministry of Education, Fuzhou 350007, China
[4] Ottawa Research and Development Centre, Agriculture and Agri-Food Canada, Ottawa, ON K1A 0C6, Canada; Jiali.Shang@canada.ca (J.S.); Jiangui.Liu@canada.ca (J.L.); Budong.Qian@canada.ca (B.Q.); Qi.Jing@canada.ca (Q.J.); dong.taifeng@gmail.com (T.D.)
[5] College of Land Science and Technology, China Agricultural University, Beijing 100094, China; jxhuang@cau.edu.cn
[6] Faculty of Geographical Science, Beijing Normal University, Beijing 100875, China; Yulh@mail.bnu.edu.cn
*   Correspondence: mengjh@radi.ac.cn; Tel.: +86-010-6486-9473

Received: 24 August 2020; Accepted: 20 October 2020; Published: 23 October 2020

Abstract: Green leaf area index (LAI) is an important variable related to crop growth. Accurate and timely information on LAI is essential for developing suitable field management strategies to mitigate risk and boost yield. Several remote sensing (RS) based methods have been recently developed to estimate LAI at the regional scale. However, the performance of these methods tends to be affected by the quality of RS data, especially when time-series LAI are required. For crop LAI estimation, supplementary growth information from crop model is helpful to address this issue. In this study, we focus on the regional-scale LAI estimations of spring maize for the entire growth season. Using time-series multispectral RS data acquired by an unmanned aerial vehicle (UAV) and the World Food Studies (WOFOST) crop model, three methods were applied at different crop growth stages: empirical method using vegetation index (VI), data assimilation method and hybrid method. The VI-based method and assimilation method were used to generate time-series LAI estimations for the whole crop growth season. Then, a hybrid method specially for the late-stage LAI retrieval was developed by integrating WOFOST model and data assimilation. Using field-collected LAI data in Hongxing Farm in 2014, the performances of these three methods were evaluated. At the early stage, the VI-based method ($R^2 = 0.63$, RMSE = 0.16, $n = 36$) achieved higher accuracy than the assimilation method ($R^2 = 0.54$, RMSE = 0.52, $n = 36$), whereas at the mid stage, the assimilation method ($R^2 = 0.63$, RMSE = 0.46, $n = 28$) showed higher accuracy than the VI-based method ($R^2 = 0.41$, RMSE = 0.51, $n = 28$). At the late stage, the hybrid method yielded the highest accuracy ($R^2 = 0.63$, RMSE = 0.46, $n = 29$), compared with the VI-based method ($R^2 = 0.19$, RMSE = 0.43, $n = 28$) and the assimilation method ($R^2 = 0.20$, RMSE = 0.44, $n = 29$). Based on the results above, we considered a combination of the three methods, i.e., the VI-based method for the early stage, the assimilation method for the mid stage, and the hybrid method for the late stage, as an ideal strategy for spring-maize LAI estimation for the entire growth season of 2014 in Hongxing Farm, and the accuracy of the combined method over the whole growth season is higher than that of any single method.

Keywords: crop growth; reflectance saturation; crop model; assimilation; crop growth stage; method combinations

---

1. Introduction

Leaf area index (LAI) is defined as the one-sided green leaf area per unit ground area in broadleaf canopies and as the projected needle leaf area in coniferous canopies [1,2]. Because green LAI (LAI_g) determines light interception and absorption of the crop canopy [3], estimation of crop LAI is critical for understanding the biophysical processes of crop growth that are essential for predicting crop biomass or yield [4,5].

To achieve accurate LAI estimations, a number of remote sensing (RS) based approaches have been developed over the past few decades. These approaches can be grouped into two broad categories: empirical methods and physical models [6]. Empirical methods can be further divided into parametric and non-parametric empirical methods based on whether there is an explicit relationship between selected RS bands and LAI [7]. A series of optical vegetation indices (VIs) have been developed through combinations of reflectance in two or more bands [8,9] for maize LAI estimation. For instance, a normalized difference vegetation index (NDVI) [10] calculated from the reflectance in red and near-infrared bands are commonly used for retrieving canopy biophysical properties of corn [11]. Because NDVI is highly affected by soil reflectance at low LAI and shows asymptotic saturation at high LAI [12], some more VIs were designed, such as the optimized soil adjusted vegetation index (OSAVI) [13] for considering the soil effect and the two-band enhanced vegetation index (EVI2) [14] for tackling reflectance saturation. The building of empirical methods depends on ground LAI observations which were generally made using different instruments [15–17]. Using ground LAI measurements and RS-based VIs, several empirical methods have been developed for maize LAI estimations [7,10,11]. The non-parametric empirical methods usually define the regression function between the RS information and the target variable, e.g., LAI directly [18]. Machine learning techniques, include neural network [19], support vector regression [20] and gaussian processes [21], are typical non-parametric empirical methods to generate LAI products from a variety of RS data products. Although empirical methods are effective in estimating LAI with high accuracy, they are site- and time-specific [22].

Physical models are constructed based on simulations of radiative transfer process in vegetation canopy, and therefore their utilities are not limited to a specific site or a certain time of a year as empirical methods. Physical models can be further divided into geometric optical (GO) model [23], radiative transfer (RT) model [24] and RT-GO model [25]. In these models, LAI is one of the canopy parameters that determine canopy reflectance. With the canopy reflectance acquired from RS data and other canopy parameters known, these models can be inverted to derive LAI. RT models are more suitable for crop LAI retrievals than GO models as a result, the PROSAIL model, a RT model, has been commonly used to simulate LAI for maize using RS data [26,27]. The limitations of physical models include the high cost of parameters calibration and low model running efficiency. Recently, a physical model and a machine learning technique were combined for LAI estimations [6]. For instance, the PROSAIL model and a neural network algorithm were combined to estimate maize LAI with high accuracy [28]. The machine learning algorithm can help improve the efficiency of physical models, but the high cost of parameters calibration cannot be completely avoided.

Because crop growth is a long-term process, continuous LAI information for the entire growth season is critical to calculate corresponding biomass at different growth stages and thus essential to predict the final yield. Since empirical methods and physical models require RS data as inputs for LAI estimation, the quality of RS data will influence the estimation accuracy, especially when time-series RS data are required to generate LAI estimations for different crop growth stages. In the rainy season in northeast China, clouds, shadows, and haze can strongly influence the application of multispectral RS data with high and medium spatial resolutions in time-series LAI estimation. Because plants usually develop fast in the rainy reason, the lack of good quality RS data will cause the LAI estimates to miss the critical growth stage, and thus will consequently influence biomass or yield estimation. At the late-growth stage, the appearance of more senescing leaves will limit the RS application of LAI estimations [29]. The senescing leaves are still attached to plants and can be captured by in-situ

LAI instruments. However, these leaves have reduced photosynthetic capacity and therefore present quite difference spectral behaviors from that of green leaves. The presence of senescing leaves can be at different extents at the late stage [14,30], which is a great source of uncertainty for canopy LAI estimation from RS data. The saturation of reflectance and VIs is another significant limitation for the late-growth stage. The EVI has been designed to provide improved sensitivity at high LAI [31]; however, it is still not sufficient for the spring maize.

As mentioned above, the quality issues of RS data may cause failures of directly applying RS-based empirical and physical models for LAI estimation in rainy seasons or at the late-growth stage. Hence, supplementary information should be referred to address this issue. Unlike other vegetation types, crops usually own a more regular growth period which can be useful in building mechanistic models for simulating crop growth. Crop models [32,33] are able to provide comprehensive mathematical descriptions of key crop physical and physiological processes, which cannot be achieved by RS-based methods. Therefore, it is more applicable to use crop models for LAI estimations during rainy- and late-growth seasons.

Since crop models are generally designed to simulate crop growth at the site level, several data assimilation approaches [34–36] have been developed to incorporate crop parameters estimated from RS data into crop model, aiming to extrapolate the simulations of crop models from a single site to the regional scale. Among them, the Ensemble Kalman Filter (EnKF) [37–39], particle filtering (PF) [40], and four-dimensional variational data assimilation (4DVar) [41,42] are common methods used to link crop models with RS data. Unlike RS-based LAI estimation methods, assimilation methods can tolerate time-series RS data withlong time intervals, making it possible to retrieve accurate LAI estiamtions in rainy season. Therefore, it is feasible to use crop models to fill data gaps where high-quality continuous RS data are not available. Because assimimation methods still require a small amount of RS data, it cannot address the quality issues of RS data at thelate-growth stage. Hence, further studies are needed to improve the multi-phase LAI estimates by integrating RS data and crop models.

The objective of this study is to address the quality issues of RS data for time-series LAI generation. Using the World Food Studies (WOFOST) crop model and RS data acquired by an unmanned aerial vehicle (UAV) in 2014, three methods were applied to estimate LAI for spring maize in Hongxing Farm. Based on the performances of these methods, a method combination was developed to generate LAI for the entire growth season, which includes: (1) a VI-based empirical method for early-stage LAI estimation before rainy season; (2) the EnKF assimilation method for rainy-season LAI generation; (3) the hybrid method by integrating EnKF method and WOFOST model for late-stage LAI retrieval. Complete details of the applied methods and their performances were presented in the following sections.

2. Materials and Methods

2.1. Study Area and Field Campaign

The study was conducted in an experimental field (48°08′ N, 126°57′ E, WGS84) of Hongxing Farm, located in Heilongjiang province, Northeast China. Hongxing Farm is a large state-owned farm that lies within the temperate monsoon climate zone characterized by an average annual precipitation of 548.8 mm (2014) and an average annual cumulative temperature (>10 °C) of 2293.0 °C (2014). Spring maize accounts for nearly 50% of the planting area of Hongxing Farm. The spring maize growing season extends from the beginning of May till mid-October. Another important crop in this region is soybean, which is usually rotated with spring maize. All the fields in Hongxing Farm have a unique identifier (ID). The ID of the experimental field selected for this study is 5-1-2. Plot 5-1-2 covers 11 hectares (ha), and seeded with spring maize on 18 May 2015. Figure 1 shows the location of the study area and the spatial distribution of field observation sites.

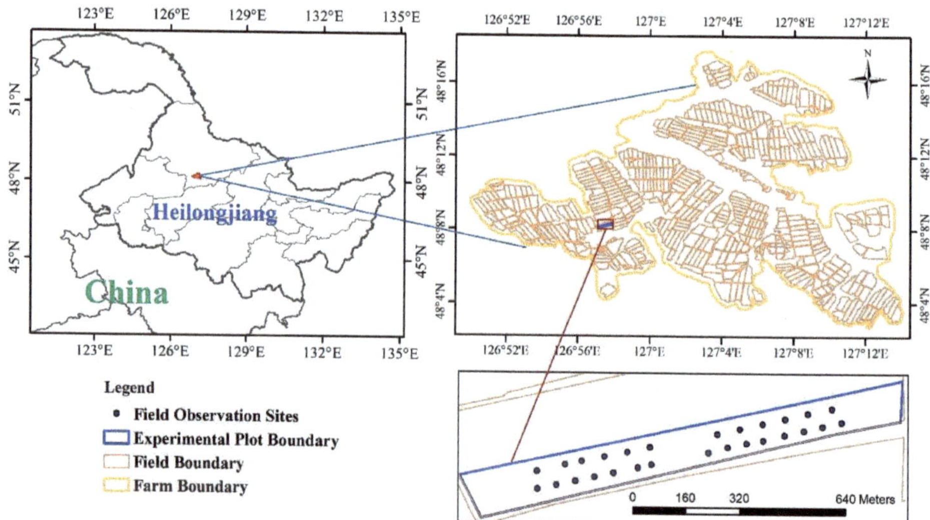

Figure 1. Locations of the study area and the field observation sites.

A total of six field campaigns were conducted in 2015 to collect data for algorithm development. During the field campaigns, LAI, basic soil available nutrient (SAN) contents, yield, and growth period were observed. For LAI, three sequential field campaigns were conducted on 29 June, 30 July and 25 August. The number of quadrats involved in these campaigns were 36, 28, and 29 respectively. The isometric sampling method was used to establish these quadrats. The distance of each quadrat is more than 100 m and the locations for 29 June are shown in Figure 1. The area of each quadrat was 4 m × 6 m. The LAI-2000 Plant Canopy Analyzer (LI-COR) [15,43] was used to measure LAI of each quadrat. Using an optical sensor, the LAI-2000 can measure the effective LAI, under the assumption of random leaf spatial distribution. Besides LAI, the instrument can also calculate the mean foliage inclination and fraction of the sky visible from beneath the canopy. In this study, a one-up-seven-down scheme was used to measure the LAI, i.e., in each quadrat, we obtained one measurement of sky light above the canopy and then seven measurements of diffuse light below the canopy. The final LAI was calculated using an LAI-2000 analyzer. Visually recognizable brown leaves in the target quadrats were eliminated before measurements so that in-situ LAI measurements only accounts for green leaves and thus can be comparable with LAI from optical RS data.

Another field campaign was conducted to collect basic SAN contents at the same locations and with the same quadrat size as the LAI campaigns, on 10 May 2015 before the application of basic fertilizer. In each quadrat, three sampling points along the diagonal were located, from which soil samples were collected. The collected samples were used to measure the available nitrogen (N), phosphorus (P), and potassium (K) contents in the lab. The mean value of the three sampling points was calculated and used as the basic SAN content of each quadrat. For crop yield, we collected the total grain weight of the plot after harvesting. The total grain weight divided by experimental plot area is used as the measured yield. The percentage impurity and water content were also recorded, and the final yield was calculated by removing the impurity and adjusting for the water content at 25%.

The key growth stages were also observed and detailed information on LAI, SAN, yield and growth stage experiments, including the sampling strategies, soil sampling procedures, and testing methods, can be found in [44,45].

2.2. Remote Sensing Data Acquired by the Unmanned Aerial System (UAS)

The RS data used in this study were acquired by a UAS. The UAS consists of three parts: an 8-rotor UAV, an MCA (Multispectral Camera Array) system, and a ground system, as shown in Figure 2. The MCA camera acquires five-band image with a resolution of 1280 × 1024 (1.3 M) pixels. The central wavelengths of the five bands are 490 nm (blue), 550 nm (green), 680 nm (red), 720 nm (red edge) and 800 nm (near-infrared). The flight height was set to 100 m above the ground level, and the overlap for the flight lines was 60% longitudinally and 40% laterally.

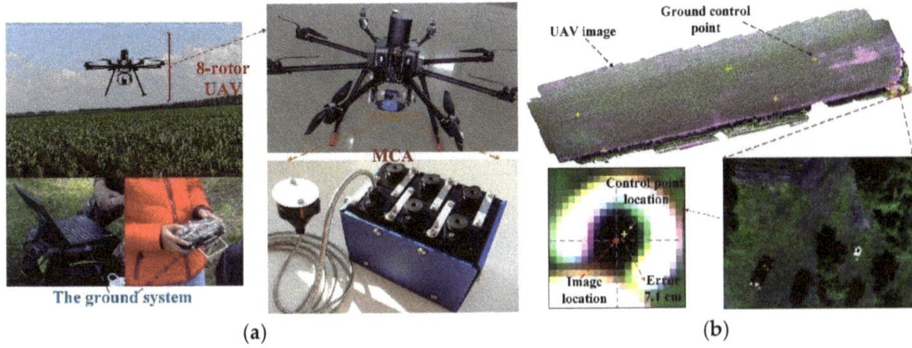

Figure 2. The components of the UAV platform (**a**) and geometric correction accuracy (**b**).

The image pre-processing includes reflectance calculation, image stitching, and geometric correction. Among the 6 MCA channels, one of them was connected to an electronic component (an e-ILS sensor) to receive incoming radiation. Using the incident solar radiation, the reflectance for the other five channels was obtained, thus the reflectance can be called at-sensor-reflectance. Because the UAV flight was conducted under clear weather condition and the fight height is 100 m, the at-sensor-reflectance can be used as the reflectance of crop canopy without the need for atmospheric correction in this study. Two software, the Agisoft PhotoScan and PIE-UAV (Pixel Information Expert), were used to generate the ALS-similar point clouds and to stitch the images using information on the flight height, camera attitude, and GPS. For geometric correction, 18 ground control points were collected within the experimental field using a hand-held Trimble GeoXH differential GPS with a mean estimated error of 0.10 m. Twelve points were used to make the geometric correction and six points were used to calculate the mean deviation error (MDE) of the corrected UAV images. The results show that the MDE was less than two pixels (0.1 m) for the June and July images and three pixels (0.15 m) for the August image. After reflectance calculation, image stitching and geometric correction, six images with a spatial resolution of 0.054 m × 0.054 m were obtained.

2.3. Estimation of LAI

The quality issues of RS data will cause failures of common RS-based empirical methods and physical models for continuous LAI estimations for the entire growth season. Hence, crop models were considered in this study to avoid the use of low quality RS data. By integrating the time-series RS data acquired by the UAS and WOFOST crop growth model, the LAI was estimated using combined methods and the processes of LAI estimation for the entire growth is shown in Figure 3.

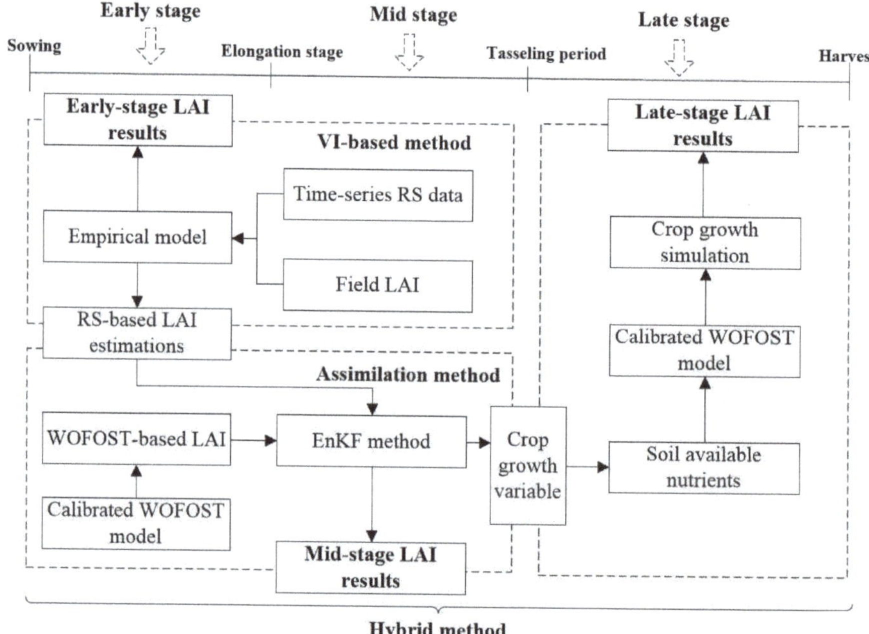

Figure 3. Processes of LAI estimation for the entire growth season using combined methods.

In Figure 3, the whole growth season was divided into three stages: the early stage, the mid stage, and the late stage. The phase from the emergence to the end of the elongation stage was considered as the early stage, and the beginning of the late stage is the tasseling period in this study. The mid stage was defined as the phase between the early and the late stage. In general, the early stage includes the period with high-quality RS time series, the mid stage can also be considered as the rainy season in this study before leaf senescence, and the late stage are characterized by reflectance saturation and leaf senescence. Based on field observations, we divided the specific time spans for the early stage (from the beginning of the growing season to 14 July) and late stage (from 12 August to the end of the growth season). The 14 July and 12 August dates were determined by calculating the middle dates of the field campaigns (30 June, 29 July and 25 August). Three methods, including the VI-based method, the assimilation method, and the hybrid method integrating crop model and the assimilation method, were applied to estimate LAI for the three stages, respectively. The details of these methods are in Sections 2.3.1–2.3.3.

2.3.1. VI-Based Method

Our previous study [44] showed that the empirical model can provide LAI estimates with higher accuracy than that of the physically based model (e.g., via PROSAIL simulation). In this study, the relationship between different VIs and ground LAI were analyzed, and the VI with the highest accuracy was selected to build a linear empirical model to estimate LAI.

To ensure the accuracy of LAI estimation, five VIs were calculated from the UAV data. Besides the commonly used *NDVI* and ratio vegetation index (*RVI*) [8], *OSAVI* was also selected to reduce the effect of soil at low LAI, and *EVI2* and modified triangular vegetation index (*MTVI2*) [9] to lower the influence of reflectance saturation. The selected VIs were calculated using the following equations:

$$NDVI = \frac{NIR - RED}{NIR + RED} \tag{1}$$

$$RVI = \frac{NIR}{RED} \tag{2}$$

$$OSAVI = \frac{1.16 \times (NIR - RED)}{NIR + RED + 0.16} \tag{3}$$

$$EVI2 = \frac{2.5 \times (NIR - RED)}{NIR + 2.4 \times RED + 1} \tag{4}$$

$$MTVI2 = \frac{1.5 \times \{1.2 * (NIR - GREEN) - 2.5 \times (RED - GREEN)\}}{\sqrt{(2 \times NIR + 1)^2 - \left(6 * NIR - 5 \times \sqrt{RED}\right) - 0.5}} \tag{5}$$

where *NIR*, *RED* and *GREEN* denote the reflectance of the near-infrared band, red band and green band, respectively. Based on the calculated VIs and ground LAI, linear functions can be built. The VI yielding the highest accuracy of LAI estimation was selected as the optimal method for the early-stage LAI estimation.

2.3.2. Data Assimilation Method

As shown in Figure 3, the WOFOST model and EnKF method was used to apply RS assimilation for mid-stage LAI simulation. As a primary member of the Wageningen crop models [46] and a core component of the Crop Growth Monitoring System (CGMS) [47], the WOFOST model can simulate daily crop physiological and ecological processes. There were two reasons to select the WOFOST model. Firstly, it can predict daily LAI by simulating CO_2 assimilation, respiration, leaf growth and dry matter formation. Secondly, following model modifications referred in [44], WOFOST can simulate LAI under nutrient-limited conditions. Compared with the water-limited LAI, the accuracy of estimated nutrient-limited LAI can be improved by eliminating the influence of SAN on crop growth.

Before WOFOST can be used for LAI simulation, its main parameters must be calibrated. In this study, three methods were used to acquire the input parameters: the SAN estimation method, field campaigns, and the FSEOPT optimization software. The method proposed in [44] was used to estimate the SAN content. Field data was also used to calibrate crop and soil water parameters. Details of the field campaign method can be found in [45]. However, there are still some core parameters cannot be calibrated using the above two methods. In this study, the FSEOPT optimization software [48] was used to calibrate the uncalibrated parameters based on field LAI observations. We selected LAI because it is the research subject of this study. The mean field LAI value of the experimental plot was calculated as the output variable. Then the calculated LAI was used to conduct optimization in the FSEOPT software. The optimization was conducted once using LAI data collected on each date of 30 June, 29 July and 25 August, respectively. Three values for each parameter were acquired after three-time FSEOPT performances and the mean value was calculated as the final calibrated parameter. The calibrated WOFOST model was used to simulate daily crop growth and output LAI results, which is necessary for the assimilation method.

In this study, the EnKF method [37] was adopted to assimilate the LAI estimated from the VI-based method into the WOFOST model to correct for daily simulated LAI results, and LAI values after data assimilation were taken as the output results for the mid stage. The EnKF method, which is based on the Monte Carlo simulation, performs a model forecast in which the state variables are propagated forward in time based on the modeled dynamics and updated using probability distribution and available observations [49]. EnKF is a major assimilation method that can be easily applied in the WOFOST model [50]. In the recursive algorithm of the Kalman Filter, the assimilation process is divided into two steps [41]: forecast and update. In the forecast step, the covariance matrix (A_c) of the state variables is calculated. If ensemble of the state variables is defined as A_t,

$$A_t = (x_1, x_2, x_3, \dots, x_N) \tag{6}$$

then A_c can be calculated as follow:

$$A_c = \left(A_t - \overline{A}\right) * \left(A_t - \overline{A}\right)^T / (N-1)$$

where $\overline{A}$ is the mean value of A_t, T represents the transpose of a vector, N is the element number in A_t.

In the update step, the state variables were updated using the observation (RS based LAI). The simulated ensemble of state variables was calculated by a nonlinear equation:

$$x_i(t) = M[x_i(t-1)] \tag{7}$$

where t is WOFOST simulation step (Day of year), M is the forecast equation.

If the observed ensemble (RS based LAI) was defined as D_t, the covariance matrix (D_c) of the observed ensemble can also be calculated by Equation (6). Then the standard analysis equation can be built to calculate the updated WOFOST simulated ensemble (A_a):

$$A_a = A_f + K(D_t - HA) \tag{8}$$

where A_a is the optimal estimated ensemble, A_f is the forecasted ensemble calculated using Equation (7), K is the Kalman gain matrix to weigh the difference between observation (RS-based LAI) and prior simulation of the model's state (WOFOST-based LAI), which was calculated by the following equation:

$$K = A_c H^T \left(H A_c H^T + D_c\right)^{-1} \tag{9}$$

where H is the parameter of observation operator. The A_a was used to replace A_t to realize the assimilation process.

Filter divergence was observed in our study. It represents a tendency of the standard EnKF to reject observations in favor of ensemble forecasting in subsequent stages, leading to incremental deviations of LAI from actual measurements. To address this issue, an expansion parameter (E) was used:

$$E = R * N * \frac{\sigma_1^2}{D * \sigma_2^2} \tag{10}$$

where R is a random number that is less than 1, D is the total number of days for the assimilation stage, N is the number of days (from 1 to D) for the current WOFOST's simulation step, σ_1^2 is the variance of RS based LAI of N, and σ_2^2 is the variance of WOFOST-based LAI of N. When σ_1^2/σ_2^2 is larger than 4 and E is equal or greater than 1, E will be used to enlarge K to eliminate filter divergence in this study.

2.3.3. The Hybrid Method

As mentioned in the introduction, both VI-based and assimilation methods cannot address the quality issues of RS data at the late-growth stage caused by senescing leaves and reflectance saturation. Hence, we designed a new strategy (the hybrid method) for the late-stage LAI estimation. In the hybrid method, the WOFOST model with and without UAV data assimilation were combined to improve the late-stage LAI estimation. The UAV data were assimilated into the WOFOST model to generate daily crop growth from the beginning of the growth season until the begin of late stage. Then the data assimilation was halted, and WOFOST was only used to simulate LAI until the end of the growth season. The processes of LAI estimation for the late stage is shown in Figure 4.

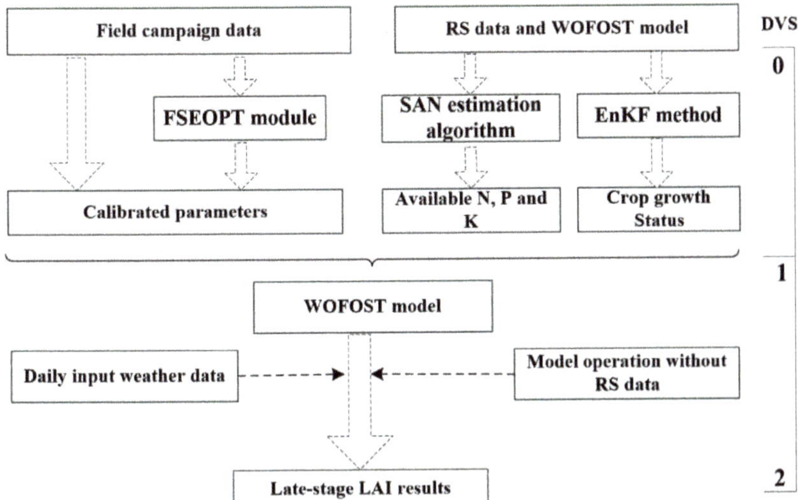

Figure 4. Processes of LAI estimation for the late stage using hybrid method (DVS: the development stage of the crop, DVS = 1 indicates the beginning of tasseling period for maize in this study).

Figure 4 indicates that the hybrid method can be considered as a specific crop model method in which the WOFOST model without using RS data was run from the end of the mid stage (DVS = 1). To enable the simulation of WOFOST, the parameters of crop growth status with DVS = 1 and RS-based SAN contents were required as initial inputs. Then, the WOFOST model was self-operated and daily LAI values were output as the late-stage results. This method was named as the hybrid method to distinguish it from the WOFOST model used in the assimilation method.

Finally, three methods were combined to generate LAI time series for the entire growth season (shown in Figure 3). Detailed LAI estimation processes are described as follows:

Step 1. For early-stage LAI estimation, the VI-based empirical method was applied. VIs were calculated from time-series RS data acquired by UAS and the empirical model was built between field LAI measurements and VIs. Then, LAI was estimated using the empirical model.

Step 2. For mid-stage LAI estimation, the RS-based LAI was assimilated into the WOFOST model to correct the daily crop growth simulations from the beginning of the growth season to the end of the mid stage to generate LAI estimations.

Step 3. For late-stage LAI simulation, crop growth was simulated using the method in Step 2 from the beginning of the growing season to the end of the mid stage. Then, we halted RS assimilation and used the WOFOST model instead to output the daily LAI.

2.4. Evaluating the Accuracies of LAI Estimations

The performance of the three LAI estimation methods was evaluated using field LAI data. The correlation coefficient (R), coefficient of determination (R^2) and *RMSE* were selected as the indices to analyze the relationship between the estimated and field-measured LAI values. The R^2 and *RMSE* were calculated as follows:

$$L_{mean} = \frac{1}{n} \sum_{k=1}^{n} L_{obs,k} \tag{11}$$

$$L_f = a \times L + b \tag{12}$$

$$R^2 = \frac{\sum_{k=1}^{n}\left(L_{f,k} - L_{mean}\right)^2}{\sum_{k=1}^{n}\left(L_{obs,k} - L_{mean}\right)^2} \tag{13}$$

$$RMSE = \sqrt{\frac{\sum_{k=1}^{n}\left(L_k - L_{obs,k}\right)^2}{n}} \tag{14}$$

where L_{obs} is the measured LAI values during the first three field campaigns introduced above, n is the number of quadrats, L_{mean} is the mean value of L_{obs}, L is the RS-based VIs (vegetation index method) or simulated LAI (WOFOST model and assimilation method), L_f is the linear fitting formula between L (independent variable) and L_{obs} (dependent variable), a and b are the coefficients.

The R^2 and $RMSE$ were both used to evaluate the performance of LAI estimation methods. Generally, the LAI estimation method with a higher R^2 corresponding to a lower $RMSE$ was selected as the ideal method combinations for time-series LAI estimation.

Furthermore, the variable coefficient (CV) was selected to examine the spatial heterogeneity of late-stage LAI estimations. The CV was calculated as follows:

$$CV = \frac{SD}{Mean} * 100\% \tag{15}$$

where SD is the standard deviation and $Mean$ is the mean value of the research data (LAI and vegetation indices in this study).

3. Results and Discussion

3.1. LAI Estimation Using the VI-Based Method

Using the linear regression model, the R values between RS-based VIs and field LAI were derived. To show the significant correlation level, a significance test was also conducted by calculating the p-value. p-value < 0.01 indicates highly significant correlation level and p-value < 0.05 for significant correlation level. The R values and significance levels are listed in Table 1. The results show that RVI provided the best LAI estimations with the highest R among all the indices on 30 June (the early-growth stage) and 29 July (the mid-growth stage). EVI2 provided LAI estimate with the highest R on 25 August (the late-growth stage). The R values on 30 June (RVI) and 29 July (RVI) reach highly significant level, reaches significant level on 25 August (EVI2).

Table 1. The R values between vegetation indices and field LAI.

Date (Month-Day)	Number of Samples	NDVI	RVI	OSAVI	EVI2	MTVI2
6-30	36	0.78 **	0.79 **	0.75 **	0.67 **	0.73 **
7-29	28	0.63 **	0.64 **	0.55 **	0.46 *	0.39 *
8-25	29	0.30	0.30	0.42 *	0.44 *	0.33

** indicates highly significant correlation, * indicates significant correlation.

Subsequently, RVI and EVI2 were selected to build the RS-based statistical models for LAI estimation. The R^2 and RMSE values of the field LAI and estimated LAI were also calculated. The statistical models and the accuracy analysis results are shown in Figure 5.

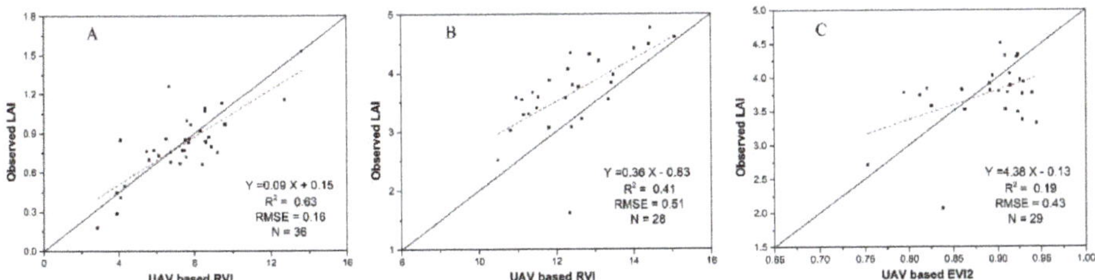

Figure 5. Established empirical vegetation index models (**A**) 30 June, (**B**) 29 July, (**C**) 25 August.

From Figure 5, we can observe that the estimation accuracy on 25 August was considerably lower than that on 30 June and 29 July. The VI saturation that occurs during the late-growth season is an important reason for the significantly reduced accuracy on 25 August. In Table 2, the CV indices of the five VIs on 30 June, 29 July and 25 August were calculated to represent the levels of VI saturation on different dates. The results (listed in Table 2) show that the CV values on 29 July and 25 August were lower than that on 30 June, which means that there is a saturation trend toward the mid and late season.

Table 2. CV values of the vegetation indices for the experimental field (%).

Time (Month-Day)	Field LAI	NDVI	RVI	OSVI	EVI2	MTVI2
6-30	31.51	10.92	30.54	13.44	20.37	20.82
7-29	17.96	1.49	9.35	2.83	6.82	4.12
8-25	12.71	1.51	4.64	2.50	5.38	4.43

To compare the levels of VI saturation at different stages, LAI estimations using the VI-based method were shown in Figure 6. The figure reveals that the relative dynamic range (the dynamic range ratio of simulated and observed LAI) was reduced on 25 August respective to 29 July and 30 June. This result suggests that although the saturation has already appeared on 30 July, it is more obvious on 25 August. Hence, the saturation is an important reason for the poor performance of VI-based empirical method for LAI estimation at the late stage.

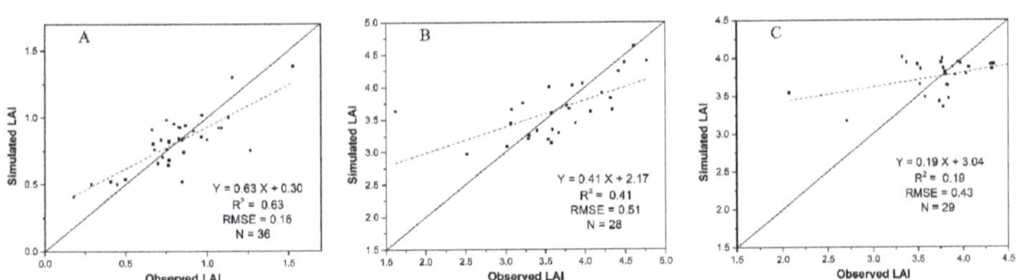

Figure 6. The accuracy of LAI estimations using the vegetation index method (**A**) 30 June, (**B**) 29 July, (**C**) 25 August.

Besides VI saturation, leaf senesce may also have contributed to the lower estimation accuracy on 25 August. At the late growth stage of spring maize, the primary receiver of dry matter from photosynthesis will be shifted to storge organs. The WOFOST' parameters FOTB (fraction of above ground dry matter to storage organs) and FLTB (fraction of above ground dry matter to leaves) can be used to demonstrate this phenomenon. The FOTB on 25 August is higher than that on 29 July, while FLTB shows the opposite pattern (25 August: FOTB = 0.69, FLTB = 0.12; 29 July FOTB = 0.20, FLTB = 0.18). Less dry matter supply will decrease leaf bioactivity and thus generate more inactive leaves. Chemical changes in inactive leaves can hardly be detected by optical RS data. In general, saturation and senescence jointly decreased the correlation between LAI and RS-based VIs.

From Figure 6A, we can find that the accuracy of VI-based method for the early-stage LAI estimation is higher than the assimilation method. Hence, the statistical model shown in Figure 5A was selected as the ideal method to provide early-stage LAI results (shown in Figure 3). Furthermore, the RS-based LAI was also required by the assimilation method, and the three statistical models shown in Figure 5 were used to generate the RS-based LAI for the entire growth season. Then the inverse distance weighted interpolation method was used to generate daily RS-based LAI for data assimilation.

Although the VI-based statistical models can provide higher accuracy LAI estimations, these models are commonly used for specific sensors and sampling conditions because the lack of physical basis. Thus, any changes in location, crop type and application years can cause failures in the original model application [51,52]. Because radiative transfer model [53] is the core algorithm of the physical models, they reflect the transfer and interaction of radiation inside the canopy; hence, the physical model methods can overcome the limitations of sensors, geographical locations, and application times. A physical model [26] should be considered before applying the LAI estimation method to a larger area with different crops for different years. Meanwhile, detailed radiation transfer and interaction simulations in physical models require more input parameters; consequently, more field experiments should be conducted before applying physical models. Because there is no significant change in environmental conditions, growth stages, and remote sensing data calibration, the VI-based statistical model was selected and re-calibration of the relationship between RS-based VI and field LAI each time is not necessary.

3.2. LAI Simulation Using the Assimilation Method

3.2.1. WOFOST Model Calibration

The results of the WOFOST model calibration are presented in this section. The input parameters of the WOFOST model include meteorological, soil, and crop parameters. In addition to the daily meteorological and fertilization data, which were provided by Hongxing Farm, other parameters also need to be calibrated in this study. Based on a sensitivity analysis (discussed in previous work [44]), 18 parameters sensitive to LAI were selected as core parameters to be calibrated using the three previously described calibration methods in this study. The values of the calibrated parameters and their calibration methods are listed in Table 3.

Table 3. Core parameter calibration results of the WOFOST model.

Parameters	Description	Original Values	Calibrated Values	Unit	Calibration Method
TSUM1	Temperature sum from emergence to anthesis	695	890	°C × d	Field campaign
TSUM2	Temperature sum from anthesis to maturity	800	710	°C × d	Field campaign
CVL	Conversion efficiency of assimilates into leaf	0.68	0.64	kg/kg	Field campaign
CVO	Conversion efficiency of assimilates into storage organ	0.67	0.81	kg/kg	Field campaign
CVR	Conversion efficiency of assimilates into root	0.69	0.70	kg/kg	Field campaign
CVS	Conversion efficiency of assimilates into stem	0.66	0.66	kg/kg	Field campaign
FRTB	Fraction of total dry matter to root	0–0.37	0–0.40	kg/kg	Field campaign
FOTB	Fraction of above ground dry matter to storage organs (DVS = 0.1–1.7)	0–1.00	0–0.74	kg/kg	Field campaign

Table 3. *Cont.*

Parameters	Description	Original Values	Calibrated Values	Unit	Calibration Method
FLTB	Fraction of above ground dry matter to leaves (DVS = 0.1–1.7)	0–0.62	0.20–0.75	kg/kg	Field campaign
FSTB	Fraction of above ground dry matter to stem (DVS = 0.1–1.7)	0–0.85	0.06–0.57	kg/kg	Field campaign
NBASE	Basic soil nitrogen content	100	40–410	mg/kg	SAN estimation method
PBASE	Basic phosphorus content	100	10–80	mg/kg	SAN estimation method
KBASE	Basic potassium content	100	20–340	mg/kg	Field campaign
SMFCF	Soil moisture content at field capacity	0.11	0.46	cm^3/cm^3	FSEOPT software
SMW	Soil moisture content at wilting point	0.04	0.20	cm^3/cm^3	FSEOPT software
SM0	Soil moisture content of saturated soil	0.39	0.570	cm^3/cm^3	FSEOPT software
RDMCR	Maximum root depth allowed by soil	10	2.4	m	FSEOPT software
SPAN	Life span of leaves growing at 35 °C	33	28	day	FSEOPT software

The accuracy of simulated crop growth by the calibrated WOFOST model was firstly evaluated using emergence time, anthesis time, maturity time and yield. These parameters were simulated for the experimental plot and compared with field observations, and the results are listed in Table 4. Comparing to observations, the crop growth, as indicated by those parameters in Table 4, simulated by the calibrated model is better than the original model.

Table 4. Performance of calibrated and un-calibrated WOFOST model.

Variable	Method	Values	Error
Emergence time	Observed	1 June	-
	Original model	23 May	−8 days
	Calibrated model	28 May	−4 days
Anthesis time	Observed results	25 July	-
	Original model	15 July	−10 days
	Calibrated model	29 July	4 days
Maturity time	Observed results	27 September	-
	Original model	22 September	−5 days
	Calibrated model	30 September	3 days
Yield (kg/ha)	Observed results	9179	-
	Original model	9607	−428
	Calibrated model	9104	75

3.2.2. LAI Simulation Using the WOFOST Model

We applied the calibrated WOFOST model to simulate LAI which is required for the assimilation process, and the model performances were also evaluated using field LAI. Results (shown in Figure 7) show that the calibrated model can simulate LAI on 30 June with high accuracy, but yielded LAI estimates with much lower accuracy on 25 July and 25 August. Calibration errors are the potential reason for the lower accuracy at the mid- and late-stages. Some core parameters of the WOFOST model can only be acquired as constant values at the site level; but in reality they may change with time. Thus, errors cannot be completely avoided using either of the three calibration methods. These errors will propagate and accumulate, hence reduce the accuracy of LAI simulation over time. The higher simulation accuracy on 30 June than the other two dates also indicates that reducing the simulation time can weaken the calibration error propagation and accumulation and help the WOFOST model provide better LAI simulation.

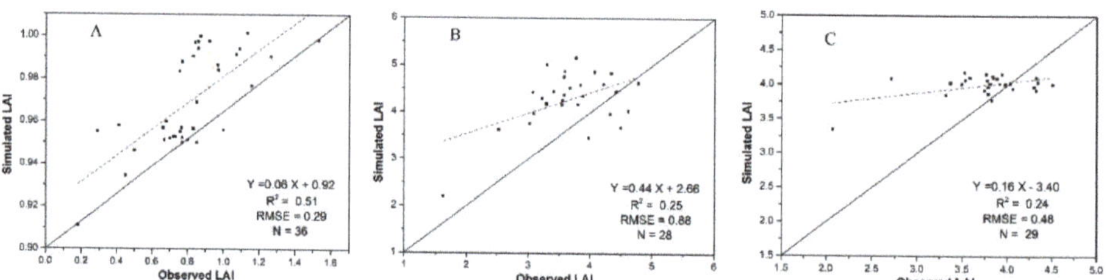

Figure 7. The LAI estimation accuracies for the calibrated WOFOST model (**A**) 30 June, (**B**) 29 July, (**C**) 25 August.

Although the single WOFOST model cannot provide LAI estimates with higher accuracy than the VI-based method and the assimilation method for all the three growth stages (shown in Figures 6–8), WOFOST was still required by the assimilation method and the hybrid method (shown in Figure 3). Errors in the WOFOST-based growth simulation will be reflected in the performance of the assimilation and the hybrid methods. Hence, improve the performance of WOFOST is beneficial for the estimation of LAI time series in this study. Because LAI is simulated and analyzed at the pixel or sampling point level, the parameters with high spatial heterogeneity should also be calibrated at the pixel level to ensure the LAI simulation accuracy. Among input parameters, soil parameters, including soil water and soil available nutrients, are the main variables that change within the experimental field in this study. Considering that soil water contents can be simulated by WOFOST model using daily precipitation data, the SAN values at the pixel level should be acquired before conducting WOFOST-based LAI simulation.

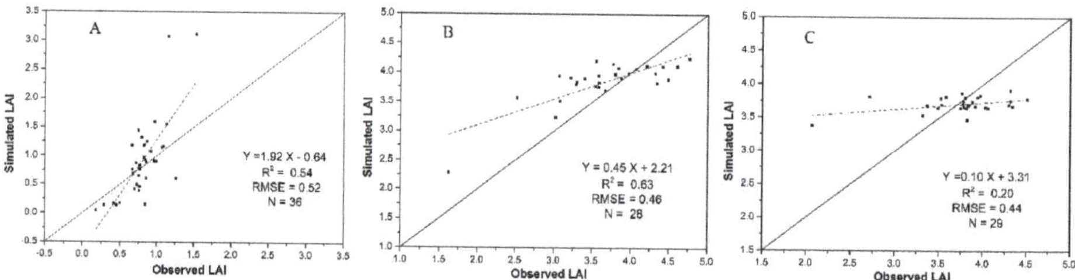

Figure 8. The LAI estimation accuracies using assimilation method (**A**) 30 June, (**B**) 29 July, (**C**) 25 August.

The SAN estimation method proposed in [44] was used to provide available N, P and K based on RS data and the WOFOST model. From the accuracy analysis results, we can find that the estimation accuracy for K ($R^2 = 0.15$, RMSE = 23.56 mg/kg, mean = 168.26 mg/kg) was significantly lower than that of N ($R^2 = 0.48$, RMSE = 18.45 mg/kg, mean value = 296.78 mg/kg) and P ($R^2 = 0.37$, RMSE = 7.05 mg/kg, mean value = 31.63 mg/kg). Optimizing the K estimation would be useful to further improve the LAI simulations. The instability of K and the lower effect of potassium ions in comparison with the other two ions on crop growth, especially on leaf growth, are the possible causes for the lower estimation accuracy using the proposed method. Adjusting the SAN estimation method to consider these factors will be the focus of future studies to further improve the K estimation accuracy.

3.2.3. LAI Simulation Using the EnKF Assimilation Method

Using the method presented in Section 3.1, LAI was estimated using time-series UAV data first, and then assimilated into the WOFOST model through the EnKF method. The calibrated model with data assimilation was used to generate LAI for 30 June, 29 July and 25 August. The R^2 and RMSE were calculated for the three LAI simulations using field LAI. The results (listed in Figure 8) show that the UAV data improved the WOFOST model's LAI simulation performance for 30 June and 29 July. However, the accuracy for 25 August did not improve. Compared with the VI-based empirical method, the assimilation method provides higher simulation accuracy for 29 July and similar accuracies for 30 June and 25 August. From Figure 8A, we see that there are two samples with higher simulated LAI (around 3). Comparing Figures 6B and 8B, we can also find that the assimilation can provide mid-stage LAI estimation with higher accuracy, thus, this method was selected as the ideal method for this stage (shown in Figure 3).

To show the improvements that both calibration and assimilation can bring in LAI simulation, the LAI trajectories were gathered from three methods including the calibrated WOFOST model, the assimilation method, and the VI-based empirical method. We selected the mean LAI value of the experimental plot and LAI's variation range (10–90%) as the indexes and generate three trajectories (shown in Figure 9). The results show that the assimilation can correct WOFOST's LAI simulation to certain extent. Because we didn't conduct UAV flight after 25 August, the trajectories end at that time (Day of year; DOY = 137).

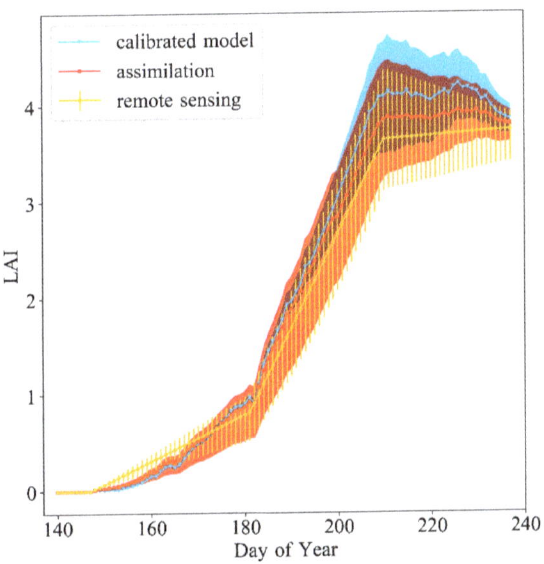

Figure 9. The seasonal LAI estimated using three methods: the calibrated WOFOST model, the assimilation method, and the VI-based empirical using RS data.

3.3. LAI Estimations Using the Hybrid Method

From above analyses, we can find that the late-stage LAI estimation accuracies of the VI-based, the WOFOST model and the assimilation methods at the late stage are much lower than those at the early and mid-stages. Model calibration, reflectance and VI saturation, and canopy changes associated with leaf aging, jointly contributed to the reduced performance to certain extents. Hence, the hybrid method was developed to improve the accuracy of AI estimation at the late stage. Using this method, LAI values on 25 August were simulated, and the accuracy was evaluated using field LAI. The analysis result, which is shown in Figure 10, shows that combining the WOFOST model with RS data assimilation provides better LAI simulations than the VI-based empirical method (Figure 5C) and the RS assimilation method (Figure 8C) alone because the problems of leaf aging and error accumulation can be avoided.

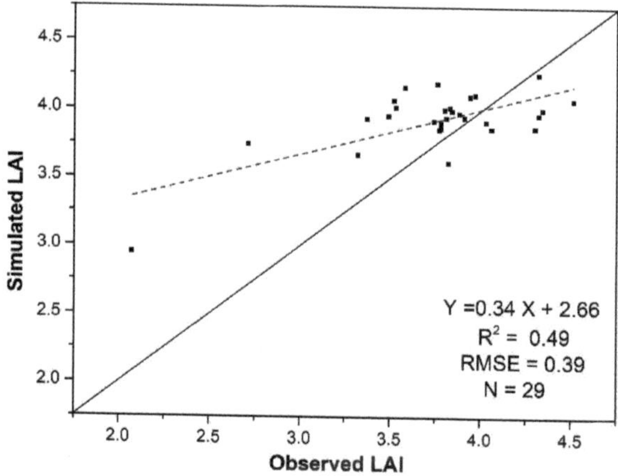

Figure 10. The LAI estimation accuracies for 25 August using the hybrid method.

The time span delimitation of different crop growth stages is the main limitations for the application of the hybrid method. The time window used for the three crop growth stages in this study influences the application times of the VI-based empirical method, assimilation and crop model without data assimilation. For instance, the beginning of the late-growth stage is the period when reflectance saturation and leaf senescence both appeared; hence, the WOFOST model was used to replace the assimilation method during that period. However, we calculated the start date from the field campaign dates, which may be inaccurate because the goal of the field campaign did not involve measuring reflectance saturation and leaf senescence. More field campaigns should be conducted to acquire the best application time for the three methods.

4. Conclusions

In this study, we applied three methods to conduct time-series LAI estimations for spring maize. Using field-measured LAI over an experimental field, the performances of these methods were investigated. Results show that the VI-based method can be used to provide LAI estimation with high accuracy before the appearance of reflectance saturation and leaf senescence. With the application of the UAV data, the accuracy of the assimilation method was improved, and this method was an ideal choice for the mid-stage LAI simulation. The hybrid method is designed to address reflectance saturation and leaf senescence issues of the VI-based method and error accumulation of the WOFOST model in the late stage, and its LAI estimation accuracy of at the late stage is higher than the other two methods. In general, the accuracy analysis results show that the combined methods, which applied the VI-based method for the early stage, the assimilation method for the mid stage, and the hybrid method for the late stage, can provide highly accurate continuous LAI estimations for the entire growth season. Meanwhile, it's also worth noting that the experiment was only conducted for one year due to the high cost and logistic requirements. The field data gathered in the experiment are insufficient to well evaluated the robustness of the applied methods. A multi-year experiments with additional crop types will be the focus of future studies to further analyze the performance of the combined methods for time-series LAI generation.

Author Contributions: J.M., J.S., Y.Q. and Z.C. conceived and designed the experiments; Z.C. applied the combined methods, performed the experiments and analyzed the data; J.L., J.H., B.Q., Q.J., T.D. and L.Y. contributed proposed the comments on method optimization and data analysis. All authors have read and agreed to the published version of the manuscript.

Funding: This research was funded by National Natural Science Foundation of China (grant number: 41871261), Fujian Provincial Department of Education science and technology project (grant number: JAT190083), and the GF6 Project (grant number: 09-Y20A05-9001-17/18).

Acknowledgments: We are grateful to DLO Winand Staring Centre, Wageningen, for providing the original WOFOST code and necessary parameter documentations, to Hongxing Farm for providing their database to calibrate the WOFOST parameters and offering us the experimental field and their laboratory for field campaigns, and to the China Scholarship Council (CSC) for providing a scholarship for Zhiqiang Cheng to pursue studies at Agriculture and Agri-Food Canada as a joint Ph.D. student.

Conflicts of Interest: The authors declare no conflict of interest.

References

1. Chen, J.; Cihlar, J. Retrieving leaf area index of boreal conifer forests using Landsat TM images. *Remote Sens. Environ.* **1996**, *55*, 153–162. [CrossRef]
2. Myneni, R.; Hoffman, S.; Knyazikhin, Y.; Privette, J.; Glassy, J.; Tian, Y.; Wang, Y.; Song, X.; Zhang, Y.; Smith, G. Global products of vegetation leaf area and fraction absorbed PAR from year one of MODIS data. *Remote Sens. Environ.* **2002**, *83*, 214–231. [CrossRef]
3. Sellers, P.; Dickinson, R.; Randall, D.; Betts, A.; Hall, F.; Berry, J.; Collatz, G.; Denning, A.; Mooney, H.; Nobre, C. Modeling the exchanges of energy, water, and carbon between continents and the atmosphere. *Science* **1997**, *275*, 502–509. [CrossRef] [PubMed]
4. Dong, T.; Liu, J.; Qian, B.; He, L.; Liu, J.; Wang, R.; Jing, Q.; Champagne, C.; McNairn, H.; Powers, J.; et al. Estimating crop biomass using leaf area index derived from Landsat 8 and Sentinel-2 data. *ISPRS J. Photogramm. Remote Sens.* **2020**, *168*, 236–250. [CrossRef]
5. Pagani, V.; Guarneri, T.; Busetto, L.; Ranghetti, L.; Boschetti, M.; Movedi, E. A high-resolution, integrated system for rice yield forecasting at district level. *Agric. Syst.* **2018**, *168*, 181–190. [CrossRef]
6. Verrelst, J.; Camps-Valls, G.; Munoz-Mari, J.; Pablo Rivera, J.; Veroustraete, F.; Clevers, J.; Moreno, J. Optical remote sensing and the retrieval of terrestrial vegetation bio-geophysical properties—A review. *ISPRS J. Photogramm. Remote Sens.* **2015**, *108*, 273–290. [CrossRef]
7. Dong, T.; Liu, J.; Shang, J.; Qian, B.; Ma, B.; Kovacs, J.; Walters, D.; Jiao, X.; Geng, X.; Shi, Y. Assessment of red-edge vegetation indices for crop leaf area index estimation. *Remote Sens. Environ.* **2019**, *222*, 133–143. [CrossRef]
8. Pearson, R.; Miller, L. Remote Mapping of Standing Crop Biomass for Estimation of the Productivity of the Shortgrass Prairie. In Proceedings of the Eighth International Symposium on Remote Sensing of Environment, Ann Arbor, MI, USA, 2–6 October 1972; pp. 7–12.
9. Jiang, Z.; Huete, A.; Didan, K.; Miura, T. Development of a two-band Enhanced Vegetation Index without a blue band. *Remote Sens. Environ.* **2008**, *112*, 3833–3845. [CrossRef]
10. Myneni, R.; Hall, F. The interpretation of spectral vegetation indexes. *IEEE Trans. Geosci. Remote Sens.* **1995**, *33*, 481–486. [CrossRef]
11. Jiang, Z.; Huete, A.; Chen, J.; Chen, Y.; Li, J.; Yan, G.; Zhang, X. Analysis of NDVI and scaled difference vegetation index retrievals of vegetation fraction. *Remote Sens. Environ.* **2006**, *101*, 366–378. [CrossRef]
12. Liu, J.; Pattey, E.; Jégo, G. Assessment of vegetation indices for regional crop green LAI estimation from Landsat images over multiple growing seasons. *Remote Sens. Environ.* **2012**, *123*, 347–358. [CrossRef]
13. Steven, M. The sensitivity of the OSAVI vegetation index to observational parameters. *Remote Sens. Environ.* **2004**, *63*, 49–60. [CrossRef]
14. Haboudane, D.; Miller, J.R.; Pattey, E.; Zarco-Tejada, P.J.; Strachan, I.B. Hyperspectral vegetation indices and novel algorithms for predicting green LAI of crop canopies: Modeling and validation in the context of precision agriculture. *Remote Sens. Environ.* **2004**, *90*, 337–352. [CrossRef]
15. Li-Cor. *LAI-2000 Plant Canopy Analyser: Instruction Manual*; LI–COR, Inc.: Lincoln, NE, USA, 1992.
16. Yu, L.; Shang, J.; Cheng, Z.; Gao, Z.; Wang, Z.; Tian, L.; Wang, D.; Che, T.; Jin, R.; Liu, J.; et al. Assessment of Cornfield LAI Retrieved from Multi-Source Satellite Data Using Continuous Field LAI Measurements Based on a Wireless Sensor Network. *Remote Sens.* **2020**, *12*, 3304. [CrossRef]
17. Breda, N. Ground-based measurements of leaf area index: A review of methods, instruments and current controversies. *J. Exp. Bot.* **2003**, *54*, 2403–2417. [CrossRef] [PubMed]

18. Manuel, C.; García-Haro, F.; Lorenzo, B.; Luigi, R.; Gilabert, M.; Amparo, M.; Gustau, C.; Fernando, C.; Mirco, B. A Critical Comparison of Remote Sensing Leaf Area Index Estimates over Rice-Cultivated Areas: From Sentinel-2 and Landsat-7/8 to MODIS, GEOV1 and EUMETSAT Polar System. *Remote Sens.* **2018**, *10*, 763.

19. Bacour, C.; Baret, F.; Beal, D.; Weiss, M.; Pavageau, K. Neural network estimation of LAI, fAPAR, fCover and LAIx (ab), from top of canopy MERIS reflectance data:Principles and validation. *Remote Sens. Environ.* **2006**, *105*, 313–325. [CrossRef]

20. Pan, J.; Yang, H.; He, W.; Xu, P. Retrieve leaf area index from HJ-CCD image based on support vector regression and physical model. In Proceedings of the SPIE—The International Society for Optical Engineering, Dresden, Germany, 23–26 September 2013.

21. Svendsen, D.; Martino, L.; Campos-Taberner, M.; García-Haro, F.; Camps-Valls, G. Joint Gaussian Processes for Biophysical Parameter Retrieval. *IEEE Trans. Geosci. Remote Sens.* **2017**, *56*, 1718–1727. [CrossRef]

22. Gurung, R.; Breidt, F.; Dutin, A.; Ogle, S. Predicting enhanced vegetation index (EVI) curves for ecosystem modeling applications. *Remote Sens. Environ.* **2009**, *113*, 2186–2193. [CrossRef]

23. Li, X.; Strahler, A. Geometric-optical bidirectional reflectance modeling of the discrete crown vegetation canopy-effect of crown shape and mutual shadowing. *IEEE Trans. Geosci. Remote Sens.* **1992**, *30*, 276–292. [CrossRef]

24. Verhoef, W. Light-scattering by leaf layers with application to canopy reflectance modeling the SAIL model. *Remote Sens. Environ.* **1984**, *16*, 125–141. [CrossRef]

25. Li, X.; Strahler, A.; Woodcock, C. A hybrid geometric optical-radiative transfer approach for modeling albedo and directional reflectance of discontinuous canopies. *IEEE Trans. Geosci. Remote Sens.* **1995**, *33*, 466–480. [CrossRef]

26. Duan, S.; Li, Z.; Wu, H.; Tang, B.; Ma, L.; Zhao, E.; Li, C. Inversion of the PROSAIL model to estimate leaf area index of maize, potato, and sunflower fields from unmanned aerial vehicle hyperspectral data. *Int. J. Appl. Earth Obs. Geoinf.* **2014**, *26*, 12–20. [CrossRef]

27. Li, H.; Chen, Z.; Jiang, Z.; Wu, W.; Ren, J.; Liu, B.; Hasi, T. Comparative analysis of GF-1, HJ-1, and Landsat-8 data for estimating the leaf area index of winter wheat. *J. Integr. Agric.* **2017**, *16*, 266–285. [CrossRef]

28. Wei, X.; Gu, X.; Meng, Q.; Yu, T.; Zhou, X.; Wei, Z.; Ji, K.; Wang, C. Leaf Area Index Estimation Using Chinese GF-1 Wide Field View Data in an Agriculture Region. *Sensors* **2017**, *17*, 1593. [CrossRef]

29. Delegido, J.; Verrelst, J.; Rivera, J.; Antonio, R.; Moreno, J. Brown and green LAI mapping through spectral indices. *Int. J. Appl. Earth Obs. Geoinf.* **2015**, *35*, 350–358. [CrossRef]

30. Fang, H.; Ye, Y.; Liu, W.; Wei, S.; Ma, L. Continuous estimation of canopy leaf area index (LAI) and clumping index over broadleaf crop fields: An investigation of the PASTIS-57 instrument and smartphone applications. *Agric. For. Meteorol.* **2018**, *253–254*, 48–61. [CrossRef]

31. Huete, A.; Didan, K.; Miura, T.; Rodriguez, E.; Gao, X.; Ferreira, L. Overview of the radiometric and biophysical performance of the MODIS vegetation indices. *Remote Sens. Environ.* **2002**, *83*, 195–213. [CrossRef]

32. Mabhaudhi, T.; Modi, A.; Beletse, Y. Parameterisation and evaluation of the FAO-AquaCrop model for a South African taro (Colocasia esculenta L. Schott) landrace. *Agric. For. Meteorol.* **2014**, *192–193*, 132–139. [CrossRef]

33. Abrahamsen, P.; Hansen, S. Daisy: An open soil-crop-atmosphere system model. *Environ. Model. Softw.* **2000**, *15*, 313–330. [CrossRef]

34. Jin, X.; Kumar, L.; Li, Z.; Feng, H.; Xu, X.; Yang, G.; Wang, J. A review of data assimilation of remote sensing and crop models. *Eur. J. Agron.* **2018**, *92*, 141–152. [CrossRef]

35. Huang, J.; Ma, H.; Su, W.; Zhang, X.; Huang, Y.; Fan, J.; Wu, W. Jointly assimilating MODIS LAI and ET products into the SWAP model for winter wheat yield estimation. *IEEE J. Sel. Top. Appl. Earth Obs. Remote Sens.* **2015**, *8*, 4060–4071. [CrossRef]

36. Wen, Z.; Huaing, J.; Li, L.; Zhang, X.; Ma, H.; Gao, X.; Huang, H.; Xu, B.; Xiao, X. Assimilating Soil Moisture Retrieved from Sentinel-1 and Sentinel-2 Data into WOFOST Model to Improve Winter Wheat Yield Estimation. *Remote Sens.* **2019**, *11*, 1618.

37. Zhao, Y.; Chen, S.; Shen, S. Assimilating remote sensing information with crop model using Ensemble Kalman Filter for improving LAI monitoring and yield estimation. *Ecol. Model.* **2013**, *270*, 30–42. [CrossRef]

38. Huang, J.; Gómez-Dans, J.; Huang, H.; Ma, H.; Wu, Q.; Lewis, P.; Liang, S.; Chen, Z.; Xue, J.; Wu, Y.; et al. Assimilation of remote sensing into crop growth models: Current status and perspectives. *Agric. For. Meteorol.* **2019**, *276–277*, 107609. [CrossRef]
39. Huang, J.; Sedano, F.; Huang, Y.; Ma, H.; Li, X.; Liang, S.; Tian, L.; Zhang, X.; Fan, J.; Wu, W. Assimilating a synthetic Kalman filter leaf area index series into the WOFOST model to improve regional winter wheat yield estimation. *Agric. For. Meteorol.* **2016**, *216*, 188–202. [CrossRef]
40. Chen, Y.; Cournède, P. Data assimilation to reduce uncertainty of crop model prediction with Convolution Particle Filtering. *Ecol. Model.* **2014**, *290*, 165–177. [CrossRef]
41. Huang, J.; Tian, L.; Liang, S.; Ma, H.; Becker-Reshef, I.; Huang, Y.; Su, W.; Zhang, X.; Zhu, D.; Wu, W. Improving winter wheat yield estimation by assimilation of the leaf area index from Landsat TM and MODIS data into the WOFOST model. *Agric. For. Meteorol.* **2015**, *204*, 106–121. [CrossRef]
42. Huang, J.; Ma, H.; Sedano, F.; Lewis, P.; Liang, S.; Wu, Q.; Su, W.; Zhang, X.; Zhu, D. Evaluation of regional estimates of winter wheat yield by assimilating three remotely sensed reflectance datasets into the coupled WOFOST–PROSAIL model. *Eur. J. Agron.* **2019**, *102*, 1–13. [CrossRef]
43. Behera, S.; Srivastava, P.; Pathre, U.; Tuli, R. An indirect method of estimating leaf area index in Jatropha curcas L. using LAI-2000 Plant Canopy Analyzer. *Agric. For. Meteorol.* **2010**, *150*, 307–311. [CrossRef]
44. Cheng, Z.; Meng, J.; Qiao, Y.; Wang, Y.; Dong, W.; Han, Y. Preliminary study of soil available nutrient simulation using a modified WOFOST model and time-series remote sensing observation. *Remote Sens.* **2018**, *10*, 64. [CrossRef]
45. Cheng, Z.; Meng, J.; Wang, Y. Improving spring maize yield estimation at field scale by assimilating time-series HJ-1 CCD data into the WOFOST model using a new method with fast algorithms. *Remote Sens.* **2016**, *8*, 303. [CrossRef]
46. Van Ittersum, M.; Leffelaar, P.; van Keulen, H.; Kropff, M.; Bastiaans, L.; Goudriaan, J. On approaches and applications of the Wageningen crop models. *Eur. J. Agron.* **2003**, *18*, 201–234. [CrossRef]
47. Van Diepen, C.A.; Wolf, J.; van Keulen, H.; Rappoldt, C. WOFOST: A simulation model of crop production. *Soil Use Manag.* **1989**, *5*, 16–24. [CrossRef]
48. Stol, W.; Rouse, D.; van Kraalingen, D.; Klepper, O. *FSEOPT A FORTRAN Program for Calibration and Uncertainty Analysis of Simulation Models*; A joint publication of Centre for Agrobiological Research (CABO-DLO) and Department of Theoretical Production Ecology, Agricultural University: Wageningen, The Netherlands, 1992; p. 24.
49. Li, Y.; Zhou, Q.; Zhou, J.; Zhang, G.; Chen, C.; Wang, J. Assimilating remote sensing information into a coupled hydrology-crop growth model to estimate regional maize yield in arid regions. *Ecol. Model.* **2014**, *291*, 15–27. [CrossRef]
50. Ma, H.; Huang, J.; Zhu, D.; Liu, J.; Su, W.; Zhang, C.; Fan, J. Estimating regional winter wheat yield by assimilation of time series of HJ-1 CCD NDVI into WOFOST–ACRM model with Ensemble Kalman Filter. *Math. Comput. Model.* **2013**, *58*, 753–764. [CrossRef]
51. Colombo, R.; Bellingeri, D.; Fasolini, D.; Marino, C. Retrieval of leaf area index in different vegetation types using high resolution satellite data. *Remote Sens. Environ.* **2003**, *86*, 120–131. [CrossRef]
52. Meroni, M.; Colombo, R.; Panigada, C. Inversion of a radiative transfer model with hyperspectral observations for LAI mapping in poplar plantations. *Remote Sens. Environ.* **2004**, *92*, 195–206. [CrossRef]
53. Canisius, F.; Fernandes, R.; Chen, J. Comparison and evaluation of Medium Resolution Imaging Spectrometer leaf area index products across a range of land use. *Remote Sens. Environ.* **2010**, *114*, 950–960. [CrossRef]

Article

Automatic Seamline Determination for Urban Image Mosaicking Based on Road Probability Map from the D-LinkNet Neural Network

Shenggu Yuan [1], Ke Yang [2,*], Xin Li [3] and Hongyue Cai [3]

[1] China Transport Telecommunications and Information Center, Beijing 100011, China; shengguyuan@whu.edu.cn
[2] Department of Systems Design Engineering, University of Waterloo, Waterloo, ON N2L 3G1, Canada
[3] Guojiao Spatial Information Technology (Beijing) Co., Ltd., Beijing 100011, China; lixin_kj@cttic.cn (X.L.); caihongyue@cttic.cn (H.C.)
*   Correspondence: ke.yang@uwaterloo.ca; Tel.: +1-226-3559610

Received: 10 February 2020; Accepted: 24 March 2020; Published: 26 March 2020

Abstract: Image mosaicking which is a process of constructing multiple orthoimages into a single seamless composite orthoimage, is one of the key steps for the production of large-scale digital orthophoto maps (DOM). Seamline determination is one of the most difficult technologies in the automatic mosaicking of orthoimages. The seamlines that follow the centerlines of roads where no significant differences exist are beneficial to improve the quality of image mosaicking. Based on this idea, this paper proposes a novel method of seamline determination based on road probability map from the D-LinkNet neural network for urban image mosaicking. This method optimizes the seamlines at both the semantic and pixel level as follows. First, the road probability map is obtained with the D-LinkNet neural network and related post processing. Second, the preferred road areas (*PRAs*) are determined by binarizing the road probability map of the overlapping area in the left and right image. The PRAs are the priority areas in which the seamlines cross. Finally, the final seamlines are determined by Dijkstra's shortest path algorithm implemented with binary min-heap at the pixel level. The experimental results of three group data sets show the advantages of the proposed method. Compared with two previous methods, the seamlines obtained by the proposed method pass through the less obvious objects and mainly follow the roads. In terms of the computational efficiency, the proposed method also has a high efficiency.

Keywords: mosaicking; urban image; seamline determination; deep learning; D-LinkNet

---

1. Introduction

Orthoimages have increasingly become a popular visualization product and planning instrument for integrating the rich information content of images with the geometric properties of maps (ground projection) and can be easily combined with additional information from geographic information systems (GIS) to create an orthoimage map [1]. However, with the development of technology, the orthoimage spatial resolution becomes higher, and the coverage area of an individual orthoimage is typically very small Thus, image mosaicking is a necessary process of constructing multiple images into a large-scale and single seamless composite image. This process has been applied in a wide variety of applications such as environmental monitoring, agricultural monitoring, and disaster management [2,3]. Orthoimages are typically orthorectified by the Digital Terrain Model (DTM) of the same geographical area. The quality of DTM directly affects the accuracy of orthorectification. Objects not contained in the DTM cannot be orthorectified correctly. Those objects would appear at different locations in the overlapping area and cause visual discontinuities in image mosaicking. An ideal seamline should avoid such objects [1,4–8]. A seamline is the line along which overlapping areas will be mosaicked. Each pixel in the final mosaicking result is represented entirely by only one orthoimage based on which side of the seamline it lies on. Seamline is also helpful when overlapping areas have significant differences in features. When mosaicking orthoimages, seamline determination is one of the most difficult technologies for compositing a single seamless orthoimages. The purpose of seamline determination is to find the seamlines with the minimal intensity and gradient differences in the overlapping area. In this paper, our work focuses on automatic seamline determination for urban image mosaicking.

In order to minimize the transition of the final mosaic image, the ideal seamline should avoid crossing obvious objects as much as possible and go along the objects which have small relief displacement. Differential expression is essential for seamline determination, which is a measure of the difference between left and right image overlap areas [1,7]. The method based on differential expression is the basic method of seamline optimization. The method first measures the difference between the overlap region images to form a difference matrix, and then uses the path search algorithm to obtain the final optimized seamline. According to the differential expression algorithm and the path search algorithm, the recently proposed seamline determination algorithms are as follows.

Milgram [9] defined the "best" seamline point for each line of the overlapping area that minimizes the sum of the gray differences between the left and right images. Afek and Brand [10] integrated global feature matching and local transformation into seamline determination. Soille [11] used the mathematical morphology and marker-controlled segmentation paradigm to determinate the seamlines. The difference (geometric and radiometric discontinuities) can be minimized if the seamlines go along salient image structures.

Kerschner [1] proposed a "two snake" method for seamline determination. The main idea is to design a double snake model, through mutual attraction of the two snake lines, and finally form a snake line to obtain an optimized seamline. The energy of the double snakes is defined based on similarity. The seamlines go along the region of maximum similarity. The criteria for regions of similarity are color similarity (hue and intensity) and texture similarity (orientation and magnitude of gradients). Wang et al. [12] proposed a seamline determination in image mosaicking using improved snakes. The integrated snake model and Bresenham algorithm was presented, which the Bresenham algorithm was used to calculation the photometric. This solves the local optimum problem that exists in the snake model to some extent, but not completely.

Ma and Sun [13] proposed a seamlines optimization for image mosaicking with airborne light detection and ranging (LiDAR). According to the raw laser scanning dataset, the high ground objects of the overlap area were identified as the obstacle. Then, the A* algorithm was used to determine the final seamlines, and the seamlines were kept away from these obstacles in the registered images. Wang and Wan [14–16] presented a seamline determination with the aid of vector roads for the first time. In this approach, firstly, with the help of the vector roads and their widths, the seamlines will go along the centerlines of roads with a large width as much as possible and avoid crossing the obvious objects. Finally, the shortest path algorithm is applied to determine the final seamline. Chen et al. [17] first used the Digital Surface Model (DSM) and Digital Terrain Model (DTM) to derive an Orthoimage Elevation Synchronous Model (OESM) that accurately reflected the pixel of each digital orthophoto image, and then obtained the final optimized seamline using the Dijkstra algorithm. Wang et al. [2] used vector building maps to determine the seamlines, which guaranteed the seamlines avoiding the crossing of buildings as much as possible. Different from the method of tracking vector roads, the seamlines determined by this method went along the middle line between buildings in order to avoid crossing the obvious objects, especially for high-rise buildings.

Using the normalized difference vegetation index (NDVI) and morphological building index (MBI), Pan et al. [18] introduced ground object classification into the seamline optimization method. In this approach, based on ground object classes, three types of areas, that is, obstacle areas, preferred areas, and general areas, are further formed. Then, each type of region is assigned a different weight to optimize obtained pixel-size costs. Finally, Dijkstra's algorithm is carried out to search the shortest path as the final seamlines based on previously determined pixel-size costs in overlapping areas.

Chon et al. [19] first sought the maximum difference by minimizing the maximum, and then used Dijkstra's algorithm to determine the final seamlines. The method, which is based on minimizing the maximum difference, measures the difference by a local region. It first calculates the Normalized Cross Correlation (*NCC*) of the central pixel in this window, and makes the contrast between the normalized correlation coefficients by exponential stretching of further expansion. Then, using the strategy of minimizing the maximum difference on the basis of the difference matrix, the seamline is prohibited from crossing the region, and finally the final optimized seamline is obtained by the Dijkstra algorithm.

Pan et al. [6] used image segmentation to determine seamlines for orthoimage mosaicking in an urban area for the first time. This method uses segmentation to improve seamline determination. Firstly, the preferred regions were selected according to the spans of objects segmented by the mean shift algorithm. Then, Dijkstra's shortest path algorithm was adopted to determinate the final seamline. Since this method was proposed, several object-based methods have been used to optimize seamlines. After that, Pan et al. [20] introduced the region change rate (RCR) into seamline optimization for orthoimage mosaicking. The change rate of the regions acquired by the mean shift segmentation algorithm were defined by the percentage of the changed pixels. This method determined the seamlines at object-level and pixel-level. Wang et al. [7] adopted watershed segmentation for seamline optimization at both the object and pixel level. Using normalized cross correlation, the obvious objects, such as buildings, were excluded from the preferred objects areas at the object level. Dijkstra's algorithm found the final seamlines in the preferred objects areas at the pixel level.

Li et al. [21,22] first adopted the graph cuts energy minimization framework to find the optimized seamlines. The image color, gradient magnitude and texture were combined in the smooth energy functions in the graph cuts energy minimization framework. The determined seamlines passed through the areas of a smooth texture, such as roads, woodlands, and green lands. Li et al. [23] proposed an automatic seamline optimization based on graph cuts in UVA image mosaicking.

Yuan et al. [24] proposed a seamline optimization based on the disparity image by the semi-global matching (SGM) algorithm. After obtaining the disparity image, the mathematical morphology method was employed to deal with the noises and small holes of the disparity image in order to determine the non-ground area. Finally, an improved greedy snake algorithm was adopted for the final seamlines. Similar to that algorithm, Pang et al. [25] introduced dense matching into seamline determination. Firstly, the SGM was used to estimate the disparity of each pixel. Next, the obstacle and non-obstacle areas were determined by a predefined threshold. Finally, Dijkstra's algorithm was adopted to optimize the final seamlines in avoiding crossing the obstacle area as much as possible.

Based on the integrated deep convolutional neural network (CNN) and graph cuts energy minimization framework, Li [26] proposed a novel algorithm to optimize seamlines for image mosaicking. Different from the previous method [22], this method defined similarity energy terms of the graph cut using the semantic classification classified by the CNN instead of using the color, gradient, or texture.

In our paper, we propose a novel method of seamline determination based on a road probability map which is extracted by the D-LinkNet neural network for urban image mosaicking. This method optimizes the seamlines at both the semantic and pixel level. Firstly, the D-LinkNet neural network is adopted to obtain the road probability map of the overlapping area in the left and right image respectively. Secondly, the preferred road areas (*PRAs*) are determined by binarizing the road probability map of the overlapping area both in the left and right image. The *PRAs* are priority areas which the seamlines cross. Finally, the final seamlines are determined by Dijkstra's shortest path algorithm implemented with binary min-heap at pixel level.

The remainder of this paper is organized as follows: Section 2 describes the proposed seamline determination for urban image mosaicking based on a road probability map from the D-LinkNet neural network, where Section 2.1 introduces road probability map generation by D-LinkNet. Section 2.2

presents the determination of *PRAs*, and Section 2.3 introduces pixel-level seamline determination; Section 3 describes the experimental results and analysis, where Section 3.1 presents the experimental data and platform and Section 3.2 presents the seamline determination results and analysis. Section 4 draws the conclusions.

2. Materials and Methods

The major difficulty issue of seamline determination is to define the differential expression of the overlapping area more accurately. A cost image is generally adopted to express the difference of the overlap image. Most seamline determination methods use the pixel-by-pixel or local regular subimages to define the differential expression, and it is difficult to measure the difference accurately [6,7]. Object recognition is considered to be helpful for differential expression. If object recognition has been solved perfectly, we can set the areas with the high differential expression highest cost, such as buildings. Then, the seamlines can be guaranteed not to go across stand-alone objects such as buildings. However, object recognition is a complicated problem [6]. With the help of a vector roads network, some approaches have used vector roads to optimize the seamlines, in which the seamline follows the centerlines of roads with a large width as much as possible and avoids crossing the obvious objects. Such seamlines are benefited to maintain the integrity of objects and improve the quality of image mosaics [14–16]. Based on this idea, we utilized a road probability map from the D-LinkNet neural network to optimize the seamline determination.

The process flow for the proposed method is shown in Figure 1. The proposed algorithm optimizes the seamlines at the semantic and pixel level. The D-LinkNet neural network is adopted to achieve the road probability map of the overlapping area both in the left and right image. At the semantic level, the *PRAs* are determined by binarizing the road probability map. In this step, most of the roads will be included in the *PRAs*. The *PRAs* are the priority areas which the seamlines cross. At the pixel level, the Dijkstra's shortest path searching algorithm is adopted to find the final seamlines.

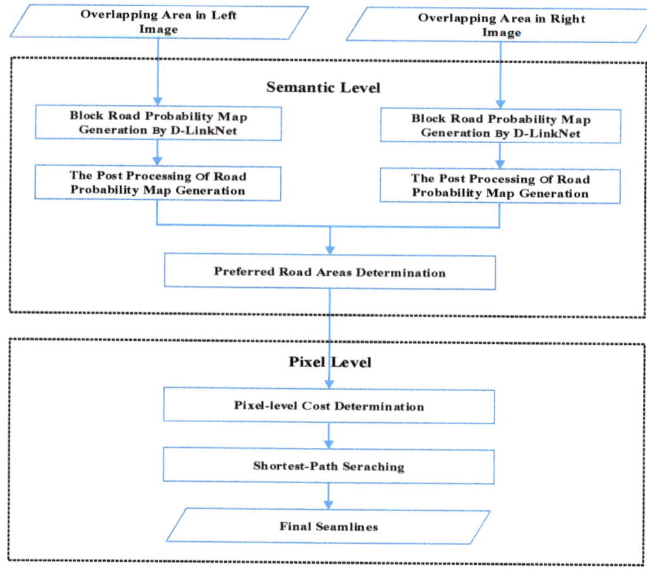

Figure 1. The process flow for the proposed method.

2.1. Road Probability Map Generation by D-LinkNet

2.1.1. Block Road Probability Map Generation by D-Linknet

Road extraction from high-resolution images is a basic application of remote sensing, which has attracted the attention of both academics and industry for a long time [27,28]. With the development of deep learning and contribution of specialized datasets from the remote sensing community, convolutional neural networks (CNNs) have been broadly used as alternatives to traditional methods for visual recognition tasks in remote sensing, including building detection [29], road segmentation [30–32], and topological map generation [33]. In our research, we focus on the main part of road extraction—road probability map generation—and model it with a convolutional neural network.

D-LinkNet [34] is adopted as a basic CNN model of the proposed method due to its excellent performance in 2018's DeepGlobe road extraction challenge and broad use as the baseline in road segmentation tasks [35]. It uses the typical encoder-decoder architecture inherited from LinkNet and adds the delated convolution part to acquire and ensemble multi-scale features to enlarge the receptive field, which is able to handle a road's properties, such as connectivity, complexity, and long span, to some extent [36]. Specifically, ResNet34 [37] is pretrained on ImageNet [38] and used as the encoder part of D-LinkNet, and several dilated convolution layers with skip connections are placed in the center part to enhance the reception ability. The decoder part uses transposed convolution [39] layers to conduct upsampling, restoring the resolution of the feature map from one that is downsampled to the original one. The architecture of the D-LinkNet is presented in Figure 2.

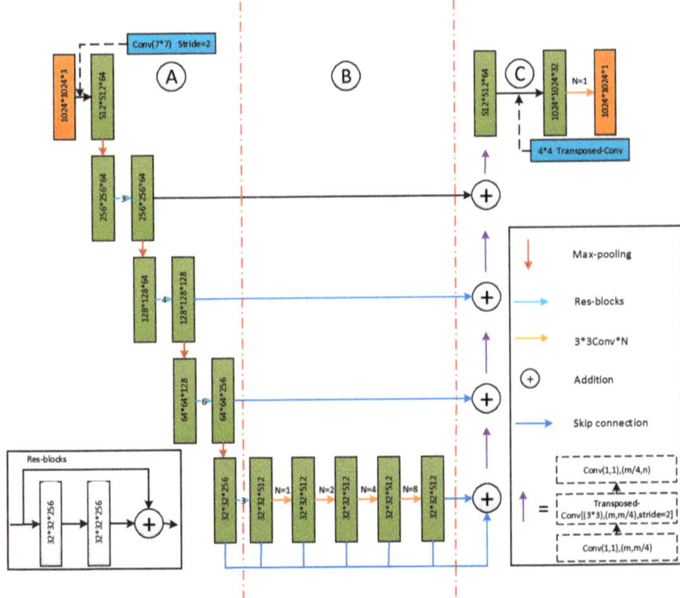

Figure 2. D-LinkNet architecture [34].

In practice, considering the large size of satellite and aerial images, clipping is necessary to generate image patches with a proper and fixed size (e.g., 1024 × 1024), to make sure that D-LinkNet implementation works under a constrained computation ability and the output of separated patches has been integrated into the final result. In addition, a small size of overlap between neighboring patches should be considered and part of the redundant output within overlap areas should be discarded, as misclassification often happens at the border pixels of a patch given that the receptive field is constrained by the edge. As for the training step of deep learning, we utilized transfer learning to accelerate the whole train process, and data augmentation of the DeepGlobe road dataset [35] to promote the network's learning ability.

Some road probability map generation by D-LinkNet is shown in Figure 3. Each size of the sample image is 1024 × 1024 pixels. The lighter the gray value of the pixel value in the image, the higher the probability that it will be recognized as a road.

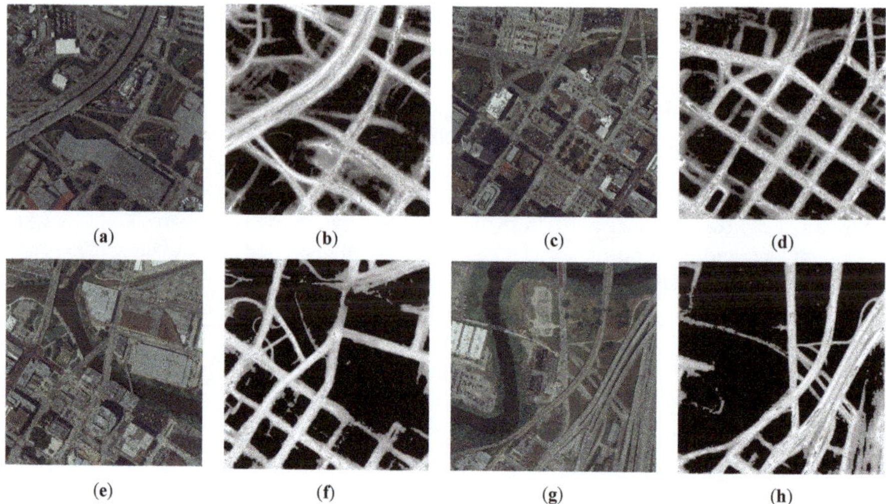

Figure 3. Road probability map generation by D-LinkNet: (**a**), (**c**), (**e**), and (**g**) are original images, (**b**), (**d**), (**f**), and (**h**) are road probability maps generation by D-LinkNet of (**a**), (**c**), (**e**), and (**g**).

2.1.2. The Post Processing of Road Probability Map Generation

When processing high-resolution remote sensing images, due to the limitations of memory and other factors, block processing is required to extract the roads using D-Link. If the block extraction results are directly stitched together, this may lead to obvious visible transitions. The stitched result is shown in Figure 4a. There are obvious visible transitions in the stitched result. Figure 4a is the result of directly stitching the image according to the size of 1024 × 1024 pixels. In order to eliminate obvious visible transitions, this paper ensures a certain overlap between adjacent blocks during blocking. The overlap provides a foundation for the subsequent elimination of obvious visible transitions. When blocking, the width of the overlapping area is 511 pixels.

Figure 4. The stitched result of the road probability map obtained by different methods: (**a**) the stitched result of directly stitching method, (**b**) the stitched result of our stitching method.

After processing according to the blocking principle to obtain the road extraction result, the post processing can be performed to eliminate obvious visible transitions on both sides of the seamline. In this paper, there are only vertical and horizontal seamlines. This kind of processing is performed on the two sides of the stitching point line by line (or row by row) within the artificially specified width range (the width of the range must be smaller than the width of overlap region). The method used is as follows:

$$OI_L^i = O_L^i + (O_L^i - O_R^i)K$$
$$OI_R^i = O_R^i + (O_R^i - O_L^i)K \tag{1}$$
$$K = i/W \quad 0 \leq i \leq W - 1$$

where O_L^i is the gray value of the pixel in the left (top) road probability map, O_R^i is the gray value of the pixel in the right(bottom) road probability map, OI_L^i is the processed gray value of the pixel in the left(top) road probability map, and OI_R^i is the processed gray value of the pixel in the right(bottom) road probability map. W is the width of the smooth area, and K is the weight [40].

In this process, the gray values of the two images to be stitched are weight averaged pixel by pixel to be used as the gray values after stitching. The weights used vary linearly and inversely within the calculation range. This process can basically eliminate the obvious difference near the stitched line, which is shown in Figure 4b. Compared with Figure 4a, there are no obvious visible transitions in Figure 4b.

2.2. Preferred Road Areas Determination

After the generation of the road probability map in the overlapping area in both the left and right image, the accuracy of detecting preferred road areas is determined by the road probability threshold for binarization. In order to estimate the road probability threshold adaptively, we used the Otsu's method [41,42] to estimate the value of the road probability threshold. Otsu's method is used to perform automatic image thresholding. The algorithm returns a single intensity threshold that separates pixels into two classes: foreground and background. This threshold is determined by minimizing the intra-class intensity variance, or equivalently, by maximizing the inter-class variance. We estimated the road probability threshold in the overlapping in the left and right image respectively. The method used is as follows Equations (2)–(4):

$$I_L(x,y) \in PRALs \begin{cases} true & P_L(x,y) >= T1 \\ false & otherwise \end{cases} \tag{2}$$

$$I_R(x,y) \in PRARs \begin{cases} true & P_R(x,y) >= T2 \\ false & otherwise \end{cases} \tag{3}$$

$$I(x,y) \in PRAs \begin{cases} true & I_L(x,y) \in PRALs \; and \; I_R(x,y) \in PRARs \\ false & otherwise \end{cases} \tag{4}$$

where $I_L(x,y)$ is the pixel in the road probability map of the overlapping area in the left image. $I_R(x,y)$ is the pixel in the road probability map of the overlapping area in the right image; $P_L(x,y)$ is the value of $I_L(x,y)$; $I_R(x,y)$ is the value of $I_R(x,y)$; $I(x,y)$ is the pixel of the overlapping area; $PRALs$ represents the preferred road areas of the overlapping area in the left image; and $PRARs$ represents the preferred road areas of the overlapping area in the right image. $PRAs$ represents the preferred road areas. $T1$ and $T2$, which are estimated by Otsu's method, are the road probability threshold of the overlapping area in the left image and right image, adaptively.

2.3. Pixel-Level Seamline Determination

2.3.1. Pixel-Level Cost Determination

After the determination of the *PRAs*, pixel-level seamline determination is used to determine the final seamlines. The two intersecting pixels of the image borders are confirmed as the start and end points [19].

Similar to Chon et al. [19], the Normalized Cross Correlation (*NCC*) is adopted to quantify the difference between the overlapping area of two images at the pixel level. Considering the efficiency, the quick *NCC* is calculated with Equation (5) using the 5×5 windows. i and j are coordinates in the image coordinate system. $I_L(i, j)$ and $I_R(i, j)$ are the gray values of the overlapping area in the left and right image at (i, j), respectively. The cost value is computed in Equation (6), which has a range of 0–1.0.

$$QNCC(x, y) = \frac{\sum\limits_{i=x-2}^{x+2} \sum\limits_{j=y-2}^{y+2} I_L(i,j)I_R(i,j) - \frac{1}{25} \sum\limits_{i=x-2}^{x+2} \sum\limits_{j=y-2}^{y+2} I_L(i,j) \sum\limits_{i=x-2}^{x+2} \sum\limits_{j=y-2}^{y+2} I_R(i,j)}{\sqrt{[\sum\limits_{i=x-2}^{x+2} \sum\limits_{j=y-2}^{y+2} I_L(i,j)^2 - \frac{1}{25}(\sum\limits_{i=x-2}^{x+2} \sum\limits_{j=y-2}^{y+2} I_L(i,j))^2][\sum\limits_{i=x-2}^{x+2} \sum\limits_{j=y-2}^{y+2} I_R(i,j)^2 - \frac{1}{25}(\sum\limits_{i=x-2}^{x+2} \sum\limits_{j=y-2}^{y+2} I_R(i,j))^2]}} \tag{5}$$

$$cost(x, y) = 0.5 - 0.5 \times QNCC(x, y) \tag{6}$$

The final pixel-level cost of pixel (x, y) in the overlapping area between images is defined as:

$$DE(x, y) = \begin{cases} w \times cost(x, y) & I(x, y) \in PRAs \\ cost(x, y) & otherwise \end{cases} \tag{7}$$

where $I(x, y)$ is the pixel of the overlapping area in the left and right image. If $I(x, y)$ belongs to *PRAs*, the cost value should be multiplied by w. w is the weight for pixels in *PRAs*, which is assigned a value much lower than 1.0. With such weight processing, this makes sure that the difference in the road area can be relatively small, so the seamlines will pass through roads as much as possible.

2.3.2. Shortest-Path Searching

After the final pixel-level cost determination, similar to Pan et al. [6], in order to minimize the difference of the seamlines, the proposed method uses the differential cost to calculate the local cost between neighboring pixels when applying Dijkstra's algorithm to search for the shortest path. The differential cost is defined in Equation (8).

$$de_{mn,pq} = |DE(m, n) - DE(p, q)| \tag{8}$$

where (m, n) and (p, q) are adjacent pixels; $DE(m, n)$ and $DE(p, q)$ are the pixel-level costs of pixels (m, n) and (p, q), respectively, which are calculated in Equation (7). Let $near(m, n)$ be the eight neighboring nodes of (m, n), $DCost(m, n)$, and $DCost(p, q)$ be the global minimum costs from the start pixel to (m, n) and (p, q), respectively. Then:

$$DCost(m, n) = \min\{de_{mn,pq} + DCost(p, q); (p, q) \in near(m, n)\} \tag{9}$$

Dijkstra's algorithm is a classic global optimization method which solves the single-source shortest-path problem for arbitrary directed graphs $G = (V, E)$ with unbounded non-negative weights [43,44]. Given a source vertex s in a weighted directed graph $G = (V, E)$ where all edges are nonnegative, the pseudo-code for Dijkstra's algorithm is presented in Algorithm 1. Dijkstra's algorithm uses a data structure for storing and querying partial solutions sorted by distance from the start. The computational complexity of the original Dijkstra's algorithm is $\Theta(|V| \cdot |V| + |E|)$, if the min-priority queue is implemented by an ordinary linked list. $|V|$ is the number of nodes in the graph and $|E|$ is the number of edges in the graph. The computational complexity depends on how to the min-priority queue is implemented. In order to improve the efficiency of Dijkstra's algorithm, similar to Wang et al. [7], the proposed method implements the min-priority queue with a binary min-heap. The pseudo-code for Dijkstra's algorithm with a binary min-heap is presented in Algorithm 2. The computational complexity of the improved Dijkstra's algorithm is $\Theta((|V|+|E|) \cdot \lg|V|)$ [7].

Algorithm 1 Dijkstra's algorithm

---

1 **Dijkstra**(G, s)
2 $dist[s] = 0$
3 **for each** vertex $v \in V$
4   **if** $v \neq s$
5          $dist[v] = \infty$
6          $pre\,[v] = $ undefined
7 $S = \varnothing$
8 $Q = V$
9 **while** $Q \neq \varnothing$ **do**
10      $u = extract\_min(Q)$
11      $S = S \cup \{u\}$
12   **for each** vertex $v \in Adj(u)$ **do**
13          $dist[v] = min(dist[v], dist[u]+w(u, v))$
14          $pre[v] = u$

Algorithm 2 Dijkstra's algorithm with Binary Min-heap

---

1 **Dijkstra_Binary_Min-heap**(G, s)
2 $dist[s] = 0$
3 **for each** vertex $v \in V$
4   **if** $v \neq s$
5          $dist[v] = \infty$
6          $pre\,[v] = $ undefined
7      $Q.add\_with\_min\text{-}priority(v, dist[v])$
8 $S = \varnothing$
9 $Q = V$
10 **while** $Q \neq \varnothing$ **do**
11      $u = extract\_min\_with\_min\text{-} priority(Q)$
12      $S = S \cup \{u\}$
13   **for each** vertex $v \in Adj(u)$ **do**
14          $dist[v] = min(dist[v], dist[u]+w(u, v))$
15          $pre[v] = u$
16          $Q.decrease\_min\text{-}priority(v, dist[v])$

---

3. Experimental Results and Analysis

3.1. Experimental Data and Platform

The experiment consists of two parts. The first part uses the D-LinkNet neural network to generate the road probability map. The second part determines the seamline based on this road probability map.

The D-LinkNet neural network was trained and tested with a single NVIDIA GeForce GTX 1080Ti using the TensorFlow library in python in Linux. The training set was composed of 6226 images from the DeepGlobe Road Extraction dataset [35] and 1000 image sets of data for manually marking roads with 0.5-m resolution remote sensing images. The image size was 1024 × 1024. Details for training a D-LinkNet-34 network are as follows: *Batchsize* = 1, *epoch* = 200, *train_best_loss* = 50, *learning_rate* = 2×10^{-4}, and the loss function defined as a mixture of binary cross entropy loss representing the error between pixels with a dice coefficient loss suitable for the error between batches based on the IOU(intersection over union) as an evaluation index. The output of the neural network is the probability of judging whether the pixel belongs to a road. It took almost 50 h to complete the network training. Once the network training is completed, the network can be applied to other datasets of different areas. After training, the error train loss is 0.18796 and the accuracy rate is 0.81204.

The proposed determination of the seamline method was implemented by C++ programming on a portable computer with four Intel(R) Core(TM) i7-6700HQ CPU at 2.60 GHz, 16.0 GB of internal memory, and a mechanical hard drive with a 1 TB capacity, a 32 MB cache, and a 7200 r/min speed for data processing. The Geospatial Data Abstraction Library (GDAL), which is a widely used open source library, was adopted to read and write a large variety of raster spatial data formats. w was the weight for pixels in *PRAs* which is shown in Equation (7). The proposed method was performed with $w = 0.001$. In our practice, the value of w was determined by our experience.

In order to verify the algorithm proposed in this paper, three sets of aerial images were selected for experiments. An overview of the three sets of data is shown in Table 1. Among them, the coverage of Dataset A is the central urban area of a large city; the coverage of Dataset B is the suburb of a big city, and the coverage of Dataset C is the center of a medium-sized city.

Table 1. Basic information of the datasets.

Dataset	Image Resolution	Imaging Size	Coverage Features
Dataset A	0.5	$3030 \times 2067 \times 3$	Central area of a big city
Dataset B	0.2	$2438 \times 4824 \times 3$	Suburb of a big city
Dataset C	0.3	$2212 \times 2693 \times 3$	Central area of a medium-sized city

3.2. Seamline Determination Results and Analysis

In order to compare the effect and efficiency of the algorithm of our proposed approach, Dijkstra's algorithm [43] and the OrthoVista method (INPHO, 2005) were selected to compare with the proposed algorithm. Dijkstra's algorithm is one of the earliest and simplest algorithms. OrthoVista is one of the most widely used professional mosaicking products in the world, which is a desktop software of INPHO's digital photogrammetric system [45].

Figure 5 illustrates the experiments of *PRAs* determination and the intermediate results for Dataset A. Figure 6 illustrates the experiments of *PRAs* determination and the intermediate results for Dataset B. Figure 7 illustrates the experiments of *PRAs* determination and the intermediate results for Dataset C. In Figures 5–7, (a) illustrates the overlapping area of the left image; (b) illustrates the overlapping area of the right image; (c) illustrates the road probability map of the overlapping area of the left image; (d) illustrates the road probability map of the overlapping area of the right image; (e) illustrates the *PRAs* of the overlapping area; and (f) illustrates the NCC cost of the overlapping area. Figures 5c, 6c, and 7c illustrate the road probability map of the overlapping area of the left image of Datasets A, B, and C, respectively. The larger the pixel value, the more likely it is to be determined as a road. Figures 5d, 6d, and 7d illustrate the road probability map of the overlapping area of the right image of Datasets A, B, and C, respectively. The larger the pixel value, the more likely it is to be determined as a road. Most of the roads in the orthoimages are extracted. In order to enhance the appearance of the road probability map, the stretch method of histogram equalize is used to adjust the value of the road probability map. Because of the differences between the overlapping area of the left and right image, the road probability maps obtained by them are also not the same, which are illustrated in (c) and (d) of Figures 5–7. With the help of the Otsu method [41,42], the preferred road areas of the left and right image are determined by the road probability threshold for binarization, respectively. The final *PRAs* of the overlapping are defined from the preferred road areas of left and right image by Equation (4). In Figures 5e, 6e, and 7e, the white areas are the *PRAs* of the overlapping areas, which are the priority areas that the seamlines pass. Most roads of the overlapping area are included in the *PRAs*, which meets our requirement. Figures 5f, 6f, and 7f illustrate the NCC cost of the overlapping area of

Datasets A, B, and C, respectively. The greater the brightness value of a pixel, the greater the NCC cost. The NCC cost is an effective method for assessing the difference of the pixel-level.

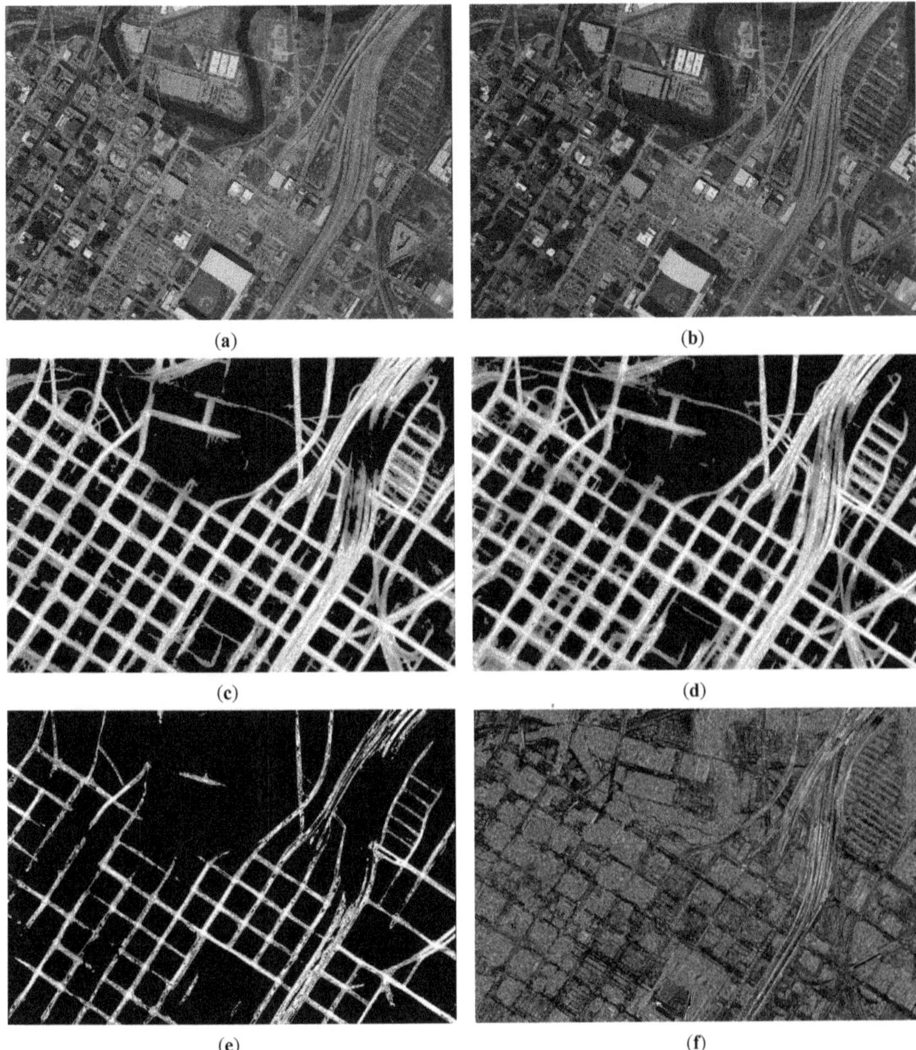

Figure 5. Experiments of preferred road areas (*PRAs*) determination for Dataset A: (**a**) the ovrlapping area of the left image, (**b**) the overlapping area of the right image, (**c**) the road probability map of the overlapping area of the left image, (**d**) the road probability map of the overlapping area of the right image, (**e**) the *PRAs* of the overlapping area, and (**f**) the Normalized Cross Correlation (*NCC*) cost of the overlapping area.

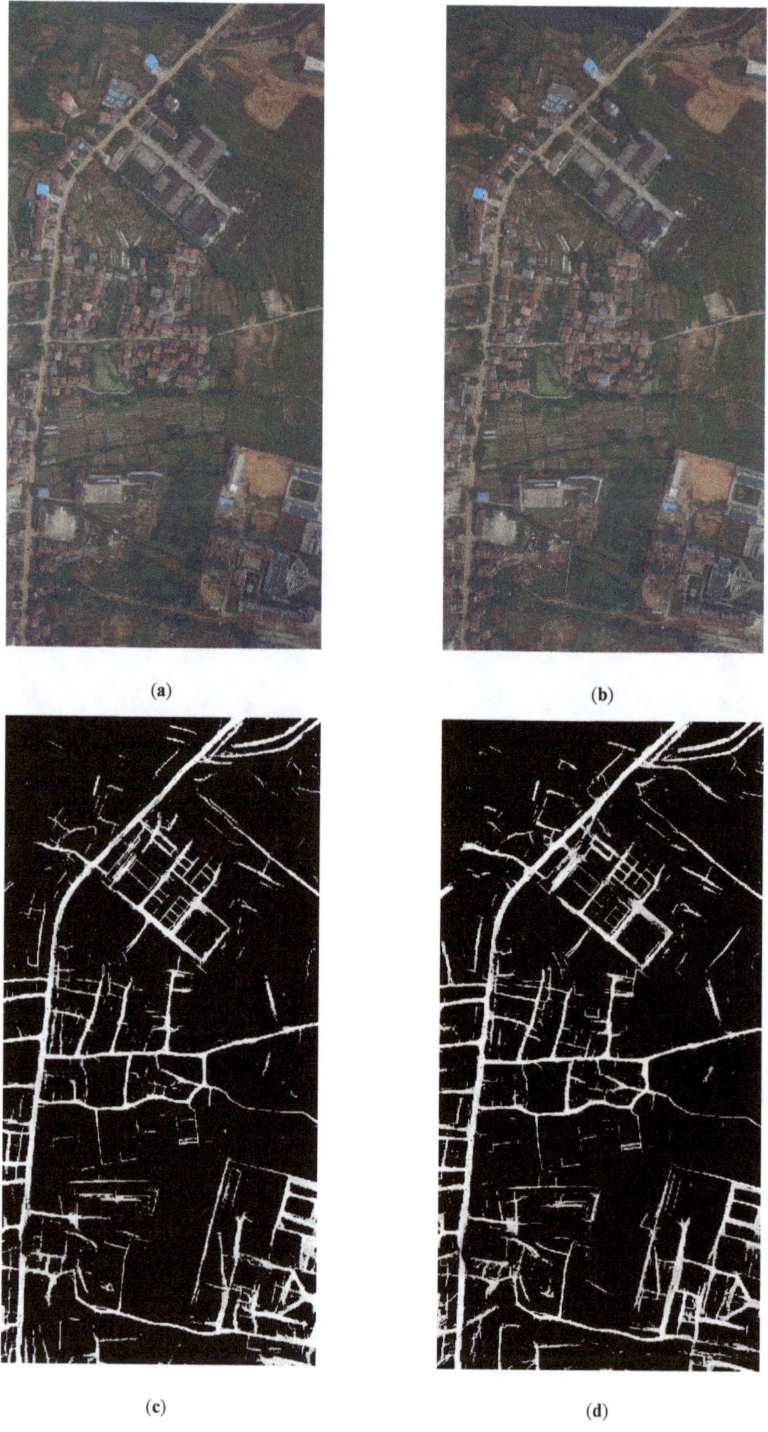

Figure 6. *Cont.*

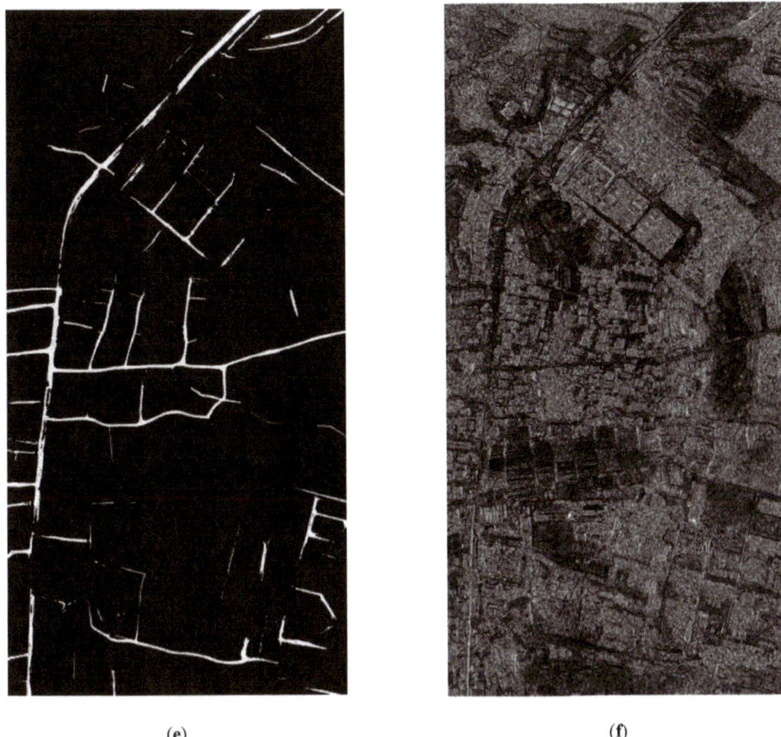

(e)

(f)

Figure 6. Experiments of preferred road areas (*PRAs*) determination for Dataset B: (**a**) the overlapping area of the left image, (**b**) the overlapping area of the right image, (**c**) the road probability map of the overlapping area of the left image, (**d**) the road probability map of the overlapping area of the right image, (**e**) the *PRAs* of the overlapping area, and (**f**) the Normalized Cross Correlation (*NCC*)cost of the overlapping area.

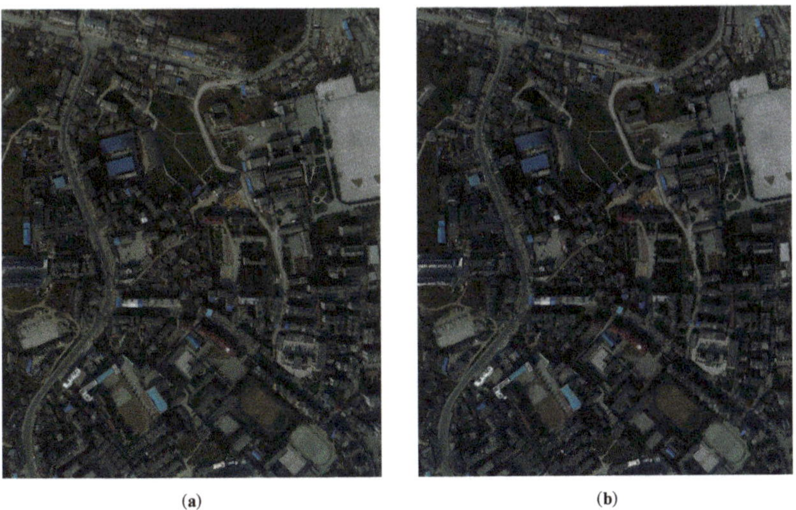

(a)

(b)

Figure 7. *Cont.*

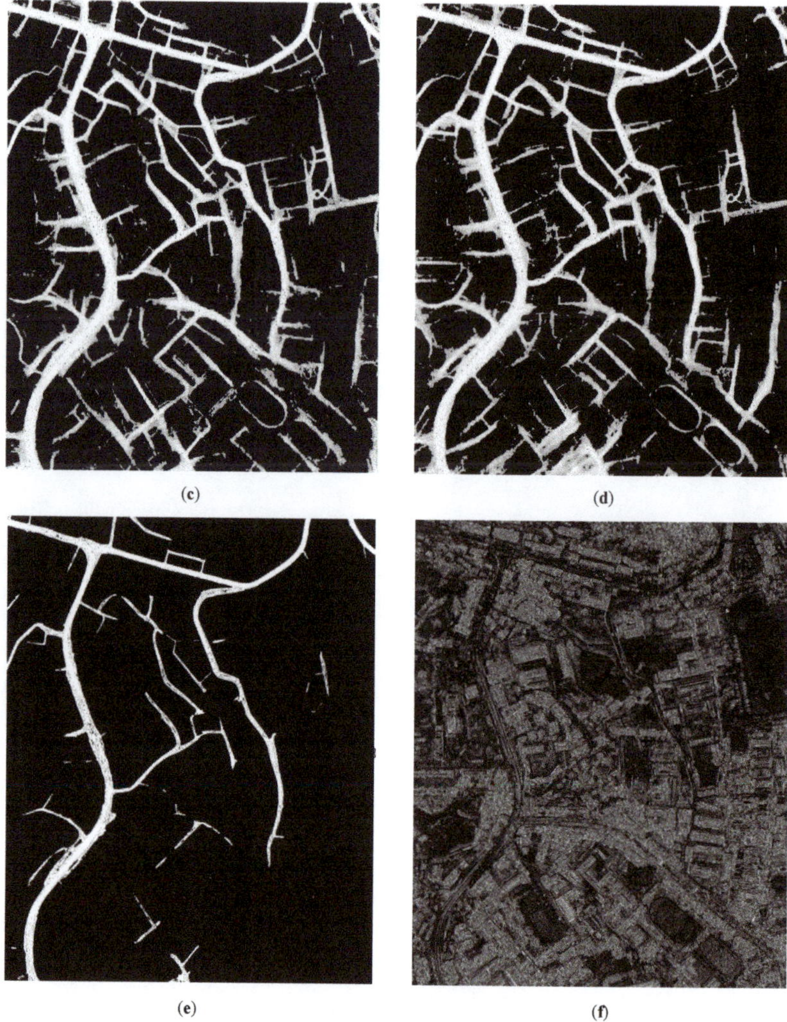

Figure 7. Experiments of preferred road areas (*PRAs*) determination for Dataset C: (**a**) the overlapping area of the left image, (**b**) the overlapping area of the right image, (**c**) the road probability map of the overlapping area of the left image, (**d**) the road probability map of the overlapping area of the right image, (**e**) the *PRAs* of the overlapping area, and (**f**) the Normalized Cross Correlation (*NCC*)cost of the overlapping area.

Figure 8 illustrates the experiments of seamline determination of the three different methods for Dataset A. Figure 9 illustrates the experiments of seamline determination of the three different methods for Dataset B. Figure 10 illustrates the experiments of seamline determination of the three different methods for Dataset C. The three methods were tested without a down-sampling strategy.

In Figures 8–10, (a), (c), and (e) illustrate the seamlines determined using Dijkstra's algorithm, OrthoVista and the proposed algorithm, respectively; (b), (d), and (f) illustrate the details of the white boxes in (a), (c), and (e), respectively. The ideal seamline should avoid crossing obvious objects, such as buildings, as much as possible and go along the objects which have small relief displacement, such as a road, river, grass, or bare land [6]. From the seamline detection results of the three data sets and especially the details of the white boxes, compared with the two previous methods, the seamlines determined by the proposed method mainly go along the roads where no significant differences exist.

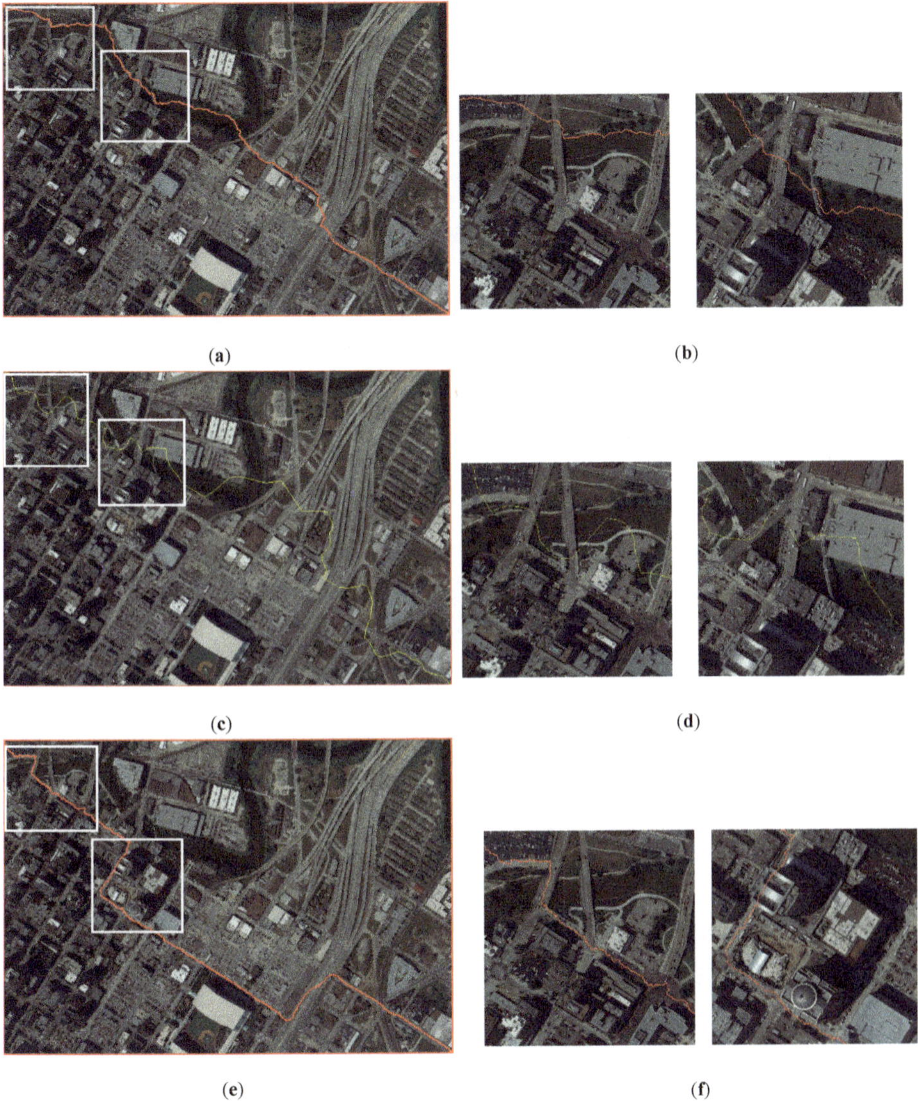

Figure 8. Experiments of seamline determination for Dataset A: (**a**) seamline determined by Dijkstra's algorithm, (**b**) details of the white box in (**a**), (**c**) seamline determined by OrthoVista, (**d**) details of the white box in (**c**), (**e**) seamline determined by the proposed method, and (**f**) details of the white box in (**e**).

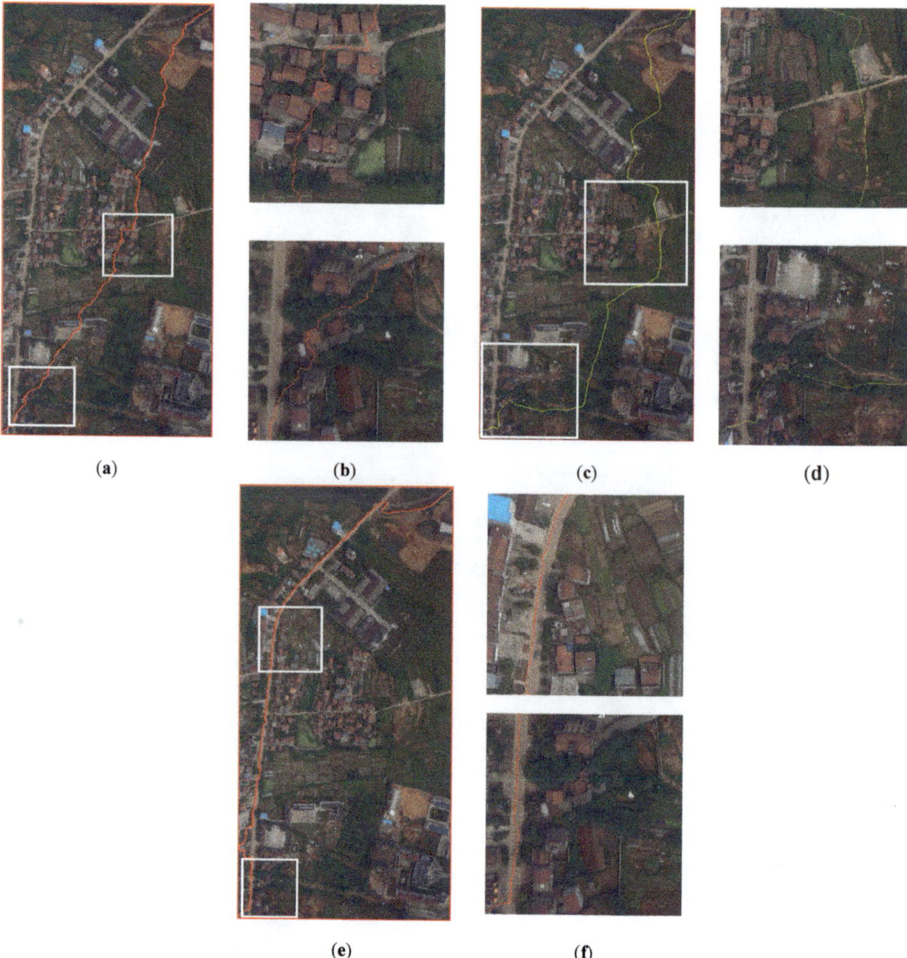

Figure 9. Experiments of seamline determination for Dataset B: (**a**) seamline determined by Dijkstra's algorithm, (**b**) details of the white box in (**a**), (**c**) seamline determined by OrthoVista, (**d**) details of the white box in (**c**), (**e**) seamline determined by the proposed method, and (**f**) details of the white box in (**e**).

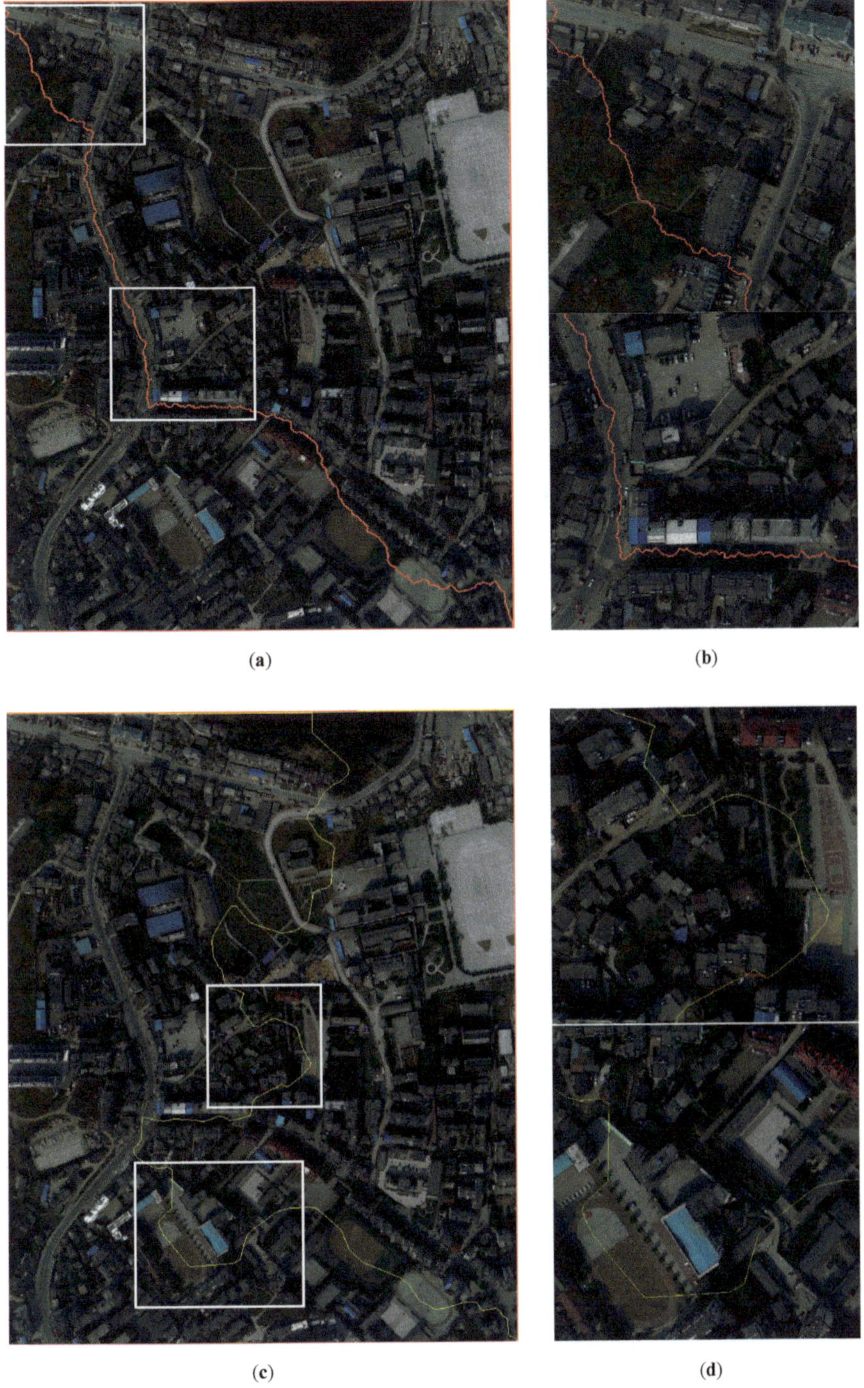

(**a**)

(**b**)

(**c**)

(**d**)

Figure 10. *Cont.*

(e)   (f)

Figure 10. Experiments of seamline determination for Dataset C: (**a**) seamline determined by Dijkstra's algorithm, (**b**) details of the white box in (**a**), (**c**) seamline determined by OrthoVista, (**d**) details of the white box in (**c**), (**e**) seamline determined by the proposed method, and (**f**) details of the white box in (**e**).

In many related studies, it is difficult to find a general method for automated quantitative assessment of the seamline quality. Therefore, similar to the evaluation method in other relevant studies [2,6,7,25], the quantitative index applied in the proposed method is the number of times that seamlines cross obvious objects. The seamlines which have a smaller number of times are considered ideal seamlines. In order to compare the efficiency of the different algorithms fairly, all the algorithms were implemented without a down-sampling strategy and parallel computing strategy. The comparison results of the three different methods in the three groups of test data are shown in Table 2.

Table 2. Comparison of previous methods with the proposed method.

Dataset	Method	Number of Obvious Objects Passed Through	Processing Time (s)
	Dijkstra's	5	329.770
1	OrthoVista	9	13.000
	Proposed	1	21.387+Δ
	Dijkstra's	6	1033.976
2	OrthoVista	6	21.000
	Proposed	0	38.353+Δ
	Dijkstra's	2	568.329
3	OrthoVista	11	9.000
	Proposed	2	8.612+Δ

The coverage area of Dataset A is located in the central urban area of large cities, for which the spatial resolution is 0.5 m. There are many high-rise buildings, overpasses, residential areas, and other buildings in the image. Such objects are the objects which the ideal seamlines should bypass. The seamline determination by Dijkstra's algorithm crossed five obvious objects, which is shown in Figure 8a. It crossed four bridges and an overpass, which is illustrated in Figure 8b. The seamline determination by the OrthoVista algorithm crossed nine obvious objects, which is shown in Figure 8c. Figure 8d shows that the seamline crossed several bridges and a buildings. Seamline determination by the proposed algorithm crossed one building, which is shown in Figure 8e. The details of the one crossed building are illustrated in Figure 8f. As shown in Figure 5e, during the optimization of the seamline at the semantic level, the proposed algorithms almost included the roads in *PRAs*. By giving the *PRAs* area a smaller weight, which is shown in Equation (7), the seamline mainly passes through the *PRAs*, and the final seamline is mainly along the roads, bypassing most of the areas with large relief displacement.

The coverage area of Dataset B is located in the suburbs area of large cities, for which the spatial resolution is 0.2 m. There is much agricultural land, woodland, and bare land in the image. There are some original villages and factories in it. A main road runs from north to south. The seamline determination by Dijkstra's algorithm crossed six obvious objects, which is shown in Figure 9a. The details of the crossed buildings are illustrated in Figure 9b. The seamline determination by the OrthoVista algorithm crossed six obvious objects, which is shown in Figure 9c. The details of the crossed buildings are illustrated in Figure 9d. Seamline determination by the proposed algorithm crossed no obvious objects and was along the north–south main road, which is shown in Figure 9e. The details of the crossed buildings are illustrated in Figure 9f. As shown in Figure 6e, during the optimization of the seamline at the semantic level, the proposed algorithms almost included the roads in *PRAs*. By giving the *PRAs* area a smaller weight, which is shown in Equation (7), the seamline mainly passes through the *PRAs*, and the final seamline is mainly along the roads, bypassing most of the areas with large relief displacement.

The coverage area of Dataset C is located in the central urban area of a medium-sized city, for which the spatial resolution is 0.5 m. In the image, in addition to buildings and residential areas, there is much agricultural land and woodland. The seamline determination by Dijkstra's algorithm crossed two obvious objects, which is shown in Figure 10a. It crossed two buildings, which is illustrated in Figure 10b. The seamline determination by the OrthoVista algorithm crossed 11 obvious objects, which is shown in Figure 10c. Figure 10d shows that the seamline crossed several buildings. Seamline determination by the proposed algorithm crossed two obvious objects, which is shown in Figure 10e. The details of the two crossed buildings are illustrated in Figure 10f. As shown in Figure 7e, during optimization of the seamline at the semantic level, the proposed algorithms almost included the roads in *PRAs*. By giving the *PRAs* area a smaller weight, which is shown in Equation (7), the seamline mainly passes through the *PRAs*, and the final seamline is mainly along the roads, bypassing most of the areas with large relief displacement.

In summary, due to the use of D-LinkNet, the roads were almost extracted and the *PRAs* were determined, which are shown in Figures 5e, 6e, and 7e. The seamline obtained by our method had the best result and passed through the less obvious objects and mainly went along the roads.

In practice, efficiency has to be taken into consideration. The processing time are recorded in the fourth column of Table 2. This shows that the seamlines determined by Dijkstra's algorithm took around 644.025 s on average. OrthoVista's method was better. It took around 14.333 s on average. The seamlines determined by the proposed method took around 22.784 +Δs on average. This processing time for the proposed method consists of two parts. The first part is the time required for extracting the road probability map using D-LinkNet, and the second part is the time required for seamline optimization. Δ represents the time required for training the network and extracting the road probability map. It took around 50 h. Regardless Δ, the proposed and OrthoVista methods are at the same level. Compared with Dijkstra's and OrthoVista's methods, Δ of the proposed method includes the network training time and road probability map extraction time. The network training time accounted for most of Δ. Although network training takes some time, once processed completely, it can be used to extract the road probability map for other data.

Figure 11 shows the experiments of seamline determination for Dataset A with different values of w. A quantitative comparison of seamlines determined by the proposed method for Dataset A with different values of w was conducted, as shown in Table 3. w is the weight for pixels in *PRAs*, which is shown in Equation (7). With such weight processing, this makes sure that the difference in the road area can be relatively small, so the seamlines will pass through roads as much as possible. In our paper, we suggested setting w to 0.001. According to Table 3, an acceptable seamlines determination result can be obtained by setting w to 0.001 for Dataset A.

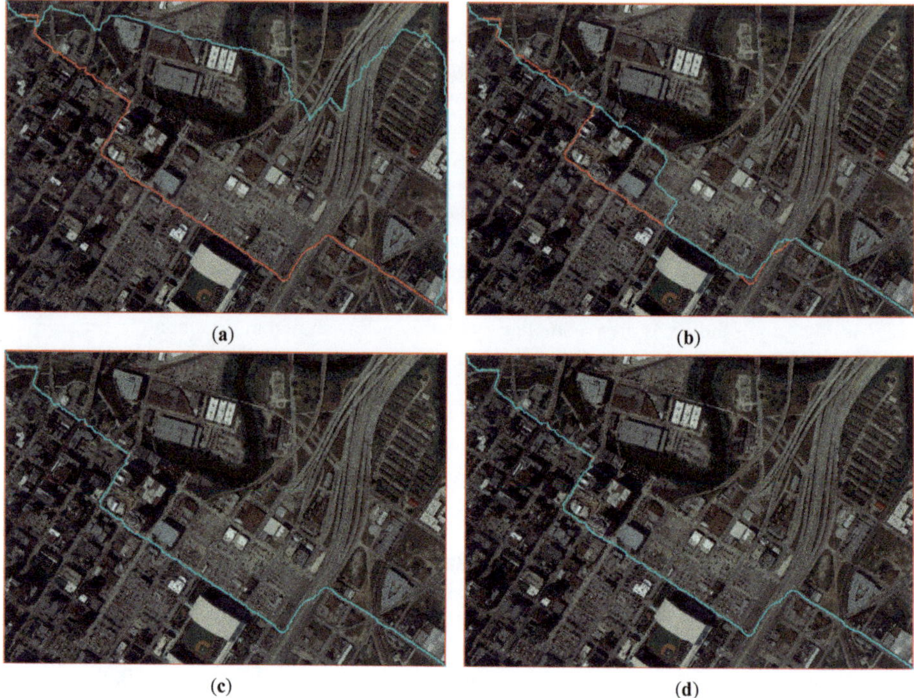

Figure 11. Experiments of seamline determination for Dataset A with different values of w: (**a**) w is 0.001 (red line) and w is 0.1 (cyan line), (**b**) w is 0.001 (red line) and w is 0.01 (cyan line), (**c**) w is 0.001 (red line) and w is 0.0001 (cyan line), and (**d**) w is 0.001 (red line) and w is 0.00001 (cyan line).

Table 3. Comparison of the proposed method with different values of w for Dataset A.

The Values of w	Number of Obvious Objects Passed Through	Processing Time (s)
0.1	11	24.522+Δ
0.01	4	18.602+Δ
0.001	1	21.387+Δ
0.0001	1	25.461+Δ
0.00001	1	25.491+Δ

4. Conclusions

In this paper, an automatic seamline determination method was presented for urban image mosaicking based on road probability map from the D-LinkNet neural network. The road probability map was used to improve the seamline determination. This method optimizes the seamlines at both the semantic and pixel level. At the semantic level, the *PRAs* are determined by binarizing the road probability map. In this step, most of the roads are included in the *PRAs*. At the pixel level, Dijkstra's algorithm is adopted to find the final seamlines. To improve the efficiency, the minimum heap is adopted to store the graph in the form of adjacency lists and extract the minimum efficiently [7]. Three group data sets of aerial orthoimages with different ground resolutions located in different cities were used to test and validate the proposed method in this paper. The comparative experimental results show the advantages of the proposed method. Compared with two previous methods, the seamline obtained by the proposed method had the best result in that it passed through the less obvious objects and mainly followed the roads. In terms of the computational efficiency, the proposed method also has a high efficiency. Moreover, the proposed method can easily be applied to the seamlines network determination framework easily [3,46,47].

Nevertheless, the proposed method may be improved in the future as follows: (1) Road probability map generation by D-LinkNet may have a significant influence on the final seamline determination. Therefore, more training samples should be made to better train the D-LinkNet neural network. (2) The proposed algorithm can be applied to the seamline network determination framework [3,46,47] to construct a single seamless composite image automatically.

Author Contributions: Conceptualization, S.Y.; methodology, S.Y. and K.Y.; software, S.Y., K.Y., and X.L.; writing—original draft preparation, S.Y., K.Y., X.L., and H.C. All authors have read and agreed to the published version of the manuscript.

Funding: This research was funded by the National Natural Science Foundation of China, grant number 41901388 and China Transport Telecommunications and Information Center Reserve Project, grant number 2018CB07. The APC was funded by the National Natural Science Foundation of China, grant number 41901388.

Conflicts of Interest: The authors declare no conflicts of interest.

References

1. Kerschner, M. Seamline detection in colour orthoimage mosaicking by use of twin snakes. *ISPRS J. Photogramm. Remote Sens.* **2001**, *56*, 53–64. [CrossRef]
2. Wang, D.; Cao, W.; Xin, X.; Shao, P.; Brolly, M.; Xiao, J.; Wan, Y.; Zhang, Y. Using vector building maps to aid in generating seams for low-attitude aerial orthoimage mosaicking: Advantages in avoiding the crossing of buildings. *ISPRS J. Photogramm. Remote Sens.* **2017**, *125*, 207–224. [CrossRef]
3. Pan, J.; Wang, M.; Ma, D.; Zhou, Q.; Li, J. Seamline network refinement based on area Voronoi diagrams 1with overlap. *IEEE Trans. Geosci. Remote Sens.* **2014**, *52*, 1658–1666. [CrossRef]
4. He, J.; Sun, M.; Chen, Q.; Zhang, Z. An improved approach for generating globally consistent seamline networks for aerial image mosaicking. *Int. J. Remote Sens.* **2019**, *40*, 859–882. [CrossRef]
5. Zheng, M.; Xiong, X.; Zhu, J. A novel orthoimage mosaic method using a weighted a* algorithm—Implementation and evaluation. *ISPRS J. Photogramm. Remote Sens.* **2018**, *138*, 30–46. [CrossRef]
6. Pan, J.; Zhou, Q.; Wang, M. Seamline determination based on segmentation for urban image mosaicking. *IEEE Trans. Geosci. Remote Sens. Lett.* **2014**, *11*, 1335–1339. [CrossRef]
7. Wang, M.; Yuan, S.; Pan, J.; Fang, L.; Zhou, Q.; Yang, G. Seamline determination for high resolution orthoimage mosaicking using watershed segmentation. *Photogramm. Eng. Remote Sens.* **2016**, *82*, 121–133. [CrossRef]
8. Dong, Q.; Liu, J. Seamline Determination Based on PKGC Segmentation for Remote Sensing Image Mosaicking. *Sensors* **2017**, *17*, 1721. [CrossRef]
9. Milgram, D.L. Computer methods for creating photomosaics. *IEEE Trans. Comput.* **1975**, *24*, 1113–1119. [CrossRef]
10. Afek, Y.; Brand, A. Mosaicking of orthorectified aerial images. *Photogramm. Eng. Remote Sens.* **1998**, *64*, 115–124.
11. Soille, P. Morphological image compositing. *IEEE Trans. Pattern Anal. Mach. Intell.* **2006**, *28*, 673–683. [CrossRef] [PubMed]

12. Wang, L.; Ai, H.; Zhang, L. Automated seamline detection in orthophoto mosaicking using improved snakes. In Proceedings of the International Conference on Information Engineering and Computer Science (ICIECS), Wuhan, China, 25–26 December 2010; pp. 1–4.

13. Ma, H.; Sun, J. Intelligent optimization of seam-line finding for orthophoto mosaicking with LiDAR point clouds. *J. Zhejiang Univ. Sci. C* **2011**, *12*, 417–429. [CrossRef]

14. Wang, D.; Wan, Y.; Xiao, J.; Lai, X.; Huang, W.; Xu, J. Aerial image mosaicking with the aid of vector roads. *Photogramm. Eng. Remote Sens.* **2012**, *78*, 1141–1150. [CrossRef]

15. Wan, Y.; Wang, D.; Xiao, J.; Wang, X.; Yu, Y.; Xu, J. Tracking of vector roads for the determination of seams in aerial image mosaics. *IEEE Geosci. Remote Sens. Lett.* **2012**, *9*, 328–332. [CrossRef]

16. Wan, Y.; Wang, D.; Xiao, J.; Lai, X.; Xu, J. Automatic determination of seamlines for aerial image mosaicking based on vector roads alone. *ISPRS J. Photogramm. Remote Sens.* **2013**, *76*, 1–10. [CrossRef]

17. Chen, Q.; Sun, M.; Hu, X.; Zhang, Z. Automatic seamline network generation for urban orthophoto mosaicking with the use of a digital surface Model. *Remote Sens.* **2014**, *6*, 12334–12359. [CrossRef]

18. Pan, J.; Yuan, S.; Li, J.; Wu, B. Seamline optimization based on ground objects classes for orthoimage mosaicking. *Remote Sens. Lett.* **2017**, *8*, 280–289. [CrossRef]

19. Chon, J.; Kim, H.; Lin, C.S. Seam-line determination for image mosaicking: A technique minimizing the maximum local mismatch and the global cost. *ISPRS J. Photogramm. Remote Sens.* **2010**, *65*, 86–92. [CrossRef]

20. Pan, J.; Wang, M.; Li, J.; Yuan, S.; Hu, F. Region change rate-driven seamline determination method. *ISPRS J. Photogramm. Remote Sens.* **2015**, *105*, 141–154. [CrossRef]

21. Li, L.; Yao, J.; Lu, X.; Tu, J.; Shan, J. Optimal seamline detection for multiple image mosaicking via graph cuts. *ISPRS J. Photogramm. Remote Sens.* **2016**, *113*, 1–16. [CrossRef]

22. Li, L.; Yao, J.; Shi, S.; Yuan, S.; Zhang, Y.; Li, J. Superpixel-based optimal seamline detection in the gradient domain via graph cuts for orthoimage mosaicking. *Int. J. Remote Sens.* **2018**, *39*, 3908–3925. [CrossRef]

23. Li, M.; Li, D.; Guo, B.; Li, L.; Wu, T.; Zhang, W. Automatic seam-line detection in UAV remote sensing image mosaicking by use of graph cuts. *ISPRS Int. J. Geo-Inf.* **2018**, *9*, 361. [CrossRef]

24. Yuan, X.; Duan, M.; Cao, J. A seam line detection algorithm for orthophoto mosaicking based on disparity image. *Acta Geod. Cartogr. Sin.* **2015**, *44*, 877–883.

25. Pang, S.; Sun, M.; Hu, X.; Zhang, Z. SGM-based seamline determination for urban orthophoto mosaicking. *ISPRS J. Photogramm. Remote Sens.* **2016**, *112*, 1–12. [CrossRef]

26. Li, L.; Yao, J.; Liu, Y.; Yuan, W.; Shi, S.; Yuan, S. Optimal seamline detection for orthoimage mosaicking by combining deep convolutional neural network and graph cuts. *Remote Sens.* **2017**, *9*, 701. [CrossRef]

27. Tupin, F.; Maitre, H.; Mangin, J.F.; Nicolas, J.M.; Pechersky, E. Detection of linear features in SAR images: Application to road network extraction. *IEEE Trans. Geosci. Remote Sens.* **1998**, *36*, 434–453. [CrossRef]

28. Zhou, T.; Sun, C.; Fu, H. Road information extraction from high-resolution remote sensing images based on road reconstruction. *Remote Sens.* **2019**, *11*, 79. [CrossRef]

29. Sun, L.; Tang, Y.; Zhang, L. Rural building detection in high-resolution imagery based on a two-stage cnn model. *IEEE Geosci. Remote Sens. Lett.* **2017**, *14*, 1998–2002. [CrossRef]

30. Zhang, Z.; Liu, Q.; Wang, Y. Road extraction by deep residual U-Net. *IEEE Geosci. Remote Sens. Lett.* **2018**, *15*, 749–753. [CrossRef]

31. Wei, Y.; Wang, Z.; Xu, M. Road structure refined CNN for road extraction in aerial image. *IEEE Geosci. Remote Sens. Lett.* **2017**, *14*, 709–713. [CrossRef]

32. Lyu, Y.; Huang, X. Road segmentation using CNN with GRU. *arXiv* **2018**, arXiv:1804.05164.

33. Ma, J.; Zhao, J. Robust topological navigation via convolutional neural network feature and sharpness measure. *IEEE Access* **2017**, *5*, 20707–20715. [CrossRef]

34. Zhou, L.; Zhang, C.; Wu, M. D-LinkNet: LinkNet with pretrained encoder and dilated convolution for high resolution satellite imagery road extraction. In Proceedings of the 2018 IEEE/CVF Conference on Computer Vision and Pattern Recognition Workshops (CVPRW), Salt Lake City, UT, USA, 18–22 June 2018; pp. 192–196.

35. Demir, I.; Koperski, K.; Lindenbaum, D.; Pang, G.; Huang, J.; Basu, S.; Hughes, F.; Tuia, D.; Raska, R. Deepglobe 2018: A challenge to parse the earth through satellite images. In Proceedings of the 2018 IEEE/CVF Conference on Computer Vision and Pattern Recognition Workshops (CVPRW), Salt Lake City, UT, USA, 18–22 June 2018; pp. 172–181.

36. Chaurasia, A.; Culurciello, E. LinkNet: Exploiting encoder representations for efficient semantic segmentation. *arXiv* **2017**, arXiv:1707.03718.

37. He, K.; Zhang, X.; Ren, S.; Sun, J. Deep residual learning for image recognition. In Proceedings of the IEEE Conference on Computer Vision and Pattern Recognition, Las Vegas Blvd, Las Vegas, NV, USA, 27–30 June 2016; pp. 770–778.
38. Deng, J.; Dong, W.; Socher, R.; Li, L.; Li, K.; Fei-Fei, L. Imagenet: A large-scale hierarchical image database. In Proceedings of the IEEE Conference on Computer Vision and Pattern Recognition, Miami, FL, USA, 20–25 June 2009; pp. 248–255.
39. Chen, C.; Tian, X.; Xiong, Z.; Wu, F. UDNet: Up-down network for compact and efficient feature representation in image super-resolution. In Proceedings of the 2017 IEEE International Conference on Computer Vision Workshop (ICCVW), Venice, Italy, 22–29 October 2017; pp. 1069–1076.
40. Wen, H.; Zhou, J. An improved algorithm for image mosaic. In Proceedings of the 2008 International Symposium on Information Science and Engieering (ISISE 2008), Shanghai, China, 20–22 December 2008; pp. 497–500.
41. Otsu, N. A threshold selection method from gray-level histograms. *Automatica* **1975**, *11*, 23–27. [CrossRef]
42. Sahoo, P.; Soltani, S.; Wong, A. A Survey of Thresholding Techniques. *Comput. Vis. Graph. Image Process.* **1988**, *41*, 233–260. [CrossRef]
43. Dijkstra, E.W. A note on two problems in connexion with graphs. *Numer. Math.* **1959**, *1*, 269–271. [CrossRef]
44. Cormen, T.; Leiserson, C.; Rivest, R.; Stein, C. *Introduction to Algorithms*; MIT Press: Cambridge, MA, USA, 2001.
45. Welcome to Amigo Optima. Available online: http://www.amigooptima.com/inpho/ortho-vista.php/ (accessed on 2 December 2019).
46. Mills, S.; McLeod, P. Global seamline networks for orthomosaic generation via local search. *ISPRS J. Photogramm. Remote Sens.* **2013**, *75*, 101–111. [CrossRef]
47. Pan, J.; Wang, M.; Li, D.; Li, J. Automatic generation of seamline network using area voronio diagrams with overlap. *IEEE Trans. Geosci. Remote Sens.* **2009**, *47*, 1737–1744. [CrossRef]

 *sensors*

Article

Satellite Laser Ranging for Retrieval of the Local Values of the Love h_2 and Shida l_2 Numbers for the Australian ILRS Stations

Marcin Jagoda [1,*], Miłosława Rutkowska [1], Paweł Lejba [2], Jacek Katzer [3], Romuald Obuchovski [4] and Dominykas Šlikas [4]

[1]   Faculty of Civil Engineering, Environmental and Geodetic Sciences, Koszalin University of Technology, Śniadeckich 2, 75-453 Koszalin, Poland; miloslawa.rutkowska@tu.koszalin.pl

[2]   Space Research Centre, Polish Academy of Sciences, Bartycka 18A, 00-716 Warsaw, Poland; plejba@cbk.poznan.pl

[3]   Faculty of Geoengineering, University of Warmia and Mazury in Olsztyn, Prawocheńskiego 15, 10-720 Olsztyn, Poland; jacek.katzer@uwm.edu.pl

[4]   Department of Geodesy and Cadastre, Vilnius Gediminas Technical University, Saulėtekio 11, LT-10233 Vilnius, Lithuania; romuald.obuchovski@vgtu.lt (R.O.); dominykas.slikas@vgtu.lt (D.Š.)

*   Correspondence: marcin.jagoda@tu.koszalin.pl

Received: 26 October 2020; Accepted: 28 November 2020; Published: 30 November 2020

Abstract: This paper deals with the analysis of local Love and Shida numbers (parameters h_2 and l_2) values of the Australian Yarragadee and Mount Stromlo satellite laser ranging (SLR) stations. The research was conducted based on data from the Medium Earth Orbit (MEO) satellites, LAGEOS-1 and LAGEOS-2, and Low Earth Orbit (LEO) satellites, STELLA and STARLETTE. Data from a 60-month time interval, from 01.01.2014 to 01.01.2019, was used. In the first research stage, the Love and Shida numbers values were determined separately from observations of each satellite; the obtained values of h_2, l_2 exhibit a high degree of compliance, and the differences do not exceed formal error values. At this stage, we found that it was not possible to determine l_2 from the data of STELLA and STARLETTE. In the second research stage, we combined the satellite observations of MEO (LAGEOS-1+LAGEOS-2) and LEO (STELLA+STARLETTE) and redefined the h_2, l_2 parameters. The final values were adopted, and further analyses were made based on the values obtained from the combined observations. For the Yarragadee station, local $h_2 = 0.5756 \pm 0.0005$ and $l_2 = 0.0751 \pm 0.0002$ values were obtained from LAGEOS-1 + LAGEOS-2 and $h_2 = 0.5742 \pm 0.0015$ were obtained from STELLA+STARLETTE data. For the Mount Stromlo station, we obtained the local $h_2 = 0.5601 \pm 0.0006$ and $l_2 = 0.0637 \pm 0.0003$ values from LAGEOS-1+LAGEOS-2 and $h_2 = 0.5618 \pm 0.0017$ from STELLA + STARLETTE. We found discrepancies between the local parameters determined for the Yarragadee and Mount Stromlo stations and the commonly used values of the h_2, l_2 parameters averaged for the whole Earth (so-called global nominal parameters). The sequential equalization method was used for the analysis, which allowed to determine the minimum time interval necessary to obtain stable h_2, l_2 values. It turned out to be about 50 months. Additionally, we investigated the impact of the use of local values of the Love/Shida numbers on the determination of the Yarragadee and Mount Stromlo station coordinates. We proposed to determine the stations (X, Y, Z) coordinates in International Terrestrial Reference Frame 2014 (ITRF2014) in two computational versions: using global nominal h_2, l_2 values and local h_2, l_2 values calculated during this research. We found that the use of the local values of the h_2, l_2 parameters in the process of determining the stations coordinates influences the result.

Keywords: Love/Shida numbers; satellite laser ranging (SLR); Yarragadee station; Mount Stromlo station; LAGEOS; STELLA; STARLETTE satellites; SLR stations coordinates; ITRF2014

1. Introduction

There are different kinds of external forces acting on the Earth which cause its gradual changes; for this reason, our planet needs constant monitoring. One of these forces is tidal forces, which are reflected, among other things, in the displacement of Earth's masses and, consequently, in changes in the position of points on the Earth's surface. To describe the flexible reaction of the Earth to tidal stresses, the concept of so-called Love (h, k) and Shida (l) numbers were introduced. These are tidal parameters whose detailed description was presented in the fundamental works of A.E.H. Love "Some problems of geodynamics" [1] and T. Shida and M. Matsoyama "Note of Hecker's observations" [2].

The Earth's dynamics is currently studied using satellite measurement techniques, including the satellite laser ranging (SLR) technique [3]. In our earlier research programme, e.g., [4–6], we have successfully demonstrated that the SLR technique makes it possible to determine tidal parameters with very high accuracy, it was also indicated in [7]. Other satellite measuring techniques can also be used for such purposes, e.g., the VLBI technique [8,9]; satellite altimetry [10,11]. All these publications are focused on determining the global values of tidal parameters averaged over the whole Earth.

Due to the heterogeneous structure of our planet, it is reasonable that the reaction to tidal stresses is not the same for the whole Earth. With this in mind, we have launched a research programme to analyze the local tidal parameters. The research carried out so far focused on the Baltic Sea region [12,13], where local tidal parameters for the SLR stations from Poland and Latvia were analyzed based on data from the LAGEOS-1 and LAGEOS-2 satellites. In this study, we made an attempt to determine and analyze local values of tidal parameters for two Australian SLR stations: Yarragadee (no. 70900513, approx. 29° S, 115° E) and Mount Stromlo (no. 78259001, approx. 35° S, 149° E). In addition, we assessed the impact of their use on the determination of the coordinates of these stations. These tasks constitute the main research objective of this work. Estimation of the minimum time interval ensuring the stability of the determination and the assessment of the possibility of determining local tidal parameters from the data of the LEO satellites STELLA and STARLETTE constitute the intermediary purpose of this study.

The data provided by the Australian Yarragadee and Mount Stromlo stations are extremely important for geodynamic research. The global SLR network consists of 38 stations, of which only eight are located in the Southern Hemisphere; two of them on the Australian continent. Their location is shown in Figure 1.

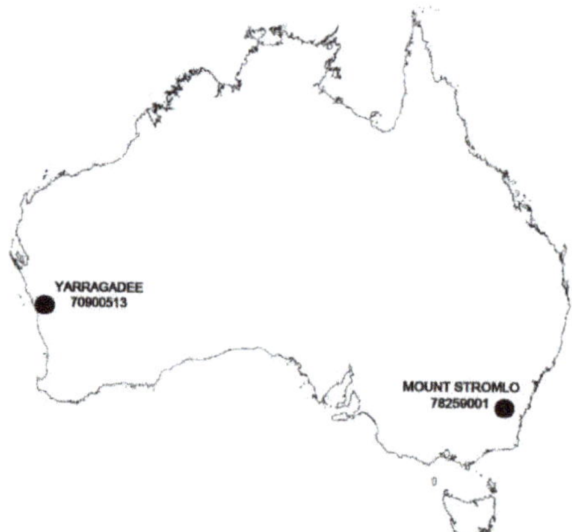

Figure 1. Location of the Australian satellite laser ranging (SLR) stations.

The Yarragadee station is located in Western Australia, near the city of Dongara. The Mount Stromlo satellite laser ranging observatory is located in the south-eastern part of the continent, near the city of Geraldton. These stations are part of the Western Pacific Laser Tracking Network and contribute to the International Laser Ranging Service (ILRS). These are some of the best stations in terms of accuracy and number of observations. The data they collect plays a very important role, in synergy with other geodetic techniques, in defining International Terrestrial Reference Frame (ITRF) and determining Earth Orientation Parameters (EOP).

The basis of the satellite laser ranging technique (SLR) is the measurement of two-way time of light pulses flight between a station and a satellite fitted with retroreflectors. The distance measured to the satellite must be adjusted to accommodate the effects of a speed of light decrease and the difference between the straight and curved paths of ray. Furthermore, it must take into consideration the distance from the retroreflector to the satellite mass center, influence of the satellite motion, the Earth rotation and relativistic effects [14]. In its simplified form, the equation of laser observation is as follows [14]:

$$\rho = \frac{C\Delta t}{2} \tag{1}$$

where ρ is the distance between a station and a satellite, Δt is the two way time interval of light pulses flight between a station and a satellite, and C is the speed of light.

Typical, geodetic SLR satellites are sphere-shaped, covered with retroreflectors and can be divided into two main groups: Medium Earth Orbit (MEO) satellites; e.g., LAGEOS-1 (Perige = 5860 km), LAGEOS-2 (Perige = 5620 km); and Low Earth Orbit (LEO) satellites, e.g., STELLA (Perige = 804 km) and STARLETTE (H = 812 km). The data from these satellites is widely used in geodynamic research; e.g., to determine stations coordinates [15–17], to study the gravitational field of the Earth [18], to determine Earth Orientation Parameters [19–22], or to study the tidal phenomenon [23,24]. In this work, we used the data of the LAGEOS-1, LAGEOS-2, STELLA, and STARLETTE satellites to determine the local values of the tidal parameters and coordinates of the Australian SLR Yarragadee and Mount Stromlo stations. A detailed description of the SLR technique can be found in the works [25,26], while a wide range of applications of laser satellites in geodynamic research has been presented in [27–29].

The gravitational impact of the Moon, the Sun, and the Solar System planets on the Earth's surface results in the creation of earth and ocean tides. The tidal forces cause the displacement of earth and ocean masses. A detailed description of the tide phenomenon and its mathematical basis can be found in fundamental work "The tides of the planet Earth" by P. Melchior [30]. These changes in the distribution of the Earth's masses related to tides are expressed by movements of observation stations, as described by Equation (2) given in [31]:

$$\Delta X = \sum_{j=2}^{3}\left[\frac{GM_j}{GM_E}\frac{a_e^4}{d_j^3}\right]\left\{\left[3l_2\left(\hat{R}_j\hat{r}_{sta}\right)\right]\overline{X}_j + \left[3\left(\frac{h_2}{2}-l_2\right)\left(\hat{R}_j\hat{r}_{sta}\right)^2 - \frac{h_2}{2}\right]\overline{X}_{sta}\right\}$$

$$\Delta Y = \sum_{j=2}^{3}\left[\frac{GM_j}{GM_E}\frac{a_e^4}{d_j^3}\right]\left\{\left[3l_2\left(\hat{R}_j\hat{r}_{sta}\right)\right]\overline{Y}_j + \left[3\left(\frac{h_2}{2}-l_2\right)\left(\hat{R}_j\hat{r}_{sta}\right)^2 - \frac{h_2}{2}\right]\overline{Y}_{sta}\right\} \tag{2}$$

$$\Delta Z = \sum_{j=2}^{3}\left[\frac{GM_j}{GM_E}\frac{a_e^4}{d_j^3}\right]\left\{\left[3l_2\left(\hat{R}_j\hat{r}_{sta}\right)\right]\overline{Z}_j + \left[3\left(\frac{h_2}{2}-l_2\right)\left(\hat{R}_j\hat{r}_{sta}\right)^2 - \frac{h_2}{2}\right]\overline{Z}_{sta}\right\}$$

where

GM_j—gravitational parameter for the Moon ($j = 2$) or the Sun ($j = 3$),

GM_E—gravitational parameter for the Earth,

a_e—equatorial radius,

d_j—distance to the Moon ($j = 2$) or Sun ($j = 3$),

$\hat{R}_j$—the unit vector from the geocenter to the Moon ($j = 2$) or Sun ($j = 3$),

$\hat{r}_{sta}$—the unit vector from the geocenter to the station,

$(\overline{X}_j, \overline{Y}_j, \overline{Z}_j)$—the Cartesian components of the unit vector $\hat{R}_j$,

$(\overline{X}_{sta}, \overline{Y}_{sta}, \overline{Z}_{sta})$—the Cartesian components of the unit vector $\hat{r}$,

h_2, l_2—second degree of Love and Shida numbers.

In Equation (2), there are the tidal parameters h_2 and l_2 (Love and Shida numbers for the second-degree tides). The former refers to the radial tidal displacement of the station, the latter to the horizontal displacement, it is described in [32]. The tidal parameters are a measure of the flexible Earth's response to stresses created by tidal forces. If we assume that the Earth is a rigid body, then no elastic deformation takes place, and h, l are both 0. If we assume another extreme case in which the Earth is not just elastic but rather a liquid body, then the Love and Shida numbers are both equal to 1. Thus, for a rigid Earth $h = 0$, $l = 0$, for a liquid Earth $h = 1$, $l = 1$ and for an elastic Earth they take intermediate values: $0 < h < 1$, $0 < l < 1$. According to International Earth Rotation and Reference Systems Service (IERS) Conventions (IERS Technical Note No. 36) [33], the Earth's global (so called nominal) averaged values of the Love and Shida numbers for the second degree tides are $h_2 = 0.6078$, $l_2 = 0.0847$.

2. Materials and Methods

To determine the local tidal parameters of the Australian Yarragadee and Mount Stromlo stations, we used observation data in the form of normal points of these stations collected for the LAGEOS-1, LAGEOS-2, STELLA, and STARLETTE satellites for the 5-year interval from 01.01.2014 to 01.01.2019. For the Yarragadee station, these were respectively: 57,299 LAGEOS-1 normal points, 58,133 LAGEOS-2 normal points, 38,031 STELLA normal points, 90,953 STARLETTE normal points; for the Mount Stromlo station: 25,249 LAGEOS-1 normal points, 25,962 LAGEOS-2 normal points, 21,654 STELLA normal points, 50,172 STARLETTE normal points. The method of creating normal points from SLR measurements is described in [34]. The data from the analyzed period were used to create 7-day orbital arcs. In total, 260 orbital arcs were obtained for each of the satellites. Satellite orbits were determined using the Cowell Numerical Integration method as described in detail in [31], using standard procedures, force models and constants recommended by the International Earth Rotation and Reference Systems Service (IERS) [33] and International Laser Ranging Service (ILRS) [35]. RMS values of the post-fit residuals, calculated from formula (3), were used as the satellites orbits accuracy determination [5]:

$$\text{RMS of the post-fit residuals} = \sqrt{\frac{\sum\limits_{i=1}^{n}(O_i - C_i)^2}{n-1}} \tag{3}$$

where i denotes successive number of normal points, $(O_i - C_i)$ is the SLR observation minus the computed distance from the station to the satellite. The following values were obtained: RMS of the post-fit residuals: RMS(LAGEOS-1) = 1.02 cm, RMS(LAGEOS-2) = 1.01 cm, RMS(STELLA) = 1.98 cm, RMS(STARLETTE) = 1.87 cm.

To determine the local tidal parameters h_2, l_2 values of the Yarragadee and Mount Stromlo stations and their coordinates, an observation Equation (4) was formulated and solved using the Bayesian least square method, a detailed description of this procedure is given in [31]. The local tidal parameters and coordinates were determined independently for both of the analyzed stations.

$$(O_i - C_i) = -\left\{ \sum_{j=1}^{n} \frac{\partial C_i}{\partial \varepsilon_j} d\varepsilon_j + \frac{\partial C_i}{\partial h_2} dh_2 + \frac{\partial C_i}{\partial l_2} dl_2 \right\} + dO_i \tag{4}$$

where

j—number of adjusted parameters (satellite position and velocity, empirical accelerations, and the station position),

$d\varepsilon_j$—corrections to the j-th parameter,

dh_2, dl_2—corrections for Love number h_2 and for Shida number l_2,

dO_i—error of observation associated with the i-th measurement.

Given in Equation (4) the $\frac{\partial C_i}{\partial h_2}$, $\frac{\partial C_i}{\partial l_2}$ quantities are calculated by differentiating Equation (2) and are expressed as follows [31]:

$$\frac{\partial C_i}{\partial h_2} = \frac{\partial C_i}{\partial X_{sta}}\frac{\partial X_{sta}}{\partial h_2} + \frac{\partial C_i}{\partial Y_{sta}}\frac{\partial Y_{sta}}{\partial h_2} + \frac{\partial C_i}{\partial Z_{sta}}\frac{\partial Z_{sta}}{\partial h_2} \tag{5}$$

$$\frac{\partial C_i}{\partial l_2} = \frac{\partial C_i}{\partial X_{sta}}\frac{\partial X_{sta}}{\partial l_2} + \frac{\partial C_i}{\partial Y_{sta}}\frac{\partial Y_{sta}}{\partial l_2} + \frac{\partial C_i}{\partial Z_{sta}}\frac{\partial Z_{sta}}{\partial l_2} \tag{6}$$

where

$$\frac{\partial X_{sta}}{\partial h_2} = \sum_{j=2}^{3} \frac{GM_j}{GM_E}\frac{a_e^4}{d_j^3}\left[\frac{3}{2a_e}(\hat{R}_j\hat{r}_{st})^2 - \tfrac{1}{3}\right]X_{sta},$$

$$\frac{\partial Y_{sta}}{\partial h_2} = \sum_{j=2}^{3} \frac{GM_j}{GM_E}\frac{a_e^4}{d_j^3}\left[\frac{3}{2a_e}(\hat{R}_j\hat{r}_{st})^2 - \tfrac{1}{3}\right]Y_{sta}, \tag{7}$$

$$\frac{\partial Z_{sta}}{\partial h_2} = \sum_{j=2}^{3} \frac{GM_j}{GM_E}\frac{a_e^4}{d_j^3}\left[\frac{3}{2a_e}(\hat{R}_j\hat{r}_{st})^2 - \tfrac{1}{3}\right]Z_{sta},$$

$$\frac{\partial X_{sta}}{\partial l_2} = 3\sum_{j=2}^{3} \frac{GM_j}{GM_E}\frac{a_e^4}{d_j^3}\left[3(\hat{r}_{st}\hat{R}_j)\overline{X}_j - \tfrac{1}{a_e}(\hat{R}_j\hat{r}_{st})^2 X_{sta}\right],$$

$$\frac{\partial Y_{sta}}{\partial l_2} = 3\sum_{j=2}^{3} \frac{GM_j}{GM_E}\frac{a_e^4}{d_j^3}\left[3(\hat{r}_{st}\hat{R}_j)\overline{Y}_j - \tfrac{1}{a_e}(\hat{R}_j\hat{r}_{st})^2 Y_{sta}\right], \tag{8}$$

$$\frac{\partial Z_{sta}}{\partial l_2} = 3\sum_{j=2}^{3} \frac{GM_j}{GM_E}\frac{a_e^4}{d_j^3}\left[3(\hat{r}_{st}\hat{R}_j)\overline{Z}_j - \tfrac{1}{a_e}(\hat{R}_j\hat{r}_{st})^2 Z_{sta}\right].$$

The sequential method was used to determine local tidal parameters. In the first step, the h_2 and l_2 parameters were determined separately from each orbital arc (arc1, arc2, arc3, … , arc260). The following steps consisted in adding subsequent arcs to the calculations, one after another, following the scheme: arc1 + arc2, arc1 + arc2 + arc3, … , arc1 + arc2 + … + arc260. In each subsequent step, h_2 and l_2 parameters were re-computed. The values given in IERS Technical Note No. 36 [33] were taken as priori values ($h_2 = 0.6078$ and $l_2 = 0.0847$). In the first calculation stage, the local tidal parameters were determined separately from LAGEOS-1, LAGEOS-2, STELLA, and STARLETTE data, then data from LAGEOS-1 and LAGEOS-2 and STELLA and STARLETTE were pooled (LAGEOS-1+LAGEOS-2 and STELLA+STARLETTE) and re-computed to increase the accuracy and stability of the solutions. The final values were adopted and further analyses were made based on the values obtained from the combined observations of 260 orbital arcs.

Additionally, the coordinates of the Yarragadee and Mount Stromlo stations were determined in course of the analysis. These coordinates were calculated from the Equation (4). The determination method of the stations' coordinates from the SLR data was set out in detail in [14,36]. The coordinates of the Yarragadee and Mount Stromlo stations were determined from the LAGEOS-1 + LAGEOS-2 data with a presumptive assumption of the stations coordinates in the ITRF2014 reference frame [37]. The adjustment was performed in two calculation versions. As regards the first one, Yarragadee and Mount Stromlo stations coordinates were calculated using of the global nominal values (recommended in the IERS Conventions [33]) of tidal parameters. In the second one, Yarragadee and Mount Stromlo stations coordinates were estimated using the local values of tidal parameters calculated in this present paper. The impact of the application of different values of tidal parameters on the determination of these stations coordinates was then investigated.

The GEODYN II NASA GSFC software [31] was used for all the calculations related to the determination of satellite orbits, local tidal parameters and coordinates of Yarragadee and Mount Stromlo SLR stations.

3. Results and Discussion

In this paper, we present the results of the determination of the local values of tidal parameters h_2, l_2 for the Australian SLR stations Yarragadee and Mount Stromlo, and their coordinates in the ITRF2014 reference frame [37]. The first stage of the research included determining the local tidal parameters separately from the data of each of the satellites: LAGEOS-1, LAGEOS-2, STELLA, and STARLETTE. The obtained h_2, l_2 values show a high degree of consistency, and the differences do not exceed formal error values (please refer to Table 1). At this stage, we found that it was not possible to determine l_2 from STELLA and STARLETTE data. Then, to increase the accuracy and stability of the solutions, we pooled the data of the individual satellite groups, LAGEOS-1+LAGEOS-2 and STELLA+STARLETTE, and re-computed the local tidal parameters. We did not determine the l_2 parameter from STELLA+STARLETTE data. The values obtained in this way were assumed final and subjected to further analysis. The final estimated values of the local tidal parameters for the Yarragadee and Mount Stromlo stations are given in Table 1, whereas the results of the sequential determination method are shown in Figures 2–7. For clarity and readability of the figures, we present results orbital arcs combined in groups of ten (arcs 1–10, 1–20, 1–30, ... , 1–260).

Table 1. Local tidal parameters h_2, l_2 for Yarragadee and Mount Stromlo SLR stations.

SLR Data	Yarragadee (No. 70900513)		Mount Stromlo (No. 78259001)	
	h_2	l_2	h_2	l_2
LAGEOS-1	0.5764 ± 0.0007	0.0744 ± 0.0004	0.5616 ± 0.0009	0.0646 ± 0.0005
LAGEOS-2	0.5758 ± 0.0007	0.0748 ± 0.0004	0.5609 ± 0.0009	0.0650 ± 0.0005
LAGEOS-1+LAGEOS-2	0.5756 ± 0.0005	0.0751 ± 0.0002	0.5601 ± 0.0006	0.0637 ± 0.0003
STELLA	0.5741 ± 0.0022	0.0334 ± 0.0014 (unacceptable value)	0.5622 ± 0.0026	0.0212 ± 0.0020 (unacceptable value)
STARLETTE	0.5750 ± 0.0019	0.1785 ± 0.0013 (unacceptable value)	0.5604 ± 0.0022	0.0093 ± 0.0018 (unacceptable value)
STELLA+STARLETTE	0.5742 ± 0.0015	not estimated	0.5618 ± 0.0017	not estimated

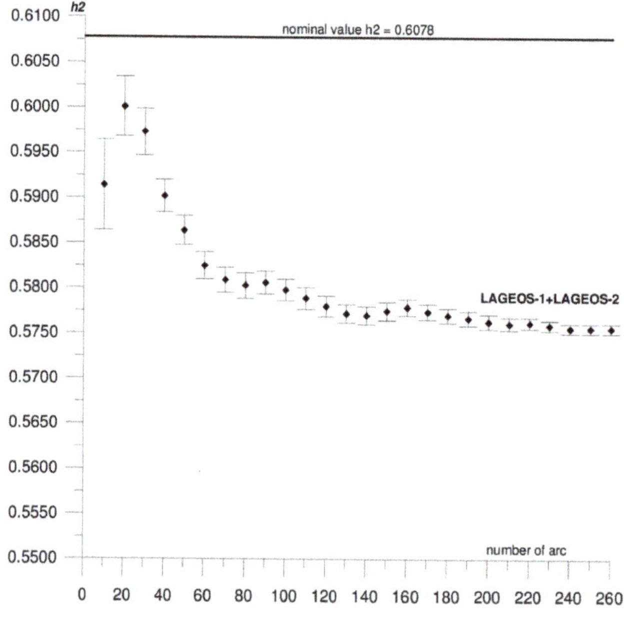

Figure 2. Sequential solution for the Yarragadee local h_2 parameter based on LAGEOS-1+LAGEOS-2 data.

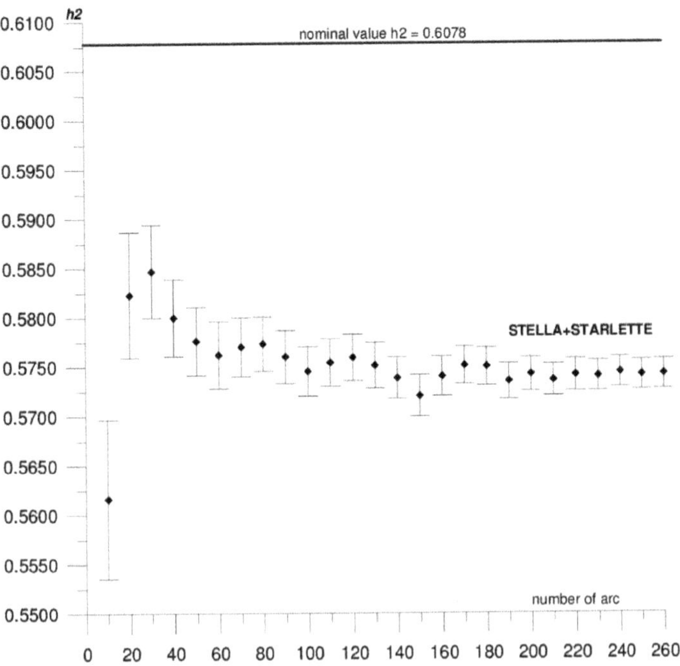

Figure 3. Sequential solution for the Yarragadee local h_2 parameter based on STELLA+STARLETTE data.

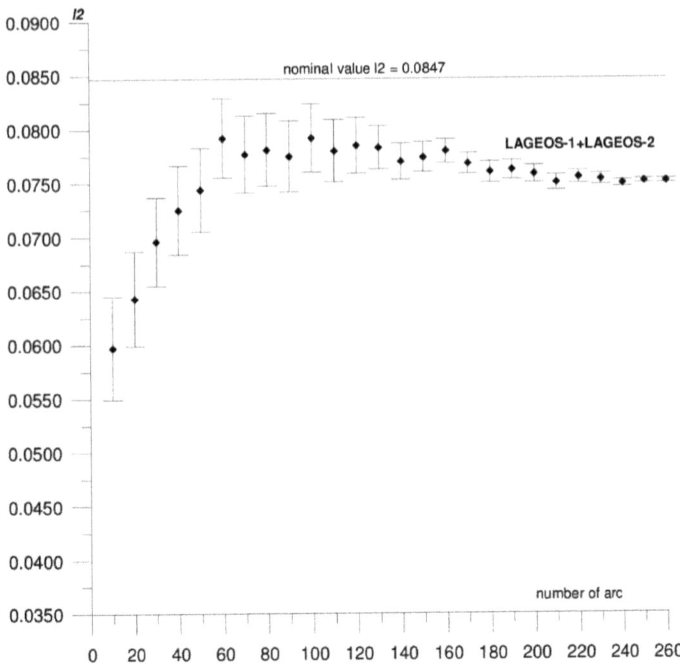

Figure 4. Sequential solution for the Yarragadee local l_2 parameter based on LAGEOS-1+LAGEOS-2 data.

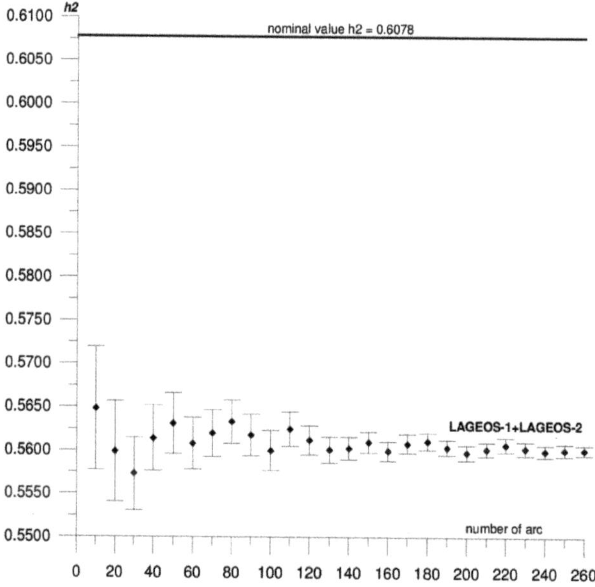

Figure 5. Sequential solution for the Mount Stromlo local h_2 parameter based on LAGEOS-1+LAGEOS-2 data.

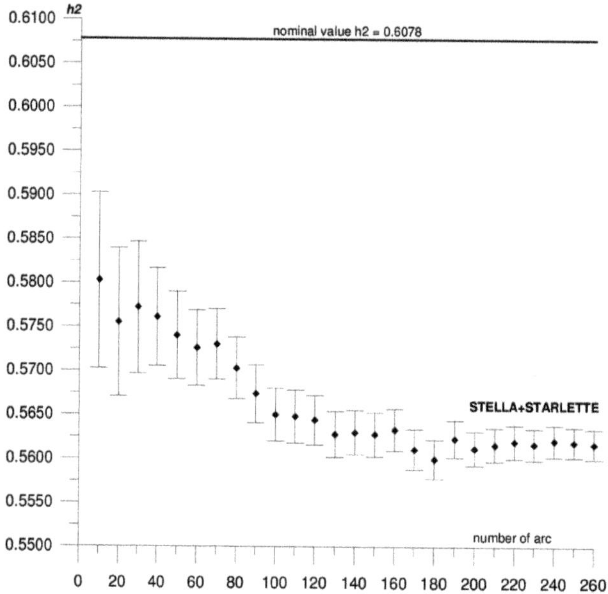

Figure 6. Sequential solution for the Mount Stromlo local h_2 parameter based on STELLA+STARLETTE data.

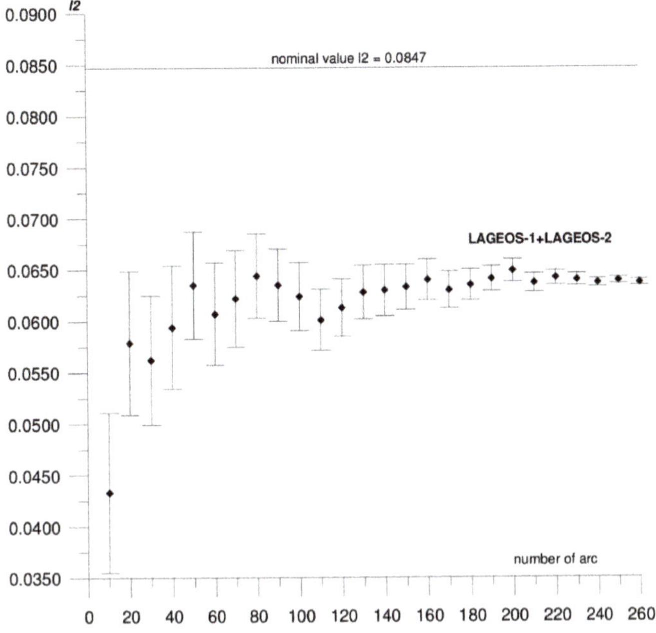

Figure 7. Sequential solution for the Mount Stromlo local l_2 parameter based on LAGEOS-1+LAGEOS-2 data.

In the first step of the sequential method, the h_2, l_2 parameters were determined from two orbital arcs. The computed values significantly deviate from the final ones. Adding arcs in weekly cycles (up to 260) allows the observation of a slowly emerging stability approaching the final h_2, l_2 values determined from the 260 arcs. The values of formal errors of the determined parameters also asymptotically approach their final values. The process of achieving stability varies across parameters and stations. For the Yarragadee station, for the h_2 parameter, the designation stability (understood as the repeatability of the results obtained for subsequently added arcs down to the level of formal error) for LAGEOS-1+LAGEOS-2 data (Figure 2) emerges at about 200 arcs. The situation is similar for the determination from STELLA+STARLETTE data (Figure 3). The l_2 parameter (Figure 4) exhibits a lower degree of determination stability, achieved after about 230 arcs. In turn, for the Mount Stromlo station, the stability of the h_2 parameter determination was achieved for about 190 LAGEOS-1+LAGEOS-2 arcs (Figure 5) and 200 STELLA+STARLETTE arcs (Figure 6). The determination stability of the l_2 parameter for the Mount Stromlo station is similar to that of Yarragadee station, and was achieved after about 230 arcs (Figure 7). It proves that the number of arcs needed to determine local tidal parameters of these stations is about 200, which corresponds to about a 50-month interval (seven-day orbital arcs). For next added arcs, the estimated parameters values vary less than the formal error value.

Figures 2–4 show the results of the sequential solution for the Yarragadee station local tidal parameters. The values of the h_2 and l_2 numbers for this station, determined from LAGEOS-1+LAGEOS-2 data are 0.5756 ± 0.0005 and 0.0751±0.0002, respectively, and differ from the global values h_2 and l_2 given in IERS Technical Note No. 36 [33] by 0.0322 (5%) and 0.0096 (11%), respectively. A similar value of the h_2 parameter was obtained from the data of the STELLA+STARLETTE satellites: h_2 = 0.5742 ± 0.0015 (the difference with respect to the global value is 0.0336, i.e., about 6%). The l_2 parameter was not determined due to an unacceptable value and large error obtained when independently determining from the STELLA and SRTARLETTE data (see Table 1). Jagoda and Rutkowska [5], where global values of tidal parameters determined from LEO satellites data were analyzed from January 2005 to July 2007, present similar findings. The values of horizontal displacement of Earth masses in effect of tidal forces which are described by Shida l_2 number are significantly lower and harder to be measured than radial displacements which are expressed by Love h_2 number. This can potentially affect a determination of l_2 parameter from the LEO satellites data.

In general, the results of determining h_2 for the Yarragadee station from STELLA+STARLETTE data are very similar to those from LAGEOS-1+LAGEOS-2, with the difference being 0.0014, i.e., in the range of formal error.

In turn, the formal error in h_2 designation is three times greater for STELLA+STARLETTE, which is due to the impaired orbit designation of these satellites. The LEO satellites STELLA and STARLETTE move in the lower, dense layers of the atmosphere (at an altitude of about 800 km), and therefore their orbits are determined with greater errors than those of LAGEOS satellites. For the STELLA and STARLETTE satellites in [38] authors obtained mean RMS values of the post-fit residuals from 1.30 cm to 1.87 cm depending on the Earth gravity field model used. In another paper [39], mean RMS values of the post-fit STELLA/STARLETTE were given from 1.87 cm to 2.90 cm depending on the frequency of estimation of empirical acceleration parameters. In [40] these were 3.11 cm for STELLA and 2.40 cm for STARLETTE. In this paper, the mean RMS values of the post-fit STELLA and STARLETTE were 1.98 cm and 1.87 cm, respectively. In turn, the LAGEOS satellite orbits at an altitude of about 6000 km are determined with an accuracy of about 1 cm, and RMS values of the post-fit of this order were obtained, e.g., in [6,39]. The mean RMS values of the post-fit residuals for LAGEOS-1 and LAGEOS-2 obtained in this analysis are 1.02 cm and 1.01 cm, respectively.

Figures 5–7 depict the results of sequential solution for Mount Stromlo station local tidal parameters. The values of the tidal parameters for this station determined from the LAGEOS-1+LAGEOS-2 data are $h_2 = 0.5601 \pm 0.0006$, $l_2 = 0.0637 \pm 0.0003$. The differences with respect to global values are 0.0477 (8%) for h_2 and 0.021 (25%) for l_2. The value of the h_2 parameter for this station determined from the STELLA+STARLETTE data is 0.5618 ± 0.0017, the difference from the nominal value is 0.046, that is about 7%. Similarly, as in the case of the Yarragadee station, the l_2 parameter from combined LEO satellites data was not determined. There is a high degree of conformity between the h_2 values obtained from LAGEOS-1+LAGEOS-2 and STELLA+STARLETTE data, with the difference being 0.0017 and not exceeding the formal error level. Similar to the Yarragadee station, the formal error of the h_2 parameter is higher for the LEO satellites; about three times in this case.

The comparison of Love/Shida numbers for the Yarragadee and Mount Stromlo stations shows that they differ by 0.0155 ± 0.0005 (LAGEOS-1+LAGEOS-2 data) and 0.0124 ± 0.0015 (STELLA+STARLETTE data) for the h_2 number and 0.0114 ± 0.0002 (LAGEOS-1+LAGEOS-2 data) for the l_2 number. These differences exceed the formal error level. So far, no similar studies have been carried out for SLR stations from Australia, so it is impossible to relate the results obtained to the work of other researchers. However, data for two European SLR stations from the Baltic Sea region are available: Borowiec (no. 78113802) and Riga (no. 18844401). Jagoda and Rutkowska [12] were found that the local tidal parameters for the Borowiec station determined from the LAGEOS satellites data in the 01.01.2009–01.01.2019 interval are $h_2 = 0.7308 \pm 0.0008$ and $l_2 = 0.1226 \pm 0.0003$. In another paper Jagoda and Rutkowska [13], were obtained $h_2 = 0.6891 \pm 0.0009$ and $l_2 = 0.1043 \pm 0.0004$ for the Riga station from the LAGEOS satellites data in the 01.01.2004–01.01.2019 interval. Significant differences can be found when comparing the results obtained in this work to the results for the European stations Borowiec and Riga. These are the largest for Mount Stromlo and Borowiec stations: about 23% for the h_2 parameter and about 48% for the l_2 parameter.

The differences between the global and local tidal parameters may be influenced by the geological structure and physical factors of the observation site, in this case Australia. The Australian continent is located in the eastern part of the Indo–Australian lithosphere plateau. The greater part of Australia is occupied by the Precambrian Craton called the Australian Craton, which is adjacent to the structure of the Flinders Ranges and the Barrier Ranges, and the structure of the Great Dividing Range [41]. The Yarragadee station is located within the Australian Craton on the so-called Perth Basin. The Perth

Basin is filled mainly with continental Permian sediments which lie directly on crystalline rocks. The rocks of the Perth Basin sedimentary cover are mainly sandy-loam and marine sediments of Triassic and Cretaceous periods [42,43]. The eastern part of Australia where the Mount Stromlo station is located, is occupied by the of the Great Dividing Range. The Palaeozoic structures of the Great Dividing Range were created as a result of subduction processes on the border between Panthalassa and the Gondwana Craton, on the fringe of which the Australian continent was situated [42]. The mountain range created as a result of these processes is characterised by a varied structure and geological history. In the western part, the structures formed in the Neoproterozoic era region dominate. In the central part, the main phases of the tectonic movements, magmatism, and metamorphism were at work from the early Ordovician to the lowest Devonian. In the eastern part, on the rocks of the older Palaeozoic there are thick sediments of the Devonian and the lower Carboniferous as well as the Permian. The tectonic movements lasted here from the Carboniferous. They were accompanied by lava outflows, the covers of which amount to many thousands of square kilometers of the area. In numerous places, after the fold movements, tectonic subsidences and grabens were formed [42,43].

In addition we have studied the impact of adjusted local tidal parameters h_2, l_2 values on the determination of the Yarragadee and Mount Stromlo SLR stations coordinates in the ITRF2014 reference frame [37]. The test consisted in determining the X, Y, Z coordinates of Yarragadee and Mount Stromlo stations in ITRF2014 in two computational versions. In the first computational version, the coordinates were determined using the nominal global values $h_2 = 0.6078$, $l_2 = 0.0847$ [33]. The second version consisted in determining the coordinates using the proposed in this analysis local tidal parameters $h_2 = 0.5756$, $l_2 = 0.0751$ for the Yarragadee station and $h_2 = 0.5601$, $l_2 = 0.0637$ for the Mount Stromlo station. Table 2 presents the test results.

Table 2. The X, Y, Z coordinates of the Yarragadee and Mount Stromlo SLR stations estimated in two calculation versions.

X, Y, Z (m) ITRF2014	X, Y, Z (m) Estimated Version 1 (Using the Nominal Global Values of h_2, l_2)	X, Y, Z (m) Estimated Version 2 (Using Local Values of h_2, l_2 Proposed in this Paper)	Version 1 Minus Version 2 (m)
YARRAGADEE (no. 70900513)			
−2389007.5340	−2389007.5204 ± 0.0022	−2389007.5171 ± 0.0022	−0.0033
5043329.4474	5043329.4418 ± 0.0019	5043329.4377 ± 0.0019	0.0041
−3078524.2232	−3078524.1935 ± 0.0017	−3078524.1883 ± 0.0017	−0.0052
MOUNT STROMLO (no. 78259001)			
−4467064.7778	−4467064.7519 ± 0.0021	−4467064.7481 ± 0.0019	−0.0038
2683034.8865	2683034.8632 ± 0.0017	2683034.8582 ± 0.0017	0.0050
−3667007.3186	−3667007.3331 ± 0.0016	−3667007.3386 ± 0.0016	0.0055

The use of local tidal parameters h_2, l_2 values instead of global nominal h_2, l_2 values affects the result of the coordinate determination. The Z coordinate seems to be the most affected one, with the difference between version 1 and 2 being 0.0055 m and −0.0052 m for Mount Stromlo and Yarragadee stations, respectively. The smallest difference was observed for the X coordinate: −0.0033 m for Yarragadee and −0.0038 m for Mount Stromlo. The Y component differed by 0.0041 m (Yarragadee) and 0.0050 m (Mount Stromlo). In [12], in a similar test performed for the Borowiec station, the same order of differences was obtained ($\Delta X = -0.0035$ m, $\Delta Y = 0.0033$ m, $\Delta Z = 0.0042$ m) as for the Yarragadee and Mount Stromlo stations. However, in [13] describing the Riga station, these discrepancies are larger, namely, $\Delta X = 0.0044$ m, $\Delta Y = -0.0047$ m, $\Delta Z = 0.0069$ m.

Similar results of determining the coordinates of the Yarragadee and Mount Stromlo stations in ITRF2014 system were obtained in the paper [17], where the authors proposed a kinematic method to estimate the coordinates of SLR stations by using the Global Navigation Satellite System (GNSS) technique onboard a low Earth orbiting (LEO) satellite. They applied SLR and GNSS observations of the GRACE-A satellite from January to December 2012. They found that the GRACE-A satellite, as a connection between the SLR and GNSS techniques, allowed the accurate estimation of SLR stations positions with the high agreement with the ITRF2014 system.

In another paper [22], the author used the STARLETTE, LAGEOS-1 and LAGEOS-2 data over a 14-year period (1993–2007) for determination and analysis of SLR stations coordinates in ITRF2005 system [44]. The author pointed out a good agreement of the estimated coordinates with respect to the values given in ITRF2005. However, in both of these studies the influence of the application of different values of h_2, l_2 parameters on the results of determining the SLR stations coordinates was not investigated.

4. Conclusions

Based on the results obtained in the considered case studies, the following conclusions can be drawn:

- There are discrepancies observed between the determined local tidal parameters h_2, l_2 for the Yarragadee and Mount Stromlo stations and the commonly used values of the h_2, l_2 parameters averaged for the whole Earth. This may be influenced by the geological structure and physical factors of the observation site. In order to confirm this, detailed geophysical analyses should be carried out. This goes beyond the scope of this work, suggesting at the same time the need for further studies in this field.
- The use of local tidal parameters values in the process of determining the stations coordinates influences the result.
- Local tidal parameters h_2, l_2 are better determined from the LAGEOS-1 and LAGEOS-2 data than from the STELLA and STARLETTE. However, the results obtained from the LEO satellites indicate that data from these satellites can be used for the determination of local tidal parameters. They can be used for stations with a low number of observations from the LAGEOS satellites.
- It is not possible to determine the l_2 parameter for the Yarragadee and Mount Stromlo stations from STELLA and STARLETTE data. The values of horizontal displacement of Earth masses which are described by the l_2 parameter are significantly lower and harder to be measured than radial displacements which are expressed by the h_2 parameter. This can potentially affect a determination of l_2 parameter from STELLA and STARLETTE data.
- The time interval adopted in the analysis is sufficient to determine the h_2 and l_2 local parameters. The results stabilize after about 200 orbital arcs, which corresponds to about 50 months from the 60-month interval adopted in the analysis.

Author Contributions: Conceptualization, M.J., P.L., and M.R.; methodology, M.J.; software, M.R.; validation, M.J., P.L., and M.R.; formal analysis, M.J., P.L., and M.R.; investigation, M.J., P.L., M.R., D.Š., and R.O.; resources, M.J., R.O., D.Š., and J.K.; data curation, M.J. and M.R.; writing—original draft preparation, M.J., P.L., and J.K.; writing—review and editing, M.J., M.R., P.L., R.O., and D.Š.; visualization, M.J., J.K., D.Š., and R.O.; supervision, M.J.; project administration, M.J.; funding acquisition, M.J. All authors have read and agreed to the published version of the manuscript.

Funding: The authors would like to express their gratitude to the National Science Center, Poland (PL—Narodowe Centrum Nauki) for the financial support for this study under Project No: 2019/03/X/ST10/01595.

Conflicts of Interest: The authors declare no conflict of interest.

References

1. Love, A.E.H. *Some Problems of Geodynamics*; Cambridge University Press: Cambridge, UK, 1911.
2. Shida, T.; Matsoyama, M. *Note of Hecker's Observations*; Kyoto Imperial University: Kyoto, Japan, 1912.
3. Tapley, B.D.; Schutz, B.E.; Eanes, R.J.; Ries, J.C.; Watkins, M.M. LAGEOS laser ranging contributions to geodynamics, geodesy, and orbital dynamics. In *Contributions of Space Geodesy to Geodynamics: Earth Dynamics. Geodynamic Series 24*; Smith, D.E., Turcotte, D.L., Eds.; American Geophysical Union: Washington, DC, USA, 1993. [CrossRef]

4. Rutkowska, M.; Jagoda, M. Estimation of the elastic Earth parameters using SLR data for the low satellites Starlette and Stella. *Acta Geophys.* **2012**, *60*, 1213–1223. [CrossRef]
5. Jagoda, M.; Rutkowska, M. Estimation of the Love and Shida numbers: H2, l2 using SLR data for the low satellites. *Adv. Space Res.* **2013**, *52*, 633–638. [CrossRef]
6. Jagoda, M.; Rutkowska, M.; Kraszewska, K.; Suchocki, C. Time changes of the potential love tidal parameters k2 and k3. *Stud. Geophys. Geod.* **2018**, *62*, 586–595. [CrossRef]
7. Wu, B.; Bibo, P.; Zhu, Y.; Hsu, H. Determination of Love numbers using Satellite Laser Ranging. *J. Geod. Soc. Jpn.* **2001**, *47*, 174–180.
8. Petrov, L. Determination of Love numbers h and l for long-period tides using VLBI. In *Viewgraphs at 14-th International Symposium on Earth Tides, August 28–September 1, 2000 in Mizusawa, Japan*; GGP Newsletter # 10: The Hague, The Netherlands, 2000.
9. Krásná, H.; Böhm, J.; Schuh, H. Tidal love and shida numbers estimated by geodetic VLBI. *J. Geodyn.* **2013**, *70*, 21–27. [CrossRef]
10. Ray, R.D.; Bettadpur, S.; Eanes, R.J.; Schrama, E.J.O. Geometrical determination of the Love number h2 at four tidal frequencies. *Geophys. Res. Lett.* **1995**, *22*, 2175–2178. [CrossRef]
11. Ray, R.D. Precise comparisons of bottom-pressure and altimetric ocean tides. *J. Geophys. Res. Oceans* **2013**, *118*, 4570–4584. [CrossRef]
12. Jagoda, M.; Rutkowska, M. Determination of the local tidal parameters for the borowiec station using satellite laser ranging data. *Stud. Geophys. Geod.* **2019**, *63*, 509–519. [CrossRef]
13. Jagoda, M.; Rutkowska, M. Estimation of the local tidal parameters h2, l2 for the Riga satellite laser ranging station based on LAGEOS data. *Est. J. Earth Sci.* **2019**, *68*, 199–205. [CrossRef]
14. Schillak, S. Analysis of the process of the determination of station coordinates by satellite laser ranging based on results of the Borowiec SLR station in 1993.5–2000.5. Part 2: Determination of the station coordinates. *Artif. Satell.* **2004**, *39*, 265–287.
15. Schillak, S.; Wnuk, E. The SLR stations coordinates determined from monthly arcs of Lageos-1 and Lageos-2 laser ranging in 1999–2001. *Adv. Space Res.* **2002**, *31*, 413–418. [CrossRef]
16. Zelensky, N.P.; Lemoine, F.G.; Chinn, D.S.; Melachroinos, S.; Beckley, B.D.; Beall, J.W.; Bordyugov, O. Estimated SLR station position and network frame sensitivity to time-varying gravity. *J. Geod.* **2014**, *88*, 517–537. [CrossRef]
17. Guo, J.; Wang, Y.; Shen, Y.; Liu, X.; Sun, Y.; Kong, Q. Estimation of SLR station coordinates by means of SLR measurements to kinematic orbit of LEO satellites. *Earth Planets Space* **2018**, *70*. [CrossRef]
18. Sośnica, K.; Thaller, D.; Jäggi, A.; Dach, R.; Beutler, G. Sensitivity of Lageos orbits to global gravity field models. *Artif. Satell.* **2012**, *47*, 47–65. [CrossRef]
19. Gourine, B. Use of Starlette and LAGEOS-1&-2 laser measurements for determination and analysis of stations coordinates and EOP time series. *Comptes Rendus Geosci.* **2012**, *344*, 319–333.
20. Gourine, B. On use of Starlette and Stella Laser measurements in determination of SLR stations coordinates and earth orientation parameters (EOP). In Proceedings of the 17th International Workshop on Laser Ranging (ILRS) At Bad Kötzing-Germany, Frankfurt, Germany, 16–20 May 2012; Volume 48, ISBN 978-3-89888-999-5.
21. Shen, Y.; Guo, J.Y.; Zhao, C.M.; Yu, X.M.; Li, J.L. Earth rotation parameter and variation during 2005–2010 solved with LAGEOS SLR data. *Geod. Geodyn.* **2015**, *6*, 55–60. [CrossRef]
22. Bloßfeld, M.; Rudenko, S.; Kehm, A.; Panafidina, N.; Müller, H.; Angermann, D.; Hugentobler, U.; Seitz, M. Consistent estimation of geodetic parameters from SLR satellite constellation measurements. *J. Geod.* **2018**, *92*, 1003–1021. [CrossRef]
23. Sośnica, K. LAGEOS sensitivity to ocean tides. *Acta Geophys.* **2014**, *63*, 1181–1203. [CrossRef]
24. Rutkowska, M.; Jagoda, M. SLR technique used for description of the Earth elasticity. *Artif. Satell.* **2015**, *50*, 127–141. [CrossRef]
25. Schillak, S. Analysis of the process of the determination of station coordinates by satellite laser ranging based on results of the Borowiec SLR station in 1993.5–2000.5. Part 1: Performance of the Satellite Laser Ranging. *Artif. Satell.* **2004**, *39*, 217–263.
26. Combrinck, L. Satellite laser ranging. In *Sciences of Geodesy—I*; Xu, G., Ed.; Springer: Berlin/Heidelberg, Germany, 2010. [CrossRef]

27. Schutz, B.E.; Cheng, M.K.; Eanes, R.J.; Shum, C.K.; Tapley, B.D. Geodynamic Results from Starlette Orbit Analysis. In *Contributions of Space Geodesy to Geodynamics: Earth Dynamics*. *Geodynamic Series 24*; Smith, D.E., Turcotte, D.L., Eds.; American Geophysical Union: Washington, DC, USA, 1993. [CrossRef]
28. Pearlman, M.; Arnold, D.; Davis, M.; Barlier, F.; Biancale, R.; Vasiliev, V.; Ciufolini, I.; Paolozzi, A.; Pavlis, E.C.; Sośnica, K.; et al. Laser geodetic satellites: A high-accuracy scientific tool. *J. Geod.* **2019**, 1–14. [CrossRef]
29. Sośnica, K. *Determination of Precise Satellite Orbits and Geodetic Parameters using Satellite Laser Ranging*; Astronomical Institute, University of Bern: Bern, Switzerland, 2014; ISBN1 8393889804. ISBN2 9788393889808.
30. Melchior, P. *The Tides of the Planet Earth*; Pergamon Press: Bruxelle, Belgium, 1978.
31. McCarthy, J.J.; Rowton, S.; Moore, D.; Pavlis, D.E.; Luthcke, S.B.; Tsaoussi, L.S. *GEODYN II System Operation Manual, 1–5*; STX System Corp: Lanham, MD, USA, 1993.
32. Mathews, P.M.; Dehant, V.; Gipson, J.M. Tidal station displacements. *J. Geoophys. Res.* **1997**, *102*, 20469–20477. [CrossRef]
33. Petit, G.; Luzum, B. *IERS Conventions. IERS Technical Note No. 36*; Verlag des Bundesamts fur Kartographie und Geodasie: Frankfurt, Germany, 2010.
34. Torrence, M.H.; Klosko, S.M.; Christodoulidis, D.C. The Construction and Testing of Normal Points at Goddard Space Flight Center. In Proceedings of the 5th International Workshop on Laser Ranging Instrumentation, Herstmonceux, UK, 10–14 September 1984; Geodetic Institute Univ: Bonn, Germany, 1984; pp. 506–511.
35. Pearlman, M.R.; Degnan, J.J.; Bosworth, J.M. The international laser ranging service. *Adv. Space Res.* **2002**, *30*, 135–143. [CrossRef]
36. Kuźmicz-Cieślak, M.; Schillak, S.; Wnuk, E. Stability of coordinates of the SLR stations on a basis of Satellite Laser Ranging. In Proceedings of the 12th International Workshop on Laser Ranging, Matera, Italy, 13–17 November 2000.
37. Altamimi, Z.; Rebischung, P.; Métivier, L.; Collilieux, X. ITRF2014: A new release of the International Terrestrial Reference Frame modeling nonlinear station motions. *J. Geophys. Res.* **2016**, *121*, 6109–6131. [CrossRef]
38. Lejba, P.; Schillak, S.; Wnuk, E. Determination of orbits and SLR stations' coordinates on the basis of laser observations of the satellites Starlette and Stella. *Adv. Space Res.* **2007**, *40*, 143–149. [CrossRef]
39. Lejba, P.; Schillak, S. Determination of station positions and velocities from laser ranging observations to Ajisai, Starlette and Stella satellites. *Adv. Space Res.* **2011**, *47*, 654–662. [CrossRef]
40. Jagoda, M.; Rutkowska, M. Estimation of the Love numbers: k2, k3 using SLR data of the LAGEOS1, LAGEOS2, STELLA and STARLETTE satellites. *Acta Geod. Geoph.* **2016**, *51*, 493–504. [CrossRef]
41. Brown, D.A.; Campbell, K.S.W.; Crook, K.A.W. *The Geological Evolution of the Australia and New Zeland*; Pergamon Press: Oxford, UK, 1968.
42. Fairbridge, R.W. (Ed.) *The Encyclopedia of World Geology Part 1*; Dowden Hutchinson & Ross Inc.: Stroudsburg, PA, USA, 1975.
43. Clarke, G.L. The geology of Australia. In *Geology. Vol IV. Encyclopedia of Life Support Systems*; Eolss Publishers Co. Ltd.: Oxford, UK, 2013.
44. Altamimi, Z.; Collilieux, X.; Legrand, J.; Garayt, B.; Boucher, C. ITRF2005: A new release of the international terrestrial reference frame based on time series of station positions and earth orientation parameters. *J. Geophys. Res.* **2007**, *112*, 1–19. [CrossRef]

Publisher's Note: MDPI stays neutral with regard to jurisdictional claims in published maps and institutional affiliations.

sensors

Article

Determination of Motion Parameters of Selected Major Tectonic Plates Based on GNSS Station Positions and Velocities in the ITRF2014

Marcin Jagoda

Faculty of Civil Engineering, Environmental and Geodetic Sciences, Koszalin University of Technology, Śniadeckich 2, 75-453 Koszalin, Poland; marcin.jagoda@tu.koszalin.pl

Abstract: Current knowledge about tectonic plate movement is widely applied in numerous scientific fields; however, questions still remain to be answered. In this study, the focus is on the determination and analysis of the parameters that describe tectonic plate movement, i.e., the position (Φ and Λ) of the rotation pole and angular rotation speed (ω). The study was based on observational material, namely the positions and velocities of the GNSS stations in the International Terrestrial Reference Frame 2014 (ITRF2014), and based on these data, the motion parameters of five major tectonic plates were determined. All calculations were performed using software based on a least squares adjustment procedure that was developed by the author. The following results were obtained: for the African plate, $\Phi = 49.15 \pm 0.10°$, $\Lambda = -80.82 \pm 0.30°$, and $\omega = 0.267 \pm 0.001°/\mathrm{Ma}$; for the Australian plate, $\Phi = 32.94 \pm 0.05°$, $\Lambda = 37.70 \pm 0.12°$, and $\omega = 0.624 \pm 0.001°/\mathrm{Ma}$; for the South American plate, $\Phi = -19.03 \pm 0.20°$, $\Lambda = -119.78 \pm 0.39°$, and $\omega = 0.117 \pm 0.001°/\mathrm{Ma}$; for the Pacific plate, $\Phi = -62.45 \pm 0.07°$, $\Lambda = 111.01 \pm 0.14°$, and $\omega = 0.667 \pm 0.001°/\mathrm{Ma}$; and for the Antarctic plate, $\Phi = 61.54 \pm 0.30°$, $\Lambda = -123.01 \pm 0.49°$, and $\omega = 0.241 \pm 0.003°/\mathrm{Ma}$. Then, the results were compared with the geological plate motion model NNR-MORVEL56 and the geodetic model ITRF2014 PMM, with good agreement. In the study, a new approach is proposed for determining plate motion parameters, namely the sequential method. This method allows one to optimize the data by determining the minimum number of stations required for a stable solution and by identifying the stations that negatively affect the quality of the solution and increase the formal errors of the determined parameters. It was found that the stability of the solutions of the Φ, Λ, and ω parameters varied depending on the parameters and the individual tectonic plates.

Keywords: GNSS stations; tectonic plate motion parameters; ITRF

Citation: Jagoda, M. Determination of Motion Parameters of Selected Major Tectonic Plates Based on GNSS Station Positions and Velocities in the ITRF2014. *Sensors* **2021**, *21*, 5342. https://doi.org/10.3390/s21165342

Academic Editor: Alfred Colpaert

Received: 6 July 2021
Accepted: 5 August 2021
Published: 7 August 2021

Publisher's Note: MDPI stays neutral with regard to jurisdictional claims in published maps and institutional affiliations.

1. Introduction

Since the beginning of the Earth, its surface has been undergoing dynamic changes, which are caused both by external forces as well as those that act inside the planet, and their effects are visible on the surface. One such dynamic change process is tectonic plate movements on the surface of the asthenosphere. Current knowledge of the motion of the plates has been applied in numerous areas of research such as environmental sciences. Processes that result from plate tectonics, such as seismic or volcanic activity, have a direct influence on the environment. An issue that has been recently discussed by McEvoy et al. [1] is the influence of the movement of the lithosphere on the safety of radioactive waste stored underground. In addition, in [2,3], the authors analysed the influence of plate tectonics on climate change. Another research area where knowledge of tectonic plate motion has been applied is geodynamics and geodesy, in particular, related to defining the Earth's reference systems [4,5]. For years, a problem for studies focused on lithospheric deformations, has been the formulation of laws that govern these phenomena, in particular, explaining the driving mechanism of this motion, and thus the dynamics of the lithosphere. A pioneer of the idea of lithospheric motion, based on the adherence of the coastal lines of both

Americas, Africa, and Europe, was A. Snider-Pellegrini [6]. However, only the arguments later provided by A. Wegener [7] contributed to the fact that the latter is considered to be the author of the foundation of the lithospheric motion theory. Currently, studies in the literature, for example, studies by [8,9], divide the lithosphere into seven major plates of various sizes (Eurasian, African, North American, South American, Australian, Pacific, and Antarctic), and several smaller ones, referred to as minor plates and microplates, which are parts of the major plates or complement them. Most of the plates are located along the western margin of the Pacific Ocean, as detailed in [10] and more detailed information about tectonic plate movement can be found, for example, in [11,12]. The 1980s was a period that witnessed dynamic development of space measurement techniques, including satellite laser ranging (SLR), doppler orbitography and radiopositioning integrated by satellite (DORIS), very long baseline interferometry (VLBI), and global navigation satellite systems (GNSS). These techniques have started to be used for precise determination of tectonic plate motion, due to the possibility of conducting observations at a global scale. Nowadays, the SLR, DORIS, VLBI, and GNSS techniques are the basis for studies on crustal movements [13], as they allow plate movements to be quantified at a level of submillimetres per year. Data from satellite systems are the basis for creating kinematic models of plate movements, referred to as plate kinematic and crustal deformation models, such as the series of models developed by H. Drewes [14–21]. Then, the obtained results have been compared with the geological models of plate movement that have been developed based on geophysical data. Examples of such models include: AMO-2 [22], NNR-NUVEL-1 [23], NNR-NUVEL-1A [24], PB2002 [25], and the newest one, NNR-MORVEL56 [26]. Conducting studies on a global scale requires a reference system that is uniform in geometrical and physical terms for the whole Earth. The International Terrestrial Reference System (ITRS) is such a reference system, and an International Terrestrial Reference Frame (ITRF) is a realization of the ITRS. To date, the International Earth Rotation and Reference System Service (IERS) [27] has realised more than ten solutions of the ITRF system, among which the currently valid one is the ITRF2014 [28]. The ITRF2014 was generated using the complete observation history of the four space techniques SLR, DORIS, VLBI, and GNSS. The corresponding international services, i.e., the International Laser Ranging Service [29], the International DORIS Service [30], the International VLBI Service [31], and the International GNSS Service [32], provided reprocessed time series of station positions and daily Earth orientation parameters (EOP). The International GNSS Service submitted time series comprising 7714 daily solutions, resulting from the second reprocessing campaign, covering the time period 1994.0–2015.1 [33]. It is worthwhile noting, here, that another realisation of the ITRF, namely the ITRF2020, should be available by the end of 2021. More details regarding specifications of the ITRF solutions can be found in [34,35] or IERS Technical Note 36 [36].

In order to meet the expectations of the users of various geodynamic, geodetic, and environmental applications, the realisations of the ITRF have been accompanied by the publication of a tectonic plate motion model, for example, APKIM2005 [19], ITRF2008 PMM [4], and ITRF2014 PMM [5]. These are kinematic models that have been created based on the position (coordinates) and velocities of the SLR, DORIS, VLBI, and GNSS stations in the given ITRF frame as one common solution (SLR + DORIS + VLBI + GNSS).

Considering the need for continuous analysis of tectonic plate movement due to the demands of contemporary geodynamics, geodesy, and the influence on the environment, this study was conducted to determine and analyse plate motion parameters based on the coordinates and velocities of SLR, DORIS, VLBI, and GNSS stations, separately, for each of these techniques. Such an approach offers the possibility to evaluate the contribution of each of these space techniques to the creation of a model of movement of specific tectonic plates, and to assess the accuracy, of each technique, for determining the plate motion parameters. In this study, a new approach is proposed for determining tectonic plate motion parameters, namely the sequential method, described in Section 2. It is successfully demonstrated that this method can be used to identify, and then to eliminate, from the calculations, the stations which, due to various reasons (e.g., being located on cracked, unstable areas of the given plate, areas of seismic activity, or on a microplate, so their movement is not consistent with the motion of the analysed plate), contribute to an increase in formal errors of the determined parameters and the effects on the determination results. The method also allows for data optimisation, i.e., specifying the minimum number of stations on the given tectonic plate for which the calculated motion parameters are stabilised (i.e., their changes do not exceed the value of formal errors). To date, studies have been carried out on the SLR [37,38], DORIS [38,39], and VLBI [38,40] techniques and, on the GNSS technique, for the Eurasian plate [41]. In this study, the subsequent stage of the conducted studies is presented. The aims of the study are: (i) to determine and analyse the motion parameters of the African, Australian, South American, Pacific, and Antarctic plates based on the coordinates and velocities of the GNSS stations in the ITRF2014; (ii) to identify the stations that increase the value of formal errors in the defined parameters and negatively affect the calculation results; and (iii) to estimate the stability of the solutions of the motion parameters of a given plate.

2. Materials and Methods

The data provided by SLR, DORIS, VLBI, and GNSS are the basis for the determination of the positions ϕ and λ of the observation stations of these techniques. The positions of the stations are subject to changes in time as a result of, among others, the movement of the tectonic plates on which they are located. According to the station positions ϕ and λ determined at time intervals Δt, the movement of the station in time $\Delta \vec{x}$ can be calculated. Then, knowing the value of the displacement $\Delta \vec{x}$ (in terms of coordinate shifts $\Delta \phi$ and $\Delta \lambda$) of the station on a given plate, we can determine three parameters that describe the movement of this plate, i.e., the geographical latitude Φ and longitude Λ of the pole of rotation Ω and the angular rotation speed ω of the given plate or the specific elements of the pole of rotation ω_x, ω_y, and ω_z around the X, Y, and Z axes (Figure 1). The relations between these values are described in Equation (1) [42]. The geometric relations between the plate motion parameters and coordinate shifts are shown in Figure 1, where X_0 is the position of the station on the initial epoch t_0, X_1 is the position of the station on the epoch t_1 (after plate moving), and point P denotes the Earth pole. Equation (1):

$$\tan \Phi = \frac{\omega_z}{\sqrt{\omega_x^2 + \omega_y^2}}$$
$$\tan \Lambda = \frac{\omega_y}{\omega_x}$$
$$\omega = \sqrt{\omega_x^2 + \omega_y^2 + \omega_z^2} \tag{1}$$
$$\omega_x = \omega \cos \Lambda \cos \Phi$$
$$\omega_y = \omega \sin \Lambda \cos \Phi$$
$$\omega_z = \omega \sin \Phi$$

According to [17], the displacement of the observational station $\Delta \vec{x} = (\vec{\Omega} \times \vec{x})\Delta t$ expressed as a function of the tectonic plate motion parameters Φ, Λ, ω ($\Delta \varphi = f(\Phi, \Lambda, \omega)$, $\Delta \lambda = f(\Phi, \Lambda, \omega)$) is described by Equation (2):

$$\Delta \varphi = \omega \cdot \Delta t \cdot \cos \Phi \cdot \sin(\lambda - \Lambda)$$
$$\Delta \lambda = \omega \cdot \Delta t \cdot (\sin \Phi - \cos(\lambda - \Lambda) \cdot \tan \varphi \cdot \cos \Phi) \tag{2}$$

Determining the motion parameters of a tectonic plate requires knowledge of the coordinates and velocities of at least two stations on that plate. This allows for the creation of two observational equations, i.e., Equation (3), and for the determination of three plate motion parameters Φ, Λ, and ω. In such a case (when the number of observational equations is higher than the number of the determined unknowns) it may be aligned with the use of the least-squares adjustment method. Assuming that more than two stations are located on a given plate allows one to evaluate the influence of the number and location of the stations on the determined motion parameters of the plate and the accuracy of their determination. A description of the least squares adjustment procedure was presented, among others, by McCarthy et al. [43], and its practical applications for the determination of plate motion parameters in [14], and in the earlier study by Jagoda et al. [40].

According to [14], the observational equations for the least squares adjustment procedure, which allow for the determination of plate motion parameters based on shifts in the station coordinates are expressed in Equation (3):

$$
\begin{aligned}
v_\varphi &= \left(\frac{\partial \Delta \varphi}{\partial \Phi}\right) d\Phi + \left(\frac{\partial \Delta \varphi}{\partial \Lambda}\right) d\Lambda + \left(\frac{\partial \Delta \varphi}{\partial \omega}\right) d\omega - \left(\Delta \varphi^{obs} - \Delta \varphi^{cal}\right) \\
v_\lambda &= \left(\frac{\partial \Delta \lambda}{\partial \Phi}\right) d\Phi + \left(\frac{\partial \Delta \lambda}{\partial \Lambda}\right) d\Lambda + \left(\frac{\partial \Delta \lambda}{\partial \omega}\right) d\omega - \left(\Delta \lambda^{obs} - \Delta \lambda^{cal}\right)
\end{aligned}
\tag{3}
$$

where *obs* and *cal* mean observed and calculated values, respectively.

The expressions given in Equation (3) are calculated based on the following relations, Equation (4) [14]:

$$
\begin{aligned}
\frac{\partial \Delta \varphi}{\partial \Phi} &= -\omega \cdot \Delta t \cdot \sin \Phi \cdot \sin(\lambda - \Lambda) \\
\frac{\partial \Delta \varphi}{\partial \Lambda} &= -\omega \cdot \Delta t \cdot \cos \Phi \cdot \cos(\lambda - \Lambda) \\
\frac{\partial \Delta \varphi}{\partial \omega} &= \Delta t \cdot \cos \Phi \cdot \sin(\lambda - \Lambda) \\
\frac{\partial \Delta \lambda}{\partial \Phi} &= \omega \cdot \Delta t \cdot \cos \Phi + \omega \cdot \Delta t \cdot \cos(\lambda - \Lambda) \cdot \tan \varphi \sin \Phi \\
\frac{\partial \Delta \lambda}{\partial \Lambda} &= -\omega \cdot \Delta t \cdot \cos \Phi \cdot \sin(\lambda - \Lambda) \cdot \tan \varphi \\
\frac{\partial \Delta \lambda}{\partial \omega} &= \Delta t \cdot (\sin \Phi - \cos(\lambda - \Lambda) \cdot \tan \varphi \cos \Phi)
\end{aligned}
\tag{4}
$$

For the purposes of this study, the sequential method was applied to determine the plate motion parameters. It involves several calculation steps. In the first step, the Φ, Λ, and ω parameters are determined based on the coordinates and velocities of two stations, here, the GNSS stations (station 1 + station 2) located on a given tectonic plate, adopted from the ITRF2014 [28] and available for users on the website http://itrf.ensg.ign.fr/ITRF_solutions/2014/ [44] (accessed on 8 June 2021).

The next steps of the sequential method consist of adding further stations to the calculations, one by one, according to the scheme: station 1 + station 2, station 1 + station 2 + station 3, station 1 + station 2 + station 3 + , . . . , station n, where n is the number of stations on the given plate as adopted in the solution. The number of stations added varies depending on the specific plate. In every subsequent step of the sequential method, the Φ, Λ, and ω parameters are calculated again. The application of the sequential method enables the obtainment of an increasingly stable adjustment, which is characterised by a decreasing formal error of the determined parameters. By increasing the number of stations, it arrives at a solution that, for a certain number of stations, is characterised by the high stability of the solution and minimum error values. A further increase in the number of stations results in the variability of the calculated motion parameter values within the limits of formal error, which is discussed in Section 3. The final values of the Φ, Λ, and ω parameters that were adopted and used in further analyses were those obtained from n stations (i.e., from the last step of the sequential method) for a given tectonic plate. The selection of the stations that are the basis for the determination of the motion parameters of specific tectonic plates should take into consideration the geophysical conditions of their location. Another factor that influences the accuracy of the solution is the geometrical configuration of the station network, which has been explained in [14,17]. Consistency with the previous determination of the Φ, Λ, and ω parameters for the SLR [37,38], DORIS [38,39], and VLBI [38,40] techniques were maintained, and the same manner of selecting stations was used. These assumptions were created based on the tests conducted during the realisation of previous studies. Thus, as in the previous studies, for example, [40,41], the stations should be located on a stable, uncracked area of the given plate, outside deformation zones. It is recommended that the stations should be distributed as evenly as possibly on the given tectonic plate. The concentration of a large number of stations in a small area does not significantly improve the accuracy and stability of the solution. The distance between the first two stations should equal approximately 60% of the distance between plate boundaries, and it should not be shorter than 50 km. The minimum timespan of observations conducted at a given station should be three years. This is required to meet the requirements of the rigid plate motion theory, which was discussed in [4]. The number of GNSS stations used in the calculations differs for specific plates: for the African plate, 25 stations were used; for the Australian plate, 20 stations were used; for the South American plate 29 stations were used; for the Antarctic plate, 13 stations were used; and finally, for the Pacific plate, 19 stations were used. As far as the Eurasian plate is concerned, which was analysed by Jagoda and Rutkowska [41], 120 GNSS stations were used (4 calculation scenarios were applied with 30 stations for each scenario). Figure 2 presents the stations that were used (black dots) and those that were rejected (red dots) from the solution and estimated positions (green stars) of the pole of rotation for particular plates (SOAM, South American; AFRC, African; PACF, Pacific; ANTC, Antarctic; AUST, Australian).

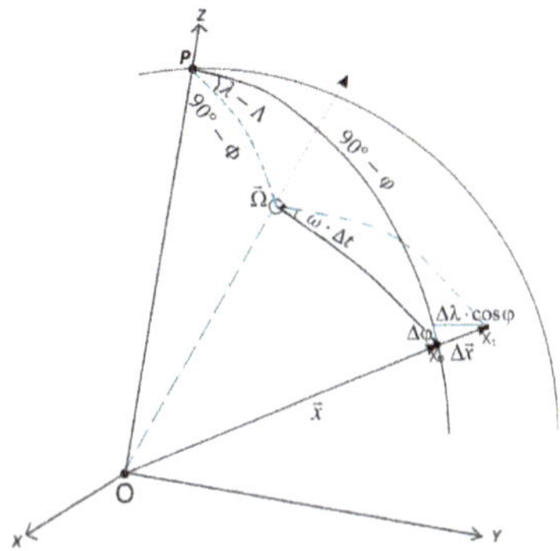

Figure 1. Relations between $\vec{\Omega}$ (Φ, Λ, and ω) and $\Delta\vec{x}$ ($\Delta\phi$ and $\Delta\lambda$).

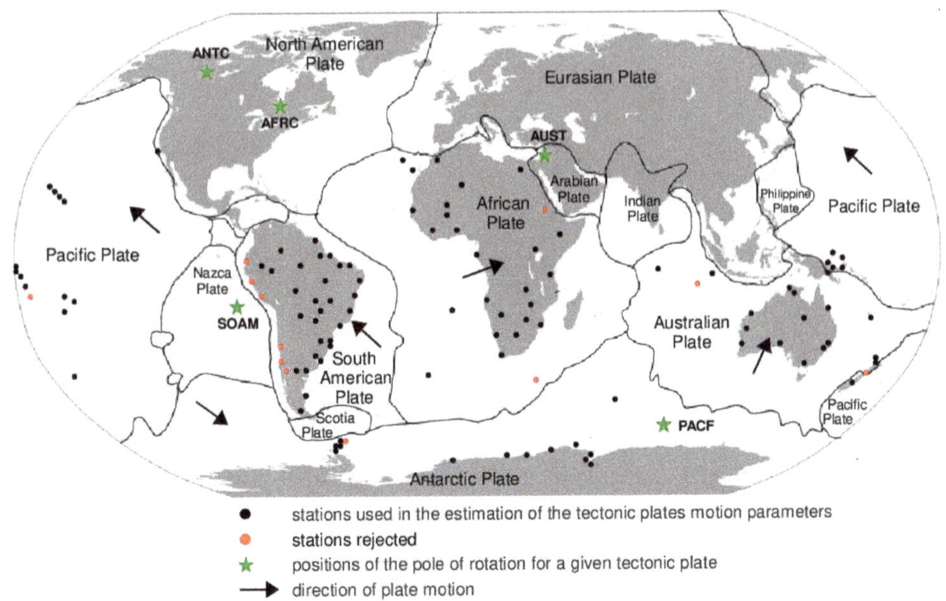

Figure 2. Distribution of the GNSS stations on tectonic plates and estimated positions of the pole of rotation for a given tectonic plate.

All calculations related to the determination of the Φ, Λ, and ω parameters were performed with use of the software in FORTRAN90, developed by the author, based on the least squares adjustment procedure. A block diagram of the adjustment of the plate motion parameters used in the author's own software was shown in an earlier study [40] and is not repeated here. The adopted weights of observations were the formal errors in the determination of shifts of individual GNSS stations given in the ITRF2014 [28].

3. Results and Discussion

The positions and velocities of the GNSS stations in the ITRF2014 [28], determined by more than 21 years of GNSS observations that covered the period 1994.0–2015.1 [33], were the basis for the determination and analysis of the values of the Φ, Λ, and ω parameters that describe the movement of the five major tectonic plates: African, Australian, South American, Pacific, and Antarctic. A study by Jagoda and Rutkowska, on the largest tectonic plate, i.e., the Eurasian plate, has already been conducted and published in the literature [41]. In addition, the last of the major plates, the North American plate, should be analysed separately due to a very large amount of observational data. The North American plate is covered with a vast number of GNSS stations, which enables a detailed analysis of the area with very high tectonic activity, located along the boundary with the Pacific plate.

The results of each step of the sequential methods for the Φ, Λ, and ω parameter values are presented in Figures 3–17 for each plate and parameter, both separately and in Appendices A–E, with the names of the stations used for the calculations. The final values of the Φ, Λ, and ω parameters and of formal errors adopted for further analyses correspond to the values from the last step of the sequential method.

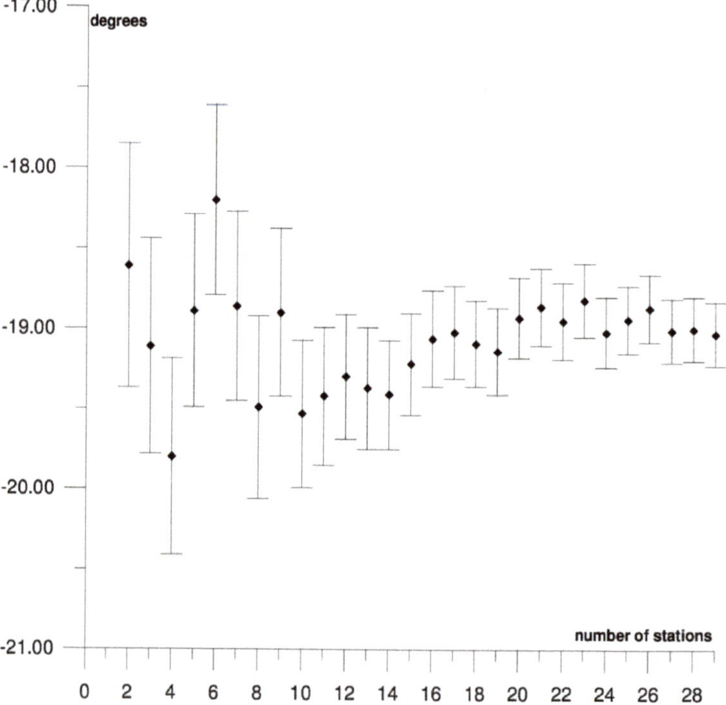

Figure 3. Results of sequential method for the Φ parameter for the South American plate.

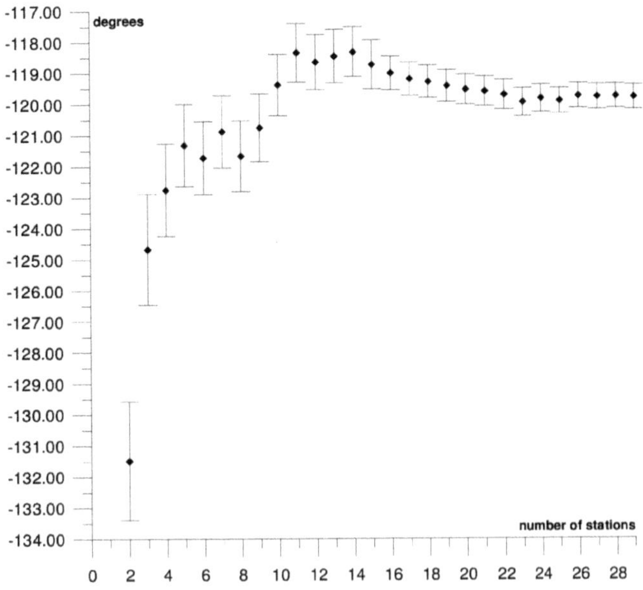

Figure 4. Results of the sequential method for the Λ parameter for the South American plate.

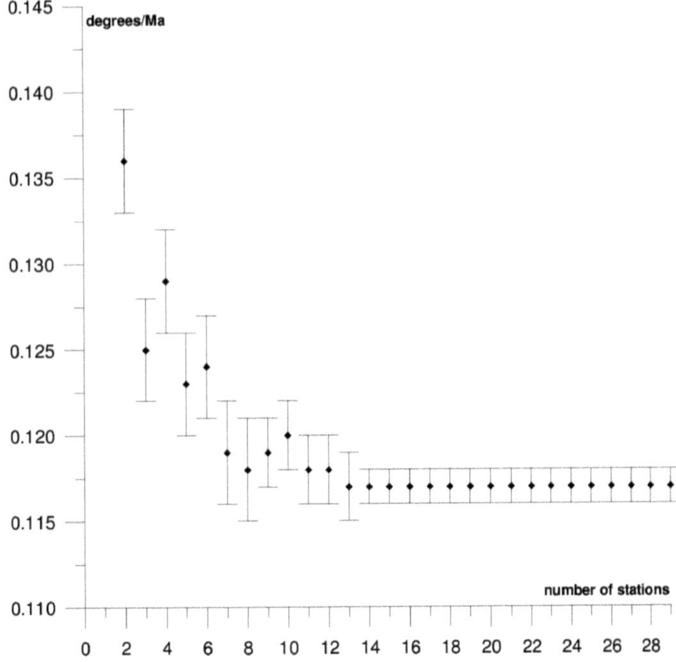

Figure 5. Results of the sequential method for the ω parameter for the South American plate.

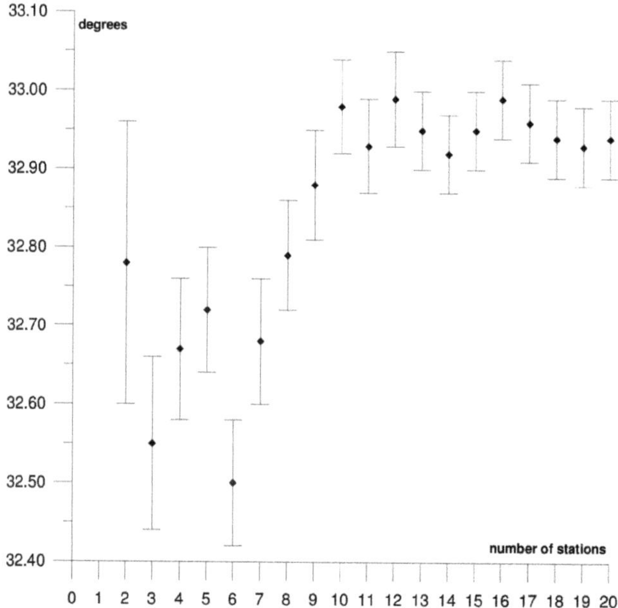

Figure 6. Results of the sequential method for the Φ parameter for the Australian plate.

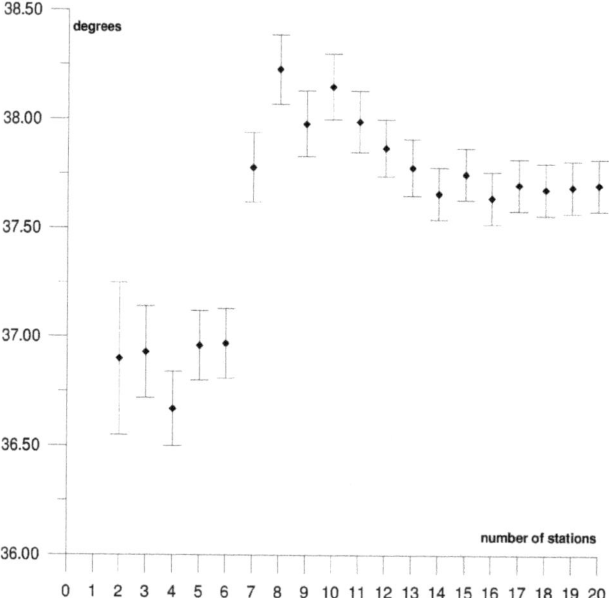

Figure 7. Results of the sequential method for the Λ parameter for the Australian plate.

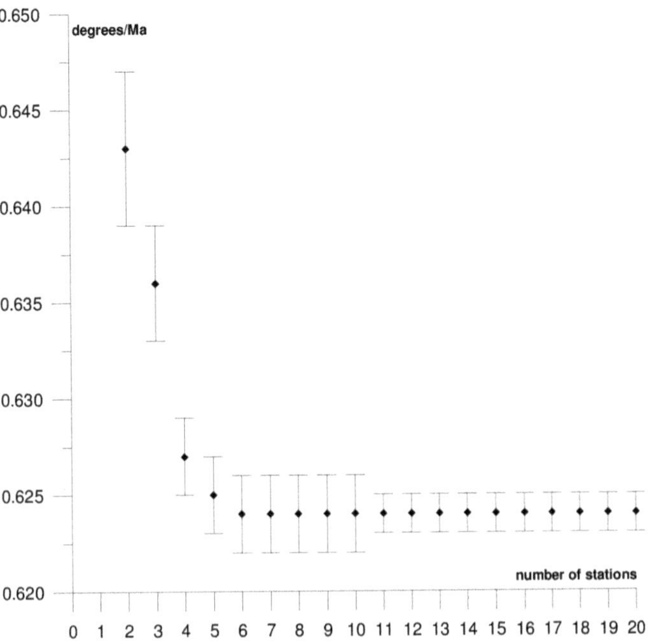

Figure 8. Results of the sequential method for the ω parameter for the Australian plate.

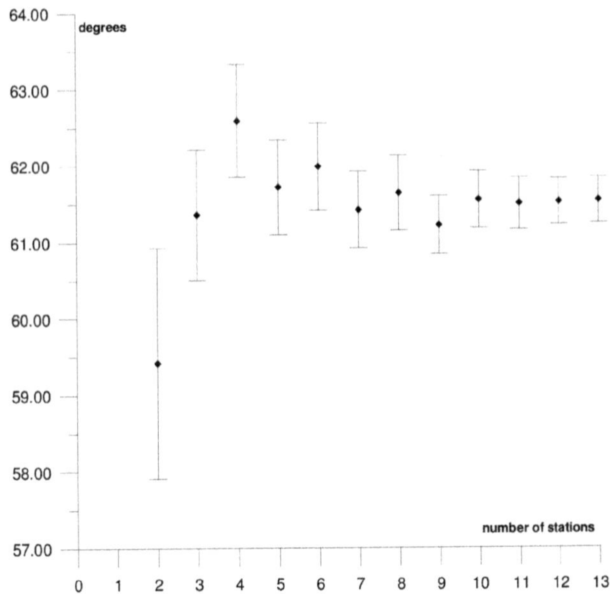

Figure 9. Results of the sequential method for the Φ parameter for the Antarctic plate.

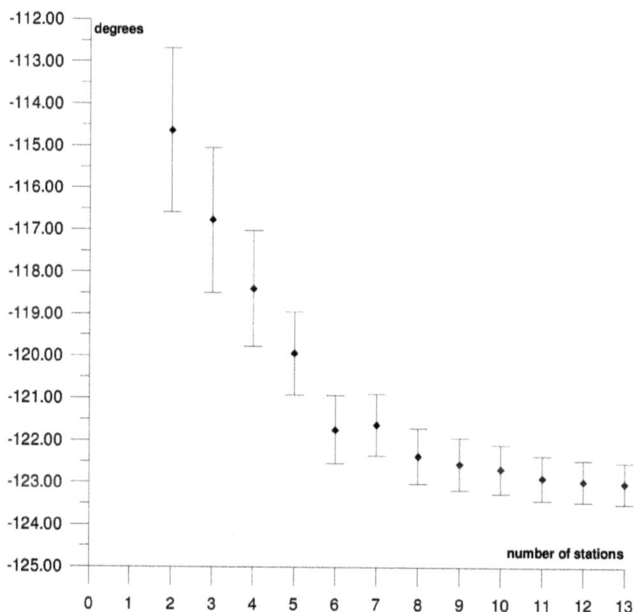

Figure 10. Results of the sequential method for the Λ parameter for the Antarctic plate.

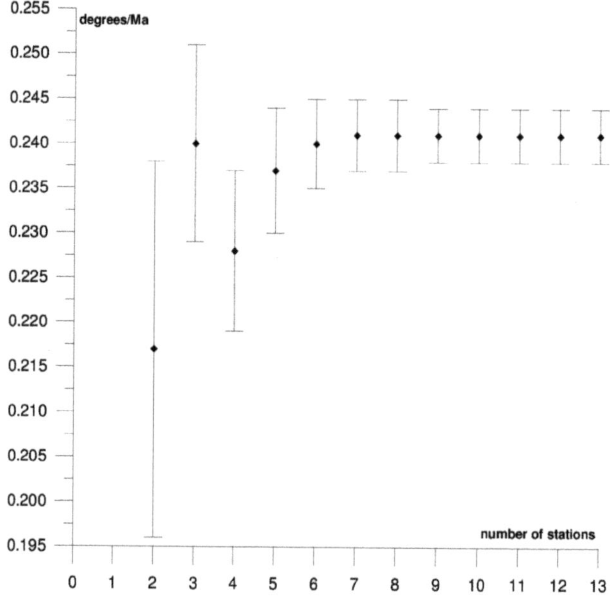

Figure 11. Results of the sequential method for the ω parameter for the Antarctic plate.

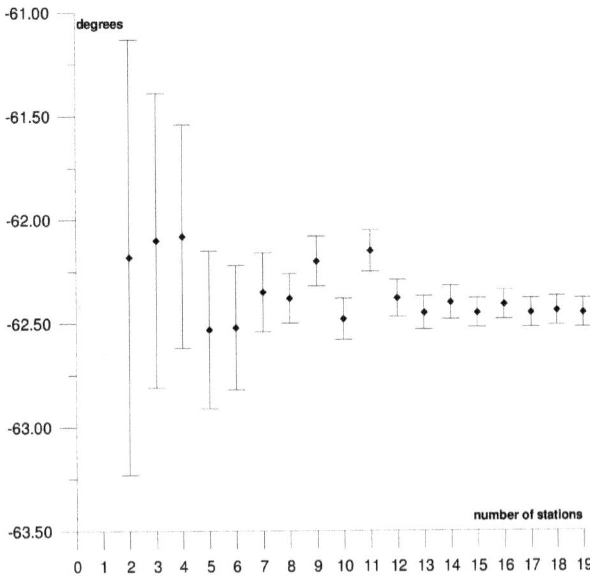

Figure 12. Results of the sequential method for the Φ parameter for the Pacific plate.

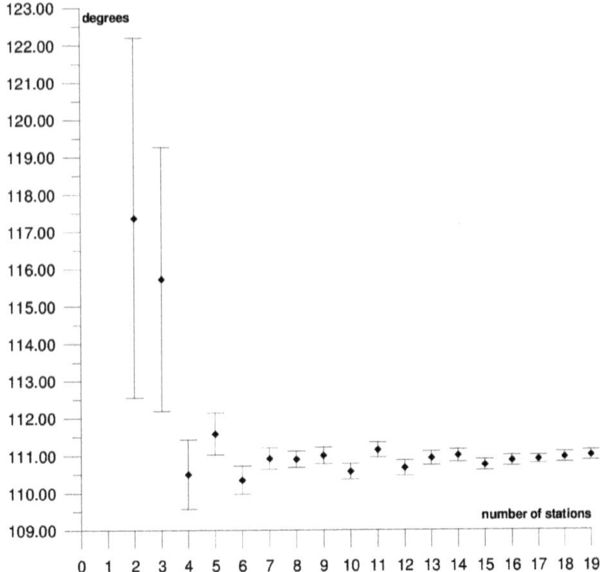

Figure 13. Results of the sequential method for the Λ parameter for the Pacific plate.

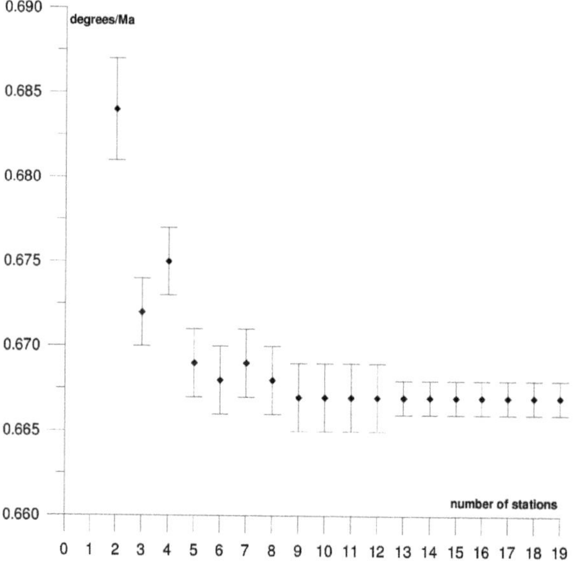

Figure 14. Results of the sequential method for the ω parameter for the Pacific plate.

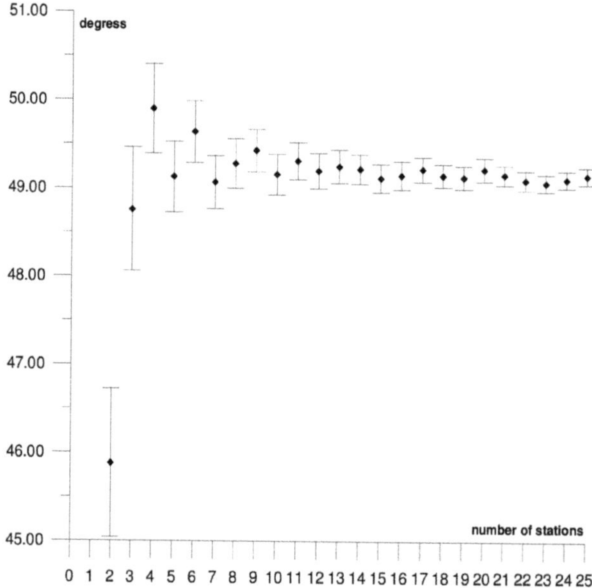

Figure 15. Results of the sequential method for the Φ parameter for the African plate.

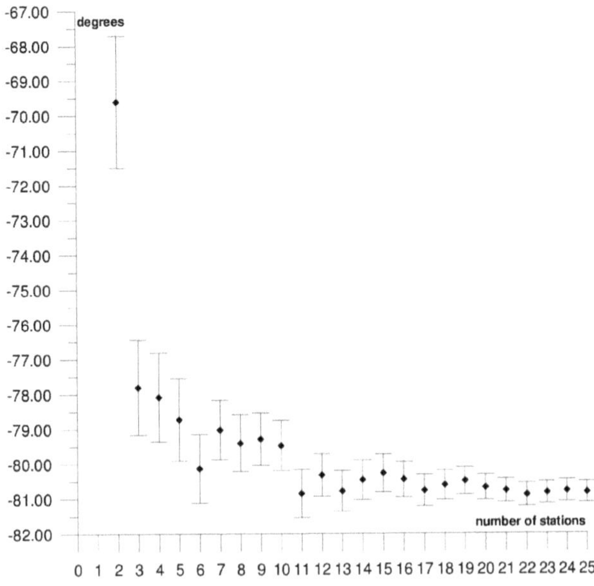

Figure 16. Results of the sequential method for the Λ parameter for the African plate.

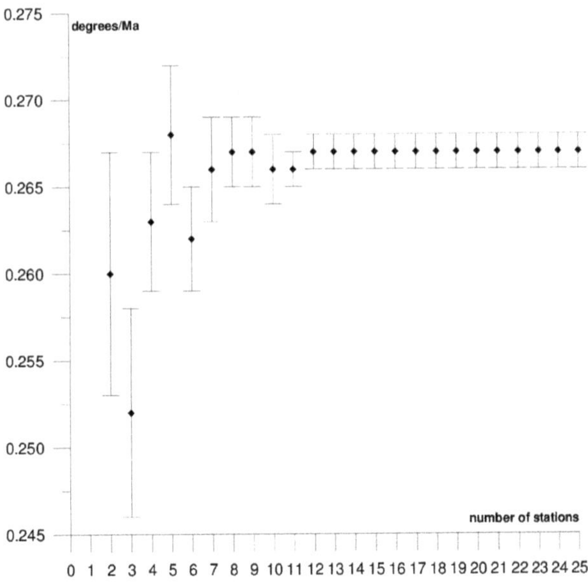

Figure 17. Results of the sequential method for the ω parameter for the African plate.

The first plate that was analysed was the South American plate, which is bordered by the African and Nazca plates as well as several smaller ones with the oceanic lithosphere. It moves northwest at a rate of approximately 3 cm/year towards the North American plate, moving continuously away from Africa. It consists of three main geological units: the South American craton, which covers the northern, eastern, and central parts of the continent; the Palaeozoic Patagonian platform in the southeast; and the Andes, an alpine mountain range that stretches along the western boundary of the plate. In total, 29 GNSS stations were used to determine the motion parameters of the South American plate. The motion parameters calculated for all stations in total are: $\Phi = -19.03 \pm 0.20°$, $\Lambda = -119.78 \pm 0.39°$, and $\omega = 0.117 \pm 0.001°/\text{Ma}$, which is presented, respectively, in Figures 3–5 and in Appendix A. The comparison of the results with the values obtained in a previous study for the DORIS technique [39] demonstrated that the consistency of results was similar to the value of formal errors, and the differences were 1.27°, 1.57°, and 0.016°/Ma for the Φ, Λ, and ω parameters, respectively. For the other techniques, i.e., SLR and VLBI, the motion parameters of this plate were not calculated due to the lack of the required number of stations.

The solution becomes stable after 10, 11, and 7 steps of the sequential method, respectively, for the Φ, Λ, and ω parameters. At that stage, the changes in the values of the calculated parameters after adding subsequent stations to the process do not exceed the formal error. Hence, one may assume that the minimum number of stations required to determine the motion parameters of this plate is approximately 13. Due to high values of formal errors and discrepancies with the final values of the determined Φ, Λ, and ω parameters (please refer to Table 1), the following stations were rejected from the solutions: Valparaiso, Antuco, Concepcion, Quito III, Arequipa, and Callao (red dots on Figure 2). These stations accumulate interseismic strain associated with locking of the Peru-Chile megathrust, and therefore are moving east relative to the South American plate [45]. The Quito III station is located near the boundaries of four smaller plates: Cocos, Caribbean, Nazca, and North Andes. Including it in the solution causes a change in the Φ, Λ, and ω parameters, respectively, by approximately 3°, 10°, and 0.01°/Ma. The other stations: Callao, Arequipa, Valparaiso, Concepcion, and Antuco are situated along the boundary of the Nazca plate. Among them, the Callao station has the highest influence on the change in motion parameters, namely parameter Φ by approximately 6°, parameter Λ by approximately 11°, and parameter ω by $-0.016°/\text{Ma}$.

Table 1. Stations rejected from the solutions.

Plate Name	Name of the Station	Φ (°)	Λ (°)	ω (°/Ma)
South American	Valparaiso	-14.47 ± 2.41	-117.08 ± 5.24	0.131 ± 0.011
	Antuco	-18.05 ± 1.06	-117.53 ± 2.54	0.110 ± 0.010
	Concepcion	-15.34 ± 2.39	-115.15 ± 5.43	0.132 ± 0.014
	Quito III	-16.09 ± 0.82	-129.78 ± 1.91	0.118 ± 0.014
	Arequipa	-16.06 ± 1.07	-127.45 ± 2.51	0.124 ± 0.008
	Callao	-13.01 ± 1.46	-130.48 ± 3.35	0.133 ± 0.012
Australian	Wellington	30.23 ± 1.03	34.14 ± 1.51	0.633 ± 0.005
	Coco Island	31.00 ± 1.27	41.08 ± 2.56	0.583 ± 0.006
Pacific	Nuku Alofa	-56.97 ± 2.15	116.27 ± 1.97	0.647 ± 0.010
Antarctic	King George Island	56.56 ± 3.40	-128.76 ± 3.07	0.183 ± 0.022
African	Awra	46.59 ± 0.47	-78.73 ± 2.02	0.250 ± 0.006
	Marion Island	50.64 ± 0.51	-84.34 ± 1.23	0.254 ± 0.005

The motion parameters of the Australian plate were determined based on the position and velocity of 20 stations. Their final values are: $\Phi = 32.94 \pm 0.05°$, $\Lambda = 37.70 \pm 0.12°$, and $\omega = 0.624 \pm 0.001°/\text{Ma}$. The results of the solution for each step of the sequential method are presented in Figures 6–8 and in Appendix B. The solutions become stable for Φ, Λ, and ω parameters, respectively, after 10, 11, and 4 steps of the sequential method. Hence, one may assume that the minimum number of stations required to determine the motion parameters of this plate is approximately 12. The Australian plate is the most tectonically stable among all of the analysed plates, it moves northeast towards the Eurasian and Pacific plate at a rate of approximately 6–7 cm/year. The greater part of the Australian plate is occupied by the Precambrian Craton, called the Australian Craton, which is adjacent to the structure of the Flinders Ranges and the Barrier Ranges, and the structure of the Great Dividing Range [46]. The tectonic stability of this plate is reflected in the formal errors of the determined motion parameters, i.e., their values are lower than those of the remaining plates. Similar findings have been observed in previous studies on the SLR and DORIS techniques [37–39]. The comparison of the results with the values obtained in an earlier study on the SLR, DORIS, and VLBI techniques, the highest compatibility of results was found with the VLBI technique [40], and the lowest one with the SLR technique [37]. The differences in the determined Φ, Λ, and ω values for the GNSS and VLBI techniques are, respectively, $0.31°$, $0.29°$, and $0.007°$/Ma, while, for the GNSS and SLR techniques, they are $1.52°$, $-1.78°$, and $0.007°$/Ma, respectively, for Φ, Λ, and ω. Two stations (Wellington and Coco Island) that disturb the result of calculating the Φ and Λ parameters by approximately $3°$ were found on the Australian plate (please refer to Table 1). Including these stations in the determination of the parameters also increases the formal errors of the parameters multiple times, therefore, these stations were rejected from the solutions. The Wellington station is located at the boundary with the Pacific plate near the so-called Alpine Fault, and the Coco Island station is located near the boundary with the Capricorn and Sunda small plates.

The next plate, the Antarctic plate is bordered by the African, South American, Australian, Pacific, Nazca, and Scotia plates. Within the plate, three main geological units can be distinguished: the Antarctic craton, which covers the eastern part of the continent; the Palaeozoic platform in the western part; and the Alpine fold zone of the Antarctic Peninsula. The Antarctic plate moves in the northeast direction (the part bordering the South American plate) and the southern direction (the part bordering the Australian plate) at a rate of about 1–1.5 cm/year. The following motion parameter values were obtained for the Antarctic plate: $\Phi = 61.54 \pm 0.30°$, $\Lambda = -123.01 \pm 0.49°$, and $\omega = 0.241 \pm 0.003°$/Ma. The results of the sequential method are depicted in Figures 9–11 and presented in Appendix C. The solution was based on 13 stations, whose positions are shown in Figure 2. The inclusion of the King George Island station located near the boundary with the Shetland microplate and the Scotia plate, in the area of the alpine fold of the Antarctic Peninsula, results in an approximately 10-fold increase in the formal errors and a change in the values of the Φ and Λ parameters by about $6°$ (Table 1); this station was not included in the calculations.

In general, the formal errors of the Φ, Λ, and ω parameters for the Antarctic plate were the highest among all of the analysed plates. The comparison of the obtained results to the solutions for the DORIS [39] and VLBI [40] techniques revealed that the compatibility was higher for DORIS. The differences in the determined values between the GNSS and DORIS techniques are $1.26°$ for Φ, $1.89°$ for Λ, and $-0.009°$/Ma for ω, while the differences between GNSS and VLBI are, respectively, $2.26°$, $4.64°$, and $0.025°$/Ma. The stability of the solution of the Φ, Λ, and ω parameters appear after five steps of the sequential method. At that stage, the changes in the values of the calculated parameters, after the addition of another station to the process, do not exceed the formal error. Hence, one may assume that the minimum number of stations required to determine the motion parameters of this plate is approximately 7.

The sequential method for the motion parameters for the Pacific plate are presented in Figures 12–14 and in Appendix D. It is the largest tectonic plate, and, within it, there are areas of very high tectonic activity, the so-called Ring of Fire. The Pacific plate moves northwest towards the Eurasian and Australian plates at a rate of approximately 6–10 cm/year. In the solution, 19 stations were included. Their locations are shown in Figure 2, and their names are listed in Appendix D. The final values of the motion parameters of the Pacific plate are: $\Phi = -62.45 \pm 0.07°$, $\Lambda = 111.01 \pm 0.14°$, and $\omega = 0.667 \pm 0.001°$/Ma. Similar values were obtained in a previous study for the SLR, DORIS, and VLBI techniques [38]. The smallest differences were noted in the comparison with the SLR technique: $0.05°$ for the Φ parameter, $2.50°$ for the Λ parameter, and the value of the ω parameter was the same.

The stability of the solution of the Φ, Λ, and ω parameters appear, respectively, after 12, 15, and 5 steps of the sequential method. Adding further stations, up to 19 stations, resulted in a change of the Φ, Λ, and ω parameters by values that did not exceed formal errors. Hence, one may assume that the minimum number of stations required to determine the motion parameters of this plate is approximately 17. The solution took into consideration the Point Reyes Lig. station located on the North American continent that, on the one hand, its movement was compatible with that of the Pacific plate and it did not have a negative effect on the solution of the Φ, Λ, and ω parameters (please refer to Appendix D). On the other hand, the Nuku Alofa station was rejected, as it caused a multi-fold increase in the values of formal errors for the Φ, Λ, and ω parameters (please refer to Table 1) and changed their values by approximately 5° (Φ and Λ) and by 0.020°/Ma (ω).

Figures 15–17 depict the results of the sequential method for the motion parameters of the African plate. It moves northeast towards the Eurasian and Arabian plates. The main geological unit is the Precambrian craton (the so–called African Megacraton) with a system of tectonic grabens, forming the East African Rift system. The sequential method was based on 25 GNSS stations, which are listed in Appendix E, and their locations are presented in Figure 2. The final values of the Φ, Λ, and ω parameters equal: $\Phi = 49.15 \pm 0.10°$, $\Lambda = -80.82 \pm 0.30°$, and $\omega = 0.267 \pm 0.001°/$Ma. The obtained values of plate motion parameters in the first few steps of calculations were significantly divergent from the final results, as shown in Appendix E. Adding further stations in the calculation process led to the stabilisation of the results, until final values were reached based on all 25 stations. The stability of the results for the Φ, Λ, and ω parameters was noted, respectively, for 10, 7, and 7 stations used in the solution. Hence, the minimum number of stations required to ensure a stable solution is approximately 12 stations. The results were rather compatible with the values obtained in previous studies for the SLR [37] and DORIS [39] techniques, and the differences are, respectively, $-1.63°$ and $-0.38°$ for Φ, $-4.00°$ and $-1.70°$ for Λ, and $-0.016°/$Ma and $0.017°/$Ma for ω. The Awra and Marion Island stations were rejected from the solution. The Awra station (located on the African continent near the boundary with the Arabian plate) contributed to a change in the Φ parameter by approximately 2.5°, Λ by approximately 2°, and ω by approximately 0.02°/Ma, while the Marion Island (located on an island near the boundary with the Antarctic plate) station changed the values by approximately 1.5° for Φ, by approximately 3.5° for Λ, and by approximately 0.01°/Ma for ω (please refer to Table 1). These stations also caused an approximately six-fold increase in the value of the formal errors in the determined parameters. The following stations are included in the solution: Addis Ababa, Mbarara, Tanzania CGPS, and Richardsbay. Although they are located on the Somalia and Lwandle small plates, they do not have a negative effect on the results of the solution (Appendix E).

Comparison with Geological Model NNR-MORVEL56 and Geodetic ITRF2014 Plate Motion Model (PMM)

Geological models are developed based on geophysical observations, such as sea floor spreading rates, earthquake slip vectors, and transform fault azimuths. The most commonly used geological model is NNR-NUVEL1A [24] that originates from NNR-NUVEL1 [23]. It has recently been replaced with a new model, namely NNR-MORVEL56 [26], which is considered to be better than NNR-NUVEL1A. The NNR-MORVEL56 model was determined from more and higher quality spreading rates and azimuths. Moreover, it excluded circum-Pacific data (earthquake slip vectors and Pacific North America spreading rates) that were biased measures of relative plate velocity, addressed in [26]. The comparison of the results of the solutions of motion parameters of the specific tectonic plates with the NNR-MORVEL56 geological model should be approached with caution. The observations that are used for creating geological models are limited to plate boundaries, where local deformations occur quite often, and therefore the obtained results are not always representative of the movement of the whole plate. Moreover, geological models provide an average movement of individual plates across a very long period of time, which may range from hundreds of thousands to millions of years, and due to this, they may not reflect the potential current speeding up or slowing down of specific tectonic plates, which, however, are recorded by very precise space techniques: SLR, DORIS, VLBI, and GNSS. Nevertheless, the ITRF2000 was defined based on the tectonic plate movement obtained from the NNR-NUVEL1A geological model. Subsequent solutions of the ITRF were adopted to the previous ones: ITRF2005 to ITRF2000, ITRF2008 to ITRF2005, and ITRF2014 to ITRF2008, but they are still indirectly linked to NNR-NUVEL1A. The question about the use of the new geological model, i.e., the NNR-MORVEL56, in future solutions of the ITRF, is still important and was discussed in [28]. To date, it has been found that the angular velocities in NNR-MORVEL56 differ significantly from those in NNR-NUVEL1A for all plates [5,26]. A detailed comparative analysis of these models was conducted by Argus et al. [26] and it will not be discussed here. However, the comparison of the motion parameters of individual plates that were determined in this study with the NNR-MORVEL56 model is presented in Table 2.

Table 2. Comparison with the NNR-MORVEL56 model.

Plate Name	NNR-MORVEL56 (1) Φ (°), Λ (°), ω (°/Ma)	This Paper (2) Φ (°), Λ (°), ω (°/Ma)	Differences (1)–(2) Φ (°), Λ (°), ω (°/Ma)
South American	-22.62 -112.83 0.109	-19.03 ± 0.20 -119.78 ± 0.39 0.117 ± 0.001	-3.59 6.95 -0.008
Australian	33.86 37.94 0.632	32.94 ± 0.05 37.70 ± 0.12 0.624 ± 0.001	0.92 0.24 0.008
Pacific	-63.58 114.70 0.651	-62.45 ± 0.07 111.01 ± 0.14 0.667 ± 0.001	-1.13 3.69 -0.016
Antarctic	65.42 -118.11 0.250	61.54 ± 0.30 -123.01 ± 0.49 0.241 ± 0.003	3.88 4.90 0.009

The calculated differences between the values of the Φ, Λ, and ω parameters for the NNR-MORVEL 56 model and the results of the solution presented in this study revealed that the most similar values were obtained for the Australian plate: the Φ parameter differed by approximately $1°$, the Λ parameter by approximately $0.2°$, and the ω parameter by $0.008°$/Ma, which corresponded to approximately 1 mm/year. A difference of approximately $1°$ for the Φ parameter was also obtained for the Pacific plate, although for the Λ parameter, the difference was higher and approximately $4°$, while the angular rotation speed, ω, differed by approximately $-0.02°$/Ma, which corresponded to approximately 2 mm/year. Regarding the South American and Antarctic plates, the difference in the values of the Φ parameter was approximately $4°$, and for the Λ parameter, approximately $7°$ (the South American plate) and approximately $5°$ (the Antarctic plate). The angular rotation speed ω differed by $-0.008°$/Ma for the South American plate and by $0.009°$/Ma for the Antarctic plate, which corresponded to approximately 1 mm/year. The NNR-MORVEL56 model does not present the results for the African plate, which distinguishes two plates within it, namely Nubia and Somalia; hence, the inability to compare the obtained results. Nubia covers approximately 95% of the surface area of the African continent and the area to the West, towards the boundary of the South American plate, while the Somalia plate covers about 5% of the continent (the Eastern part) and the area to the East to the boundary with the Australian plate and the smaller tectonic plate, i.e., the Indian plate.

The ITRF2014 PMM [5] is the geodetic model describing the motion of 11 tectonic plates (the major plates and several selected small plates). It is dedicated to the currently in force ITRF2014 frame [28]. In developing it, horizontal velocities of a subset of the ITRF2014 stations of all space techniques in a combined solution (SLR + DORIS + VLBI + GNSS) were used, localized away from plate boundaries and deforming zones. For the South American plate, it was a total of 30 stations apart from the GNSS stations also 2 DORIS ones); for the Australian plate, it was 36 stations (including five DORIS, five VLBI, and three SLR stations); for the Pacific plate, it was 18 stations (including four DORIS, three VLBI, and one SLR); for the Antarctic plate, it was seven stations (GNSS only). Similar to the NNR-MORVEL 56 model, no parameters were determined for the African plate. In the ITRF2014 PMM, the plate motion was described providing specific elements of the pole of rotation: ω_x, ω_y, ω_z, and the angular rotation speed ω. In order to compare the results, the values of geographical latitude (Φ) and longitude (Λ) of the pole of rotation determined in this study for each plate were calculated with the use of the formulas (Equation (1)) into ω_x, ω_y, and ω_z. The comparison is presented in Table 3. As it can be seen from that table, the largest differences are noted for the South American plate, whereas small differences can be noted for the Australian plate. From the analysis of separate components of the pole of rotation (ω_x, ω_y, and ω_z), it becomes evident that the largest differences for the ω_x component occur for the South American plate (-0.082 mas/year) and the Antarctic plate (-0.023 mas/year); for the ω_y component, the largest differences occur for the South American plate (0.050 mas/year) and the Australian plate (0.029 mas/year); while for the ω_z component, the largest differences occur for the Antarctic plate (-0.088 mas/year) and the Pacific plate (-0.057 mas/year). For the rotation angular velocity (ω), the biggest difference is in the case of the Antarctic plate ($-0.022°$/Ma) and the Pacific plate ($0.012°$/Ma), which corresponds to approximately 2 mm/year and 1 mm/year, respectively. For the remaining plates the differences do not exceed 1 mm/year.

Table 3. Comparison with the ITRF2014 PMM.

Plate Name	ITRF2014 PMM (1) ω_x, ω_y, ω_z (mas/yr), ω (°/Ma)	This Paper (2) ω_x, ω_y, ω_z (mas/yr), ω (°/Ma)	Differences (1)–(2) ω_x, ω_y, ω_z (mas/yr), ω (°/Ma)
South American	-0.270 ± 0.006 -0.301 ± 0.006 -0.140 ± 0.003 0.119 ± 0.001	-0.188 ± 0.003 -0.351 ± 0.003 -0.136 ± 0.002 0.117 ± 0.001	-0.082 0.050 -0.004 0.002
Australian	1.510 ± 0.004 1.182 ± 0.004 1.215 ± 0.004 0.631 ± 0.001	1.492 ± 0.004 1.153 ± 0.004 1.222 ± 0.004 0.624 ± 0.001	0.018 0.029 -0.007 0.007
Pacific	-0.409 ± 0.003 1.047 ± 0.004 -2.169 ± 0.004 0.679 ± 0.001	-0.410 ± 0.004 1.066 ± 0.004 -2.112 ± 0.004 0.667 ± 0.001	0.001 -0.019 -0.057 0.012
Antarctic	-0.248 ± 0.004 -0.324 ± 0.004 0.675 ± 0.008 0.219 ± 0.002	-0.225 ± 0.002 -0.347 ± 0.003 0.763 ± 0.021 0.241 ± 0.003	-0.023 0.023 -0.088 -0.022

In general, it can be stated that the agreement with the NNR-MORVEL56 and ITRF2014 PMM models is good for all plates.

4. Conclusions

In this study, the motion parameters, i.e., the latitude (Φ) and longitude (Λ) of the rotation pole, and the angular rotation speed (ω) of five major lithospheric plates (South American, African, Australian, Antarctic, and Pacific plates) were determined. The observational material includes the positions and velocities of 106 highly quality GNSS stations in the ITRF2014. As many as 29 of these stations are located on the South American plate, 13 stations on the Antarctic plate, 20 stations on the Australian plate, 19 stations on the

Pacific plate, and 25 stations on the African plate. The most accurate solutions of the Φ, Λ, and ω parameters were determined for the Australian and Pacific plates, while the least accurate were determined for the Antarctic plate. The comparison of the obtained results to previously conducted research on the SLR [37,38], DORIS [38,39], and VLBI [38,40] techniques revealed high compatibility. The most similar results were obtained for the comparison with the VLBI technique for the Australian plate, the SLR technique for the Pacific plate, and the DORIS technique for the African, Antarctic, and South American plates. The most accurate solutions for the Φ, Λ, and ω parameters were obtained with the GNSS technique as compared with the other techniques. Because of the dense coverage of the Earth with GNSS stations, this is the only space technique that offers the possibility to determine the motion parameters of all the major lithospheric plates. The applied sequential method allowed us to define the minimum number of stations that ensured a stable solution and to indicate the stations that negatively affected the result of the solution. The minimum number of stations that should be used to determine the Φ, Λ, and ω parameters to guarantee a stable solution differs depending on the individual plates. For the South American and Australian plates, this number is approximately 13 stations; 7 stations for the Antarctic plate; 17 stations for the Pacific plate; and 12 stations for the African plate. Including a larger number of stations in the calculations does not have a significant influence on the values of the determined Φ, Λ, and ω parameters and formal errors. A total number of 12 stations that disturbed the solutions of the Φ, Λ, and ω parameters and increased the values of formal errors were found; these stations were rejected from the solution. Six of these stations are located on the South American plate (Valparaiso, Antuco, Concepcion, Quito III, Arequipa, and Callao), two stations on the Australian (Wellington and Coco Island) and African (Awra and Marion Island) plates, and one on the Antarctic (King George Island) and Pacific (Nuku Alofa) plates. Including them in the solution leads to a change in the Φ and Λ parameters in the range of 2–11°, and of the ω parameter in the range from -0.016 to $0.058°$/Ma.

The values that were the most similar to the current geological model NNR-MORVEL56 [26] were obtained for the Australian plate (the differences do not exceed 1° for the Φ and the Λ parameters, and $0.008°$/Ma for the ω parameter). The differences for the other plates are higher, ranging from approximately 1 to 7° for the Φ and the Λ parameters, and from -0.016 to $0.008°$/Ma for the ω parameter. These differences may indicate a current slowing down or speeding up of the movement of certain tectonic plates, which are detected by the very precise GNSS space technique. Comparing the results with the geodetic model ITRF2014 PMM [5], one can find a high agreement between them, the largest differences are found for the South American plate (ω_x and ω_y components) and the Antarctic plate (ω_z component). The rotation angular velocity (ω) differs within the range from $-0.022°$/Ma (the Antarctic plate) to $0.012°$/Ma (the Pacific plate).

Funding: This research received no external funding.

Institutional Review Board Statement: Not applicable.

Informed Consent Statement: Not applicable.

Data Availability Statement: Not applicable.

Conflicts of Interest: The author declares no conflict of interest.

Appendix A

Determined plate motion parameters and their formal errors of the South American plate for the GNSS network using the sequential method.

No.	Name of the Station	$\Phi(°)$	$\Lambda(°)$	$\omega(°/\text{Ma})$
2	Brasilia + Forteleza	-18.61 ± 0.76	-131.48 ± 1.91	0.136 ± 0.003
3	2 + Buenos Aires	-19.11 ± 0.67	-124.67 ± 1.80	0.125 ± 0.003
4	3 + Rio Grande	-19.80 ± 0.61	-122.75 ± 1.49	0.129 ± 0.003
5	4 + La Plata	-18.89 ± 0.60	-121.31 ± 1.32	0.123 ± 0.003
6	5 + Bahia Blanca	-18.20 ± 0.59	-121.72 ± 1.18	0.124 ± 0.003
7	6 + Lihue Calel	-18.86 ± 0.59	-120.87 ± 1.16	0.119 ± 0.003
8	7 + Curitiba-Para	-19.49 ± 0.57	-121.66 ± 1.14	0.118 ± 0.003
9	8 + Presidente Prud	-18.90 ± 0.52	-120.75 ± 1.09	0.119 ± 0.002
10	9 + Manaus	-19.53 ± 0.46	-119.37 ± 0.99	0.120 ± 0.002
11	10 + Porto Alegre	-19.42 ± 0.43	-118.33 ± 0.95	0.118 ± 0.002
12	11 + Recif	-19.30 ± 0.39	-118.64 ± 0.89	0.118 ± 0.002
13	12 + Cananeia	-19.37 ± 0.38	-118.45 ± 0.86	0.117 ± 0.002
14	13 + Belem	-19.41 ± 0.34	-118.31 ± 0.81	0.117 ± 0.001
15	14 + Porto Velho	-19.22 ± 0.32	-118.72 ± 0.79	0.117 ± 0.001
16	15 + Macapa	-19.06 ± 0.30	-119.00 ± 0.55	0.117 ± 0.001
17	16 + Ilha Solteira	-19.02 ± 0.29	-119.19 ± 0.54	0.117 ± 0.001
18	17 + Boa Vista	-19.09 ± 0.27	-119.28 ± 0.52	0.117 ± 0.001
19	18 + Imbituba	-19.14 ± 0.27	-119.41 ± 0.51	0.117 ± 0.001
20	19 + Sao Gabriel	-18.93 ± 0.25	-119.53 ± 0.48	0.117 ± 0.001
21	20 + Sao Luis	-18.86 ± 0.24	-119.59 ± 0.47	0.117 ± 0.001
22	21 + Salvador Capita	-18.95 ± 0.24	-119.70 ± 0.47	0.117 ± 0.001
23	22 + Rio Branco	-18.82 ± 0.23	-119.94 ± 0.45	0.117 ± 0.001
24	23 + Palmas	-19.02 ± 0.22	-119.82 ± 0.44	0.117 ± 0.001
25	24 + Kourou	-18.94 ± 0.21	-119.90 ± 0.41	0.117 ± 0.001
26	25 + Yacuiba	-18.87 ± 0.21	-119.74 ± 0.40	0.117 ± 0.001
27	26 + Cuiba	-19.01 ± 0.20	-119.77 ± 0.40	0.117 ± 0.001
28	27 + Punta Arenas	-19.00 ± 0.20	-119.75 ± 0.39	0.117 ± 0.001
29	28 + Iquitos	$\mathbf{-19.03 \pm 0.20}$	$\mathbf{-119.78 \pm 0.39}$	$\mathbf{0.117 \pm 0.001}$

Appendix B

Determined plate motion parameters and their formal errors of the Australian plate for the GNSS network using the sequential method.

No.	Name of the Station	$\Phi(°)$	$\Lambda(°)$	$\omega(°/\text{Ma})$
2	Melbourne Obser + Yarragadee2	32.78 ± 0.18	36.90 ± 0.35	0.643 ± 0.004
3	2 + Christmas Islan	32.55 ± 0.11	36.93 ± 0.21	0.636 ± 0.003
4	3 + Warkworth	32.67 ± 0.09	36.67 ± 0.17	0.627 ± 0.002
5	4 + Darwin I	32.72 ± 0.08	36.96 ± 0.16	0.625 ± 0.002
6	5 + Alice Springs	32.50 ± 0.08	36.97 ± 0.16	0.624 ± 0.002
7	6 + Diego Garcia	32.68 ± 0.08	37.78 ± 0.16	0.624 ± 0.002
8	7 + Noumea	32.79 ± 0.07	38.23 ± 0.16	0.624 ± 0.002
9	8 + Karratha	32.88 ± 0.07	37.98 ± 0.15	0.624 ± 0.002
10	9 + Lae–Universit	32.98 ± 0.06	38.15 ± 0.15	0.624 ± 0.002
11	10 + Tidbinbilla1	32.93 ± 0.06	37.99 ± 0.14	0.624 ± 0.001
12	11 + Koumac	32.99 ± 0.06	37.87 ± 0.13	0.624 ± 0.001
13	12 + Sydney	32.95 ± 0.05	37.78 ± 0.13	0.624 ± 0.001
14	13 + Katherine-Nor1	32.92 ± 0.05	37.66 ± 0.12	0.624 ± 0.001
15	14 + Townsville-Ca	32.95 ± 0.05	37.75 ± 0.12	0.624 ± 0.001
16	15 + Auckland	32.99 ± 0.05	37.64 ± 0.12	0.624 ± 0.001
17	16 + Ceduna	32.96 ± 0.05	37.70 ± 0.12	0.624 ± 0.001
18	17 + Parkes	32.94 ± 0.05	37.68 ± 0.12	0.624 ± 0.001
19	18 + Perth	32.93 ± 0.05	37.69 ± 0.12	0.624 ± 0.001
20	19 + Mount Stromlo	$\mathbf{32.94 \pm 0.05}$	$\mathbf{37.70 \pm 0.12}$	$\mathbf{0.624 \pm 0.001}$

Appendix C

Determined plate motion parameters and their formal errors of the Antarctic plate for the GNSS network using the sequential method.

No.	Name of the Station	$\Phi(°)$	$\Lambda(°)$	$\omega(°/Ma)$
2	Syowa + Rothera	59.42 ± 1.51	-114.63 ± 1.95	0.217 ± 0.021
3	2 + O'Higgins	61.36 ± 0.85	-116.76 ± 1.72	0.240 ± 0.011
4	3 + Casey	62.59 ± 0.74	-118.39 ± 1.38	0.228 ± 0.009
5	4 + Vernadski	61.72 ± 0.62	-119.93 ± 0.99	0.237 ± 0.007
6	5 + Mount Fleming	61.99 ± 0.57	-121.74 ± 0.81	0.240 ± 0.005
7	6 + Ile des Petrels	61.42 ± 0.50	-121.62 ± 0.73	0.241 ± 0.004
8	7 + Mac Murdo	61.64 ± 0.49	-122.36 ± 0.66	0.241 ± 0.004
9	8 + Mawson	61.22 ± 0.38	-122.55 ± 0.62	0.241 ± 0.003
10	9 + Palmer	61.55 ± 0.37	-122.67 ± 0.58	0.241 ± 0.003
11	10 + Sanae	61.50 ± 0.34	-122.88 ± 0.53	0.241 ± 0.003
12	11 + Fishtail Point	61.52 ± 0.30	-122.96 ± 0.50	0.241 ± 0.003
13	12 + Cape Roberts	$\mathbf{61.54 \pm 0.30}$	$\mathbf{-123.01 \pm 0.49}$	$\mathbf{0.241 \pm 0.003}$

Appendix D

Determined plate motion parameters and their formal errors of the African plate for the GNSS network using the sequential method.

No.	Name of the Station	$\Phi(°)$	$\Lambda(°)$	$\omega(°/Ma)$
2	Kauai + Papeete (Tahiti)	-62.18 ± 1.05	117.37 ± 4.83	0.684 ± 0.003
3	2 + Maui I	-62.10 ± 0.71	115.73 ± 3.54	0.672 ± 0.002
4	3 + Dunedin	-62.08 ± 0.54	110.50 ± 0.93	0.675 ± 0.002
5	4 + Mauna Kea	-62.53 ± 0.38	111.58 ± 0.56	0.669 ± 0.002
6	5 + Rikitea (Ile Man)	-62.52 ± 0.30	110.35 ± 0.38	0.668 ± 0.002
7	6 + Niue Island	-62.35 ± 0.19	110.92 ± 0.29	0.669 ± 0.002
8	7 + American Samoa	-62.38 ± 0.12	110.90 ± 0.22	0.668 ± 0.002
9	8 + Apia	-62.20 ± 0.12	111.00 ± 0.22	0.667 ± 0.002
10	9 + Tubuai	-62.48 ± 0.10	110.57 ± 0.21	0.667 ± 0.002
11	10 + Futuna	-62.15 ± 0.10	111.15 ± 0.20	0.667 ± 0.002
12	11 + Point Reyes Lig	-62.38 ± 0.09	110.67 ± 0.20	0.667 ± 0.002
13	12 + Chatham Island	-62.45 ± 0.08	110.93 ± 0.18	0.667 ± 0.001
14	13 + Betio Island	-62.40 ± 0.08	111.00 ± 0.17	0.667 ± 0.001
15	14 + Kwajalein Atoll	-62.45 ± 0.07	110.75 ± 0.15	0.667 ± 0.001
16	15 + Nauru	-62.41 ± 0.07	110.87 ± 0.14	0.667 ± 0.001
17	16 + Majuro	-62.45 ± 0.07	110.90 ± 0.11	0.667 ± 0.001
18	17 + Pohnpei	-62.44 ± 0.07	110.96 ± 0.14	0.667 ± 0.001
19	18 + Honolulu	$\mathbf{-62.45 \pm 0.07}$	$\mathbf{111.01 \pm 0.14}$	$\mathbf{0.667 \pm 0.001}$

Appendix E

Determined plate motion parameters and their formal errors of the African plate for the GNSS network using the sequential method.

No.	Name of the Station	$\Phi(°)$	$\Lambda(°)$	$\omega(°/Ma)$
2	Cotonou + Lusaka	45.88 ± 0.84	-69.60 ± 1.89	0.260 ± 0.007
3	2 + Addis Ababa Uni	48.76 ± 0.70	-77.80 ± 1.36	0.252 ± 0.006
4	3 + Dakar Universit	49.90 ± 0.51	-78.08 ± 1.27	0.263 ± 0.004
5	4 + Sutherland	49.13 ± 0.40	-78.72 ± 1.18	0.268 ± 0.004
6	5 + Gough Island	49.64 ± 0.35	-80.13 ± 0.99	0.262 ± 0.003
7	6 + Gao	49.07 ± 0.30	-79.02 ± 0.85	0.266 ± 0.003
8	7 + Tamale	49.28 ± 0.28	-79.40 ± 0.82	0.267 ± 0.002
9	8 + Yamoussoukro	49.43 ± 0.24	-79.28 ± 0.75	0.267 ± 0.002
10	9 + Springbok	49.16 ± 0.23	-79.47 ± 0.72	0.266 ± 0.002
11	10 + Richardsbay	49.31 ± 0.21	-80.85 ± 0.69	0.266 ± 0.001

No.	Name of the Station	Φ(°)	Λ(°)	ω(°/Ma)
12	11 + Walvis Ba	49.20 ± 0.20	−80.32 ± 0.61	0.267 ± 0.001
13	12 + Helwan	49.25 ± 0.19	−80.78 ± 0.58	0.267 ± 0.001
14	13 + Hartebeesthoek	49.22 ± 0.17	−80.46 ± 0.57	0.267 ± 0.001
15	14 + Telecomm Centre	49.12 ± 0.16	−80.27 ± 0.55	0.267 ± 0.001
16	15 + Hermanus	49.15 ± 0.16	−80.45 ± 0.51	0.267 ± 0.001
17	16 + De Aar	49.22 ± 0.14	−80.77 ± 0.45	0.267 ± 0.001
18	17 + Ouagadougou	49.15 ± 0.13	−80.61 ± 0.42	0.267 ± 0.001
19	18 + Maspalomas	49.13 ± 0.13	−80.49 ± 0.39	0.267 ± 0.001
20	19 + Santa Cruz	49.22 ± 0.13	−80.68 ± 0.36	0.267 ± 0.001
21	20 + Saint Helena	49.16 ± 0.11	−80.77 ± 0.34	0.267 ± 0.001
22	21 + Tanzania CGPS	49.10 ± 0.11	−80.89 ± 0.33	0.267 ± 0.001
23	22 + Simonstown	49.07 ± 0.10	−80.83 ± 0.31	0.267 ± 0.001
24	23 + Mbarara	49.11 ± 0.10	−80.78 ± 0.31	0.267 ± 0.001
25	24 + Rabat	**49.15 ± 0.10**	**−80.82 ± 0.30**	**0.267 ± 0.001**

References

1. McEvoy, F.M.; Schofield, D.I.; Shaw, R.P.; Norris, S. Tectonic and climatic considerations for deep geological disposal of radioactive waste: A UK perspective. *Sci. Total Environ.* **2016**, *571*, 507–521. [CrossRef]
2. Vérard, C.; Veizer, J. On plate tectonics and ocean temperatures. *Geology* **2019**, *47*, 881–885. [CrossRef]
3. Barbuzano, J. Three times tectonics changed the climate. *EOS* **2019**, *100*. [CrossRef]
4. Altamimi, Z.; Métivier, L.; Collilieux, X. ITRF2008 plate motion model. *J. Geophys. Res. Space Phys.* **2012**, *117*, 1–14. [CrossRef]
5. Altamimi, Z.; Métivier, L.; Rebischung, P.; Rouby, H.; Collilieux, X. ITRF2014 plate motion model. *Geophys. J. Int.* **2017**, *209*, 1906–1912. [CrossRef]
6. Snider, A. *The Creation and Its Mysteries Unveiled*; Snider-Pellegrini: Paris, France, 1859; pp. 1–386.
7. Wegener, A. *Die Entstehung der Kontinente und Ozeane*; Friedr. Vieweg & Sohn Akt.-Ges.: Braunschweig, Germany, 1915.
8. Edwards, J. *Plate Tectonics and Continental Drift*; Evans Brothers Ltd.: London, UK, 2005; pp. 1–48.
9. Frisch, W.; Meschede, M.; Blakey, R. *Plate Tectonics*; Springer: Berlin/Heidelberg, Germany, 2011; pp. 1–214.
10. Bird, P. An updated digital model of plate boundaries. *Geochem. Geophys. Geosyst.* **2003**, *4*, 1–52. [CrossRef]
11. Cox, A.; Hart, R.B. *Plate Tectonics: How It Works*; John Wiley & Sons: Hoboken, NJ, USA, 1986; pp. 1–416.
12. Kearey, P. *Global Tectonics*; John Wiley & Sons Ltd.: Hoboken, NJ, USA, 2009; pp. 1–496.
13. Heflin, M. 2004. Available online: http://sideshow.jpl.nasa.gov/mbh/series.html (accessed on 8 June 2021).
14. Drewes, H. A geodetic approach for the recovery of global kinematic plate parameters. *Bull. Géodésique* **1982**, *56*, 70–79. [CrossRef]
15. Drewes, H. Models for monitoring regional geokinematics in the Alpine-Mediterranean Region. *Ann. Geophys.* **1984**, *2*, 235–238.
16. Drewes, H. Significance of kinematic plate parameters derived from actual satellite Laser ranging data. *Adv. Space Res.* **1986**, *6*, 67–70. [CrossRef]
17. Drewes, H. Global Plate Motion Parameters Derived from Actual Space Geodetic Observations. In *Global and Regional Geodynamics*; Vyskocil, P., Reigber, C., Cross, P.A., Eds.; Springer: Berlin/Heidelberg, Germany, 1990; Volume 101, pp. 30–37.
18. Drewes, H. Combination of VLBI, SLR and GPS determined station velocities for actual plate kinematic and crustal deformation models. In *Geodesy on the Move*; Springer: Berlin/Heidelberg, Germany, 1988; Volume 119, pp. 377–382.
19. Drewes, H. The actual plate kinematic and crustal deformation model APKIM2005 as basis for a non-rotating ITRF. In *Geodetic Reference Frames*; Springer: Berlin/Heidelberg, Germany, 2009; Volume 134, pp. 95–101.
20. Drewes, H.; Meisel, B. An actual plate motion and deformation model as a kinematic terrestrial reference system. *Geotechnol. Sci. Rep.* **2003**, *3*, 40–43.
21. Heidbach, O.; Drewes, H. 3-D finite element model of major tectonic processes in the Eastern Mediterranean. *Geol. Soc. Lond. Spec. Publ.* **2003**, *212*, 261–274. [CrossRef]
22. Minster, J.B.; Jordan, T.H. Present-day plate motions. *J. Geophys. Res. Space Phys.* **1978**, *83*, 5331–5354. [CrossRef]
23. Argus, D.F.; Gordon, R.G. No-net-rotation model of current plate velocities incorporating plate motion model NUVEL-1. *Geophys. Res. Lett.* **1991**, *18*, 2038–2042. [CrossRef]
24. DeMets, C.; Gordon, R.G.; Argus, D.F.; Stein, S. Effect of recent revisions to the geomagnetic reversal time scale on estimates of current plate motions. *Geophys. Res. Lett.* **1994**, *21*, 2191–2194. [CrossRef]
25. Bird, P.; Rosenstock, R.W. Kinematics of present crust and mantle flow in southern California. *Geol. Soc. Am. Bull.* **1984**, *95*, 946–957. [CrossRef]
26. Argus, D.F.; Gordon, R.G.; De Mets, C. Geologically current motion of 56 plates relative to the no-net-rotation reference frame. *Geochem. Geophys. Geosyst.* **2011**, *12*, 1–13. [CrossRef]
27. Vondrák, J.; Richter, B. International Earth Rotation and Reference Systems Service (IERS) web: www.iers.org. *J. Geod.* **2004**, *77*, 585–678. [CrossRef]

28. Altamimi, Z.; Rebischung, P.; Métivier, L.; Collilieux, X. ITRF2014: A new release of the International Terrestrial Reference Frame modeling nonlinear station motions. *J. Geophys. Res. Solid Earth* **2016**, *121*, 6109–6131. [CrossRef]
29. Pearlman, M.R.; Degnan, J.J.; Bosworth, J.M. The International Laser Ranging Service. *Adv. Space Res.* **2002**, *30*, 135–14335. [CrossRef]
30. Willis, P.; Lemoine, F.G.; Moreaux, G.; Soudarin, L.; Ferrage, P.; Ries, J.; Otten, M.; Saunier, J.; Noll, C.; Biancale, R.; et al. The International DORIS Service (IDS): Recent developments in preparation for ITRF2013. In *IAG 150 Years*; IAG Symposia Series; Rizos, C., Willis, P., Eds.; Springer: Cham, Switzerland, 2016; Volume 143, pp. 631–639. [CrossRef]
31. Nothnagel, A.; Artz, T.; Behrend, D.; Malkin, Z. International VLBI Service for geodesy and astrometry. *J. Geod.* **2017**, *91*, 711–772. [CrossRef]
32. Dow, J.M.; Neilan, R.E.; Rizos, C. The International GNSS Service in a changing landscape of Global Navigation Satellite Systems. *J. Geod.* **2009**, *83*, 191–198. [CrossRef]
33. Rebischung, P.; Altamimi, Z.; Ray, J.; Garayt, B. The IGS contribution to ITRF2014. *J. Geod.* **2016**, *90*, 611–630. [CrossRef]
34. Sillard, P.; Boucher, C. Review of algebraic constraints in terrestrial reference frame datum definition. *J. Geod.* **2001**, *75*, 63–73. [CrossRef]
35. Dermanis, A. The rank deficiency in estimation theory and the definition of reference systems. In *V Hotine-Marussi Symposium on Mathematical Geodesy*; Sans, F., Ed.; Springer: Berlin/Heidelberg, Germany, 2003; Volume 127, pp. 145–156.
36. Petit, G.; Luzum, B. *IERS Conventions*; IERS Technical Note No. 36; Verlag des Bundesamts fur Kartographie und Geodasie: Frankfurt am Main, Germany, 2010.
37. Kraszewska, K.; Jagoda, M.; Rutkowska, M. Tectonic Plate Parameters Estimated in the International Terrestrial Reference Frame ITRF2008 Based on SLR Stations. *Acta Geophys.* **2016**, *64*, 1495–1512. [CrossRef]
38. Jagoda, M.; Rutkowska, M.; Suchocki, C.; Katzer, J. Determination of the tectonic plates motion parameters based on SLR, DORIS and VLBI stations positions. *J. Appl. Geod.* **2020**, *14*, 121–131. [CrossRef]
39. Kraszewska, K.; Jagoda, M.; Rutkowska, M. Tectonic plates parameters estimated in International Terrestrial Reference Frame ITRF2008 based on DORIS stations. *Acta Geophys.* **2018**, *66*, 509–521. [CrossRef]
40. Jagoda, M.; Rutkowska, M. Use of VLBI measurement technique to determination of the tectonic plates motion parameters. *Metr. Meas. Syst.* **2020**, *27*, 151–165.
41. Jagoda, M.; Rutkowska, M. An analysis of the Eurasian tectonic plate motion parameters based on GNSS stations positions in ITRF2014. *Sensors* **2020**, *20*, 6065. [CrossRef]
42. Van Gelder, B.H.; Aardoom, L. *SLR Network Designs in View of Reliable Detection of Plate Kinematics in the East Mediterranean*; Reports of the Department of Geodesy; Delft University of Technology: Delft, The Netherlands, 1982; pp. 1–24.
43. McCarthy, J.J.; Rowton, S.; Moore, D.; Pavlis, D.E.; Luthcke, S.B.; Tsaoussi, L.S. *GEODYN II System Operation Manual, 1-5*; STX System Corp.: Lanham, MD, USA, 1993; p. 20706.
44. ITRF2014. Available online: https://itrf.ign.fr/ITRF_solutions/2014/ (accessed on 8 June 2021).
45. Argus, D.F.; Gordon, R.G.; Heflin, M.B.; Ma, C.; Eanes, R.J.; Willis, P.; Peltier, W.R.; Owen, S.E. The angular velocities of the plates and the velocity of Earth's center from space geodesy. *Geophys. J. Int.* **2010**, *180*, 913–960. [CrossRef]
46. Brown, D.A.; Campbell, K.S.W.; Crook, K.A.W. *The Geological Evolution of the Australia and New Zeland*; Pergamon Press: Oxford, UK, 1968.